AUTOMATION IN TREE FRUIT PRODUCTION: PRINCIPLES AND PRACTICE

Automation in Tree Fruit Production: Principles and Practice

Edited by

Qin Zhang

Washington State University, USA

CABI

CABI is a trading name of CAB International

CABI
Nosworthy Way
Wallingford
Oxfordshire OX10 8DE
UK

Tel: +44 (0)1491 832111
Fax: +44 (0)1491 833508
E-mail: info@cabi.org
Website: www.cabi.org

CABI
745 Atlantic Avenue
8th Floor
Boston, MA 02111
USA

Tel: +1 (617)682 9015
E-mail: cabi-nao@cabi.org

A catalogue record for this book is available from the British Library, London, UK.

Library of Congress Cataloging-in-Publication Data

Names: Zhang, Qin, 1956- editor.
Title: Automation in tree fruit production : principles and practice / edited by Qin Zhang.
Description: Boston, MA : CABI, [2017] | Includes bibliographical references and index.
Identifiers: LCCN 2017022423 (print) | LCCN 2017024618 (ebook) | ISBN 9781780648521 (ePDF) | ISBN 9781780648514 (ePub) | ISBN 9781780648507 (hbk: alk. paper)
Subjects: LCSH: Precision farming. | Farm mechanization. | Fruit trees.
Classification: LCC S494.5.P73 (ebook) | LCC S494.5.P73 A98 2017 (print) | DDC 338.1/74--dc23
LC record available at https://lccn.loc.gov/2017022423

ISBN-13: 978 1 78064 850 7

Commissioning editor: Ward Cooper
Editorial assistant: Rebecca Stubbs / Emma McCann
Production editor: Tim Kapp

Typeset by SPi, Pondicherry, India
Printed and bound in the UK by Bell & Bain Ltd, Glasgow, G46 7UQ

MIX
Paper from responsible sources
FSC® C007785

Contents

Contributors

Duke Bulanon, Northwest Nazarene University, Nampa, Idaho, USA
Thomas Burks, University of Florida, Gainesville, Florida, USA
Du Chen, China Agricultural University, Beijing, China
Joseph R. Davidson, Washington State University, Richland, Washington, USA
Reza Ehsani, University of California, Merced, California, USA
R. Karina Gallardo, Washington State University, Puyallup, Washington, USA
Markus Flury, Washington State University, Puyallup, Washington, USA
Gwen-Alyn Hoheisel, Washington State University, Prosser, Washington, USA
Manoj Karkee, Washington State University, Prosser, Washington, USA
Lav Khot, Washington State University, Prosser, Washington, USA
Afef Marzougui, Washington State University, Pullman, Washington, USA
Siddhartha Mehta, University of Florida, Gainesville, Florida, USA
Denise Neilsen, Agriculture and Agri-Food Canada, Summerland, British Columbia, Canada
Gerry Neilsen, Agriculture and Agri-Food Canada, Summerland, British Columbia, Canada
Yasin Osroosh, Washington State University, Prosser, Washington, USA
R. Troy Peters, Washington State University, Prosser, Washington, USA
Francisco Rojo, Escuela de Agronomía, Pontificia Universidad Católica de Valparaíso, Casilla 4-D, Quillota, Chile
Sindhuja Sankaran, Washington State University, Pullman, Washington, USA
Abhisesh Silwal, Washington State University, Prosser, Washington, USA
Shrinivasa Upadhyaya, University of California Davis, Davis, California, USA
Matthew Whiting, Washington State University, Prosser, Washington, USA

Caixi Zhang, Shanghai Jiaotong University, Shanghai, China
Chongyuan Zhang, Washington State University, Pullman, Washington, USA
Jingjin Zhang, Shanghai Jiao Tong University, Shanghai, China
Qin Zhang, Washington State University, Prosser, Washington, USA
David Zilberman, University of California Berkeley, Berkeley, California, USA

1 Tree Fruit Production Automation

QIN ZHANG*

Washington State University, Prosser, Washington, USA

1.1 Introduction

One solution for producing high-quality, high-yield fruit, with minimal dependence on seasonal human labor, is to create a means for automatous mechanized precision production in orchards. This involves three key technologies: agricultural automation; mechanization; and precision farming. Among them, mechanization and precision farming are at the core of a comprehensive system using automation technologies.

As one of the top-ranked engineering accomplishments of the 20th century, agricultural mechanization has made revolutionary changes in field crop production technology and made it possible to achieve high yields using minimal human labor to meet continuously growing needs for food, feed, fiber and fuel. To make machines operate efficiently, one feature of mechanized production is the uniformity of operation in a field. Even though tree fruit production is quite different from field crop production, many of the fundamental mechanization technologies for field crop production can be used directly or modified for use in tree fruit production. The uniformity of mechanized production increases efficiency at the expense of being able to respond to crop growth variabilities often caused by inter- or intra-field soil type, fertility and moisture variance.

Precision farming offers a management practice based on observing, measuring and responding to inter- and intra-field variability in crop growth, hoping to gain the highest possible either in yields or in economic returns or in both. The concept of performing field tasks precisely in response to crop growth is not new. Our ancestors exercised very small area-based, if not plant-based, precise farming practices in response to actual

* qinzhang@wsu.edu

crop growth at the location for thousands of years in the past, manually. How to effectively integrate the capability of performing uniform operations with the need for responding to the natural variations in yield potential and/or crop growth is a new challenge to be solved by mechanized precision crop production.

Invented in the early 1980s in the USA, the concept of precision agriculture divides a large field into many small management zones, allowing performance of a field operation uniformly in a specific zone according to the yield potential or actual crop growth within that zone, resulting in responsive variable operations for the entire field. The core of this technology is a responsive farming management system based on observing, measuring and responding to inter- and intra-field variability in crops. This creative and innovative technology provided famers with a functional means to practice precise responsive field operations using uniformly implemented machines in large-scale mechanized precision crop production.

Crop growth usually varies both spatially and temporally in a field and mainstream farmers are trying to maximize their profits by spending money to perform their field operations only in the right place at the right time. Enabling farmers to perform their time-sensitive site-specific mechanized precision operations requires an effective tool to support farmers making operation decisions based on both spatial and temporal crop variabilities. Because crop growth is strongly correlated to soil properties, researchers at the University of Minnesota invented a method for making precise input recommendations for fertilizers and pH corrections for fixed grid areas in a field based on soil properties sampled from corresponding grids in the 1980s. Resulting from over 30 years of development since then, technology providers have introduced yield monitoring and soil-, plant- and pest-sensing technologies, combined with the advent of global positioning systems (GPS) technology, for acquiring critical data to uncover the variabilities in crops and/or soils, and display only the information needed to support decision making in the cab for the condition. Contributing to those accomplishments, the practice of precision crop farming has been gaining ground.

Although GPS and in-cab display technologies have played an important role in many precision farming systems, precision farming does not automatically happen when a GPS unit and an in-cab display are installed. It starts with farmers gaining a basic understanding of how they could more effectively manage their resource inputs in the field corresponding to soil types, field topography and hydrology, microclimates and crop stresses, and occurs over time as they adopt new tools and management strategies using detected and verified variance of yield-affecting factors within the fields to manage inputs precisely, enabling farmers to attain all promises of the technology by crunching massive data collected from their productions, as well as those similar to theirs, using big data technology to optimize their operations.

A few core technologies available today for precision crop production include GPS, soil sampling and remote sensing, yield monitoring and mapping, automated guidance and variable-rate input controls, and big data and decision making. Among those core technologies, some could be shared by many different applications other than precision crop production. For example, GPS is a critical element of modern information infrastructure, having numerous applications affecting almost every aspect of our life from car navigation to smartphone positioning, and boosts productivity across a wide swath of the global economy, including but not limited to farming, transportation, mining, mapping and surveying, package delivery and logistical supply-chain management. Big data and decision making are also in this category.

Some of those core technologies that are specifically developed for precision crop farming are yield monitoring and mapping, automated crop scouting, tractor guidance and variable rate (VR) input controls. Monitoring site-specific yields and recording the data using harvester-mounted yield-monitoring sensors is often the first step many farmers take in adopting precision farming technologies. Yield monitoring and mapping is the process specifically created for collecting georeferenced data on crop yield and characteristics, such as moisture content, while the crop is being harvested, and graphically presenting such data to show the intra-field variation in crop yield. Instantaneous yield monitors are currently available from various technology providers. Coupled with a GPS receiver, many of those yield monitors could provide spatial coordinates for collected yield data to generate yield maps automatically. Such data recorded in a yield map could be used to compare yield variations within a field from year to year, providing farmers with the necessary information to make decisions for effective site-specific precision management of their crops.

While the yield map could provide historic yield variations within a field, farmers often also need to know the actual crop growth conditions to make adequate precision operation decisions. Satellite-, aerial-, or tractor-based crop monitoring or scouting technologies provide farmers with the capability of obtaining adequate spatial and temporal resolution of field data for various precision agriculture applications. The availability of unmanned aerial systems (UAS) for agriculture could provide farmers with much more freedom to scout the crop, and allow them to collect field data many times during the season to support near real-time soil, crop and pest management.

Aimed at increasing efficiency of crop inputs utilization, reducing the adverse impact of over-application of inputs and maximizing production profitability, variable rate application (VRA) technology plays a key role in applying the right amount of crop inputs where and when it is needed, and forms one of the fundamental technologies specifically developed for precision crop production. VRA operation is, in general, implemented using mobile crop inputs application machinery: either a seeder, a sprayer/applicator, or an irrigator, equipped with a VR input control system to change the application rate in a site-specific application. A typical VR control system often consists of a differential GPS (DGPS) receiver, a prescription map of inputs, some VR control software, and actuating controllers to make it work. A broader definition of VRA could also include targeted applications that apply the inputs accurately where the target has been detected, either at the same rate or a varying rate.

1.2 Precision and Automated Production for Tree Fruit

Tree fruit production is a highly competitive agricultural industry worldwide with growers seeking more effective methods to manage their orchard operations precisely, and automatically if possible, to remain competitive in the international market.

In principle, the precision crop production technologies developed for field crops could be adapted to tree fruit production; and precision tree fruit production is precision farming applied to enhance orchard performance by optimizing fruit yield and quality while minimizing adverse environmental impacts. It can also be accomplished by observing, measuring and responding to local variability in tree crops or soils in the hope of getting the desired yield of high-quality fruit. While it may require some adjustments or improvements to make it more fitting to tree fruit production, many of the existing precision agriculture sensing technologies developed for field crop production, including GPS, meteorological and other environmental sensors, satellite and airborne remote sensing, and geographic information systems (GIS), could easily be adopted to detect or assess the variability of factors that could influence fruit yield and quality (such as soil, topography, microclimate, plant nutrients, and water and disease stress). However, a few special sensing methods for obtaining some critical crop information that is unique in the precise management of tree fruit may be required. One example is the growing interest in measuring the amount of photosynthetic energy being absorbed by tree canopies at different times of day to support more adequate precise pruning of the trees. Another example is the use of soil, or even leaf, moisture sensors to support more efficient variable-rate irrigation to supply the trees with just the right amount of water and/or fertilizer, depending on what a particular plant needs.

Due to the morphologic and cultivation differences between tree fruit crops and field crops, some implementations in tree fruit precision farming practices could be quite different from those commonly seen in field crop farming. For example, training, pruning and thinning are some of the special and important operations only seen in tree fruit farming, requiring accurate and precise location and removal of certain branches, twigs, blossoms and/or green fruit from the trees in a precision operation. Different from conventional precision operations for field crop farming, the data required to support those operations in tree fruit farming are physical locations, sizes and orientations of objects of interest rather than the spatial or temporal variabilities of soils and plants. Therefore, different types of sensing technologies from that for conventional precision field crop farming will be needed for tree fruit farming. Meanwhile, some other implementations in tree fruit farming could be fairly similar to those of precise field crop farming in process, but require different mechanisms to fit the morphologic and cultivation features of tree crops. A few examples include irrigation, fertilization and pesticide applications. Those operations often need equipment specially designed for orchard use to perform the required precision implementations.

While harvest is an essential operation for all crops, the unpredictable distribution of fruit on trees, uneven maturity of the fruit and the high quality requirement of fruit harvested for the fresh market make mechanical fruit harvest a unique operation in precision and automated production. It is necessary to have not only adequate sensing technologies for reliably and accurately detecting the locations, sizes or even orientations of fruit on a tree, and to precisely assess multiple parameters of the fruit to determine if it is ready to be harvested, but also to have devices capable of removing the ready-to-pick fruit from the trees gently without inducing much mechanical damage to the harvested fruit. Such complicated requirements make mechanical harvesting one of the most challenging tasks in precision and automated production of tree fruit.

1.3 Special Issues for Precision and Automated Production of Tree Fruit

In general, the goal for performing precision and automated tree fruit production is the same as for field crop production, i.e. to implement an effective and sustainable management practice based on observing, measuring and responding to inter- and intra-field variability in crop growth in the hope of gaining the highest possible levels in either yield or economic returns, or both. However, it requires addressing a few specific, special challenges in production. One specific challenge is that a fruit tree is a perennial woody plant, and in general has an almost permanent trunk with a growing branch/twig structure, which could be described as a four-dimensional (4D) structure (a specific 3D trunk for individual trees, plus variations between trees and between different years). Another major challenge is the different operations required for tree fruit production as described in the previous section; and the third is the high sensitivity of the plant and the fruit to mechanical interaction. All these challenges create significant reluctance among tree fruit producers to adopt mechanization and automation technologies developed for field crops, resulting in a low level of mechanization in tree fruit production today. Some crop-specific issues need to be addressed for tree fruit automation, and this book intends to provide an overview on such crop-specific issues.

The low level of mechanization makes tree fruit production rely heavily on a large, seasonal semi-skilled workforce to perform many of the field operations such as training, pruning, thinning and harvesting, especially for fresh market fruit, creating one of the most critical labor-related risks of not having enough of the right people at the right time to perform time-sensitive field operations. However, when trying to decide on whether or not to use automated machinery and new technology in an operation, the cost will involve more than just purchasing the equipment; it will also require an initial adjustment for workers to learn to use the new technology and equipment. Not all technologies are feasible for all growers. The adoption of automation technologies for tree fruit may need to start with a techno-economic analysis, including not only the initial purchasing cost, but also the operation and maintenance costs.

Sensing is always an essential element in agricultural automation. In orchard automation, sensors are used to measure the microclimate, quantify tree-absorbed sunlight, detect fruit location in tree canopies and monitor fruit development, in addition to measuring soil properties and various crop stresses. Therefore, it can be expected that more types of different sensors will be used to obtain such additional information required in implementing tree fruit automation. While measurement could be accomplished using various types of mobile scouting platforms, it is also possible to set up wireless sensor networks permanently during the crop lifespan, due to the perennial nature of tree fruit. Scouting based on UAS (Fig. 1.1) could provide some attractive advantages over other scouting methods in orchard automation, including but not limited to: the capability of being deployed at almost any time, not limited by field accessibility as

Fig. 1.1. An example of an unmanned helicopter scouting a commercial apple orchard in the Pacific Northwest region of USA.

for ground-based scouting platforms; carrying adequate sensor(s) flying over the orchard canopy at different altitudes to get georeferenced data at either fixed or variable resolutions even within one flight; and finding a stressed area in an orchard and then going straight to the area to get more detailed data from that area. One fundamental requirement for the successful performance of an orchard automation system is that its sensing system must be able to measure the data accurately.

As in any automation system, effective orchard automation relies on the ability to make correct operation decisions. Similar to precision field crop production management, precision orchard management also aims to increase the efficiency of crop inputs utilization, reduce the adverse impact of over-application of inputs, and maximize production profitability through harvest of more and better fruit; and all the operation decisions should aim to achieve those goals. Due to necessary differences in comparison with field operations, making appropriate decisions for tree pruning and training, irrigation scheduling and pest, disease and plant nutrition management, as well as crop load management, are required to implement orchard automation processes effectively and profitably.

As high-efficiency orchard systems have been proven to produce high-value produce, there is a need to refine tree systems to develop simplified orchard systems that are productive, yield high-quality produce and, equally importantly, are amenable to facilitating the incorporation of mechanization, automation and precision horticulture strategies. This requires a transition from conventional low-density, complex orchard systems to modern planar architectures, which can be accomplished through appropriate pruning and training; this is one of the strategically critical decisions that needs to be made to utilize orchard automation effectively.

Good irrigation planning is core for precision irrigation management, and is one of the critical orchard automation practices that increases grower profitability while protecting and improving our environment. This is becoming more and more important as water becomes a limited resource due to rising urban demand and environmental restoration. Tree fruit growers often use some kind of irrigation management calculator in making their irrigation plans. Varying from calculating general design to calculating irrigation management, irrigation calculators provide growers with a practical tool to compute their irrigation needs based on their growing practices, types of soil and vegetation, and to specify their precision management plans based on the calculated needs.

Integrated pest and disease management (IPDM) is an important practice in tree fruit production for reducing dependency on pesticides in pest and disease control which helps growers to achieve the precision farming goals of improving produce safety and reducing adverse impacts on the environment. Like any other precision agriculture practice, it includes use of some means to monitor insects, pests or disease symptoms, predict pest development patterns or trends, then find methods that could effectively and economically control the pest and at the same time cause negligible impact to produce safety and minimal damage to the environment.

A fundamental concept of IPM is that a certain number of individual pests can and should be tolerated.

Aimed at improving resource-use efficiency by matching nutrient applications with physiological demand, nutrition management is an essential element in precision tree fruit farming, as it is in field crop farming. Many orchards in major tree fruit production areas are on coarse-textured soils with limited water- and nutrient-holding capacities, and fertigation is particularly useful in those operations, because it can dramatically reduce fertilizer use, which will result in reduced leaching losses and greenhouse gas emissions while improving fruit quality. Like any typical precision crop farming operation, effective precision fertigation relies on the ability to precisely determine what and how much nutrients need to be added to correct the detected nutrient deficiencies of the plant.

It is well known that the crop load will affect fruit size, fruit color, evenness of maturity, and ease of harvest as well as return bloom in the next year, making crop load management one of the most important operations in orchard management. Precise crop load management can be performed via precision pruning, precision bloom and/or green fruit thinning, and precision pollination. Figure 1.2 shows a precision bloom-thinning operation using a hand-held blossom thinner in a commercial sweet cherry orchard.

Fig. 1.2. A sweet cherry grower performing a mechanical blossom-thinning operation to remove excessive flowers in a commercial orchard in the Pacific Northwest region of the USA.

Although there are currently various orchard machines being used for field scouting, pruning and thinning, and chemical application, fresh market tree fruit harvesting still relies heavily on human pickers because machines are not yet able to offer comparable performance in removing fruit from the tree quickly and efficiently and meanwhile gently handling the fruit to prevent bruising. Aimed at creating fully automated mechanical harvesting systems that are of practical use in harvesting fresh market tree fruit, numerous research and development studies have been reported even prior to the 1980s. Despite the wide difference in specific mechanisms being used from one system to another, the investigated systems use either a shake-and-catch mechanism for mass harvesting or a pick-and-place mechanism for individual picking. In principle, shake-and-catch harvesting applies vibratory excitation to the canopy, trunk or a branch of a tree to create a detaching force on fruit to separate it from the tree and uses some type of catching device to collect the detached fruit, while pick-and-place harvesting uses an approach similar to human selective picking by first locating the fruit on the canopy, detaching it from the tree, then placing the picked fruit into a container.

Most robotic fruit harvesting systems are designed to try to mimic the precision of a human picker and the typical design often consists of a vision system for locating the fruit, a manipulator with an end-effector acting like a human arm and hand for picking the fruit, then placing it at a designated place. Moreover, a complete robotic harvesting system must provide passable mobility within the orchard, and be guided autonomously within the alleyway. To perform all these functions adequately in an autonomous way requires a comprehensive and reliable site awareness capability for the robotic system just like an experienced human worker could provide, which is still fairly challenging to achieve economically. As replacing people entirely with automated machines is not feasible at this time, mainly limited by fruit quality issues regarding harvest machines, some harvest assist solutions combining manual and automatic functions, such as auto-steered harvesting platforms, are available for growers to improve their field efficiency. Recently a prototype robotic apple picker has been developed by Abundant Robotics, Inc., a California-based company, with support from the Washington Tree Fruit Research Commission. In-orchard harvesting tests demonstrated that this prototype was capable of locating over 95% of fruit, removing around 80% of fruits under a picking speed of one fruit per second (Good Fruit Grower, 2016). Even though such a performance still does not yet quite meet the expectation of growers, it starts to make economic sense to use robotic harvesting technology. Based on current progress in robotic harvesting technology development, it is reasonable to expect that some robotic harvesting systems could become commercially available for the tree fruit industry around 2020. However, it needs to be pointed out that these performance levels are not yet achievable for other more traditional canopy structures.

1.4 Integrated Solutions to Orchard Mechanization and Automation

Mechanization and automation are the core technologies for precision management of tree fruit production, and are performed best in uniform operations, but the aforementioned issues make it difficult to achieve uniformity of operations and it is difficult to find two orchards, systems, or growers that are alike. Such a problem requires using integrated solutions to solve the challenging issues faced in orchard automation.

One level of integration could be human–machine integration which allows human orchard workers to use their skills and intelligence to maneuver machinery performing designated tasks more accurately and efficiently. For example, when mechanically harvesting fruit in an orchard, a mechanical harvesting system needs first to localize the fruit to be harvested, then position the machinery at an appropriate position to remove the target fruit. No two trees grow alike, therefore it requires a high level of intelligence, skill and experience to do the job well. A manually operated mechanical harvesting system allows orchard workers to take care of the challenging task of locating fruit and to move the machine into an appropriate position to perform effective harvesting. Human–machine integration in tree fruit mechanization requires intensive work for human orchard workers to maneuver the machinery, and the integration of electronic technology with orchard machinery, namely mechatronically enhanced orchard machinery. Mechatronically enhanced orchard machinery could elevate orchard automation to a higher level by reducing orchard workers' labor intensity, and also provide advanced functions to improve the performance of the machinery. The precision orchard sprayer is a good example of such an integration technology which uses automatic sensing systems to detect the target canopy and then controls the spraying more accurately and precisely. One study conducted by Washington State University researchers on cutworm control in vineyards showed that the use of a smart, electronically controlled target sprayer could reduce pesticide usage by 90% by applying it accurately to the targeted barrier (Kang *et al.*, 2014).

Integrating harvesting and post-harvesting tasks in an orchard harvesting system could help producers to achieve substantial benefits. Storing fruit is an expensive process in tree fruit production, and if culled fruits could be separated in-orchard during harvesting, it would help to dramatically reduce the overall cost of delivering the produce to the market. Such a functionality could be realized by equipping an on-board grading system on fruit harvesting systems. An in-orchard harvest sorting system created by the US Department of Agriculture's Agricultural Research Service (USDA ARS) researchers at Michigan State University has demonstrated the capability of sorting picked fruit into fresh, processing, and juice quality grades during harvesting which could help to save time and money by not sending lower-quality fruit to the packing house, as well as reducing the incidence of post-harvest diseases from introducing contaminated fruit (Lehnert, 2013).

As efficient orchard automation cannot be achieved by machinery/automation systems design alone, another layer of integration is the machine and orchard systems. Mechanical or robotic orchard machinery could operate at maximum productivity and efficiency in high-density blocks with narrow, uniform, accessible two-dimensional fruiting walls. Many orchards are not organized in such a way that would accommodate this requirement, and such a lack of uniformity makes these sites unsuitable for using automated or autonomous orchard machinery. One of the pioneer studies on integrating machinery and orchard systems was a robotic bulk harvesting system for apples developed by a USDA research team led by Peterson *et al.* (1999). They trained the apple trees using a Y-trellis system so that fruit grew on the side and lower branches and found them to be compatible with mechanical robotic harvesting.

The core of an integrated orchard production automation system is that the cultural practices must be suited for using mechanized operations, including field conditions, tree population and spacing, and tree canopy shape and size. Establishing favorable field conditions for machinery systems should be considered even before orchard systems are designed. Standardization of tree sizes, featured by tree height, tree shape, canopy thickness, and tree spacing within and between rows, would allow orchard machinery to operate continuously without frequent adjustment and therefore could substantially improve throughput of mechanized orchard operations and thus economic benefit. The ideal configuration of machine-friendly orchards would be a relatively uniform vertical or slightly inclined hedge fruiting wall, in a smooth and continuous tree row. Mechanical pruning allows pruning the trees uniformly for maintaining tree canopy sizes and shapes with minimal labor requirement. Properly shaped hedgerows would allow the fruit to grow on the canopy surface along the trellis wires with minimal occlusion, which could help to maximize harvesting efficiency. Uniformly pruned tree rows could also improve the canopy light exposure and enhance canopy accessibility for effective spraying, thinning and harvesting operations.

Tree fruit automation, in general, is an integration of agricultural equipment and precision agriculture management which provides tree fruit growers with an effective means to optimize their production for harvesting sufficient quality fruit. The following chapters provide a representative, snappy overview of the variety of technologies, applications, challenges and crops where automation is being developed or implemented.

References

Good Fruit Grower (2016) Mechanized Vacuum Apple Picker Demonstration. Video available at: https://www.youtube.com/watch?v=TBcWZcjXr-I (accessed 26 August 2016).

Kang, F., Li, W., Pierce, F.J. and Zhang, Q. (2014) Investigation and improvement of targeted barrier application for cutworm control in vineyards. *Transactions of the ASABE* 57(2), 381–389.

Lehnert, R. (2013) In-orchard sorting, a prototype machine divides apples into fresh, processing, and juice grades. *Good Fruit Grower* (November), 20–21.

Peterson, D.L., Bennedsen, B.S., Anger, W.C. and Wolford, S.D. (1999) A systems approach to robotic bulk harvesting of apples. *Transactions of the ASAE* 42(4), 871–876.

2 The Economics of Perennial Crops' Production Automation

R. Karina Gallardo[1]* and David Zilberman[2]

[1]*Washington State University, Puyallup, Washington, USA;* [2]*University of California Berkeley, Berkeley, California, USA*

2.1 Background

During the 20th century, technological innovations played a major role in improving agricultural productivity. Five decades ago, futurists envisioned agricultural farms that today are not that far away from that vision. In today's agricultural fields, soil sensors, drones, satellite images, efficient irrigation and mechanical harvesters, among other technologies, are a regular component of farming practices (Lusk, 2016).

It is interesting to note that mechanization devices have been readily available for grains, beans and cotton since the 1960s. However, mechanization devices for fruit and vegetables in general are lagging behind. Today, mechanical harvesters are massively used for some fruit or vegetables designed for the processing market, but the majority of fruit and vegetables for the fresh market are still reliant on manual labor (Huffman, 2012).

When plant breeder Jack Hanna and engineer Coby Lorenzen with the University of California, Davis, envisioned a machine back in the mid 1950s that could mechanically harvest tomatoes, nobody thought they would ever become successful. Countless failing prototypes and endless quantities of split tomatoes that turned into juice in the field were part of Hanna and Lorenzen's obscure days. Besides, there was no apparent need to develop a mechanical harvesting machine, because there was an abundance of cheap and efficient farm workers, many of whom came to the USA from Mexico via the Bracero program (initiated in 1942 to provide better conditions for migrant manual laborers). However, the era of abundance came to an end by 1963, when the Bracero program ended. The tomato industry increased its

* Corresponding author, email: karina_gallardo@wsu.edu

concerns that its businesses would surcease if they lost the influx of cheap labor used for tomato picking. It was then that Hanna and Lorenzen's efforts made a breakthrough. It was a convolution of circumstances: the impending need to find an alternative to the reduced availability of labor, plus the development of a new tomato variety, the 'VF-145', that could be easily de-stemmed and would better resist the severe handling of the mechanical harvester. Within 5 years, almost the entire tomato industry was planting the 'VF-145' and using the mechanical harvester (Carlisle-Cummins, 2015). Over 35 years, mechanical harvesters have reduced labor requirements of the California processing tomato industry by 92% (Huffman, 2012).

The case of the tomato mechanical harvester illustrates how technological adoption happened in the USA. One can observe that technological advancements do not entail adopting a single tool or technique; it is rather the adoption of a systems approach of technologies, which usually involves machines, improved seeds, careful tillage, fertilizer, producer use of water, and so on (Rasmussen, 1968; Sunding and Zilberman, 2001). For example, the tomato mechanical harvester was developed in parallel with the 'VF-145' tomato variety, which was more resistant to mechanical handling. The systems approach, coupled with external factors such as economic and policy changes, often influences the adoption and diffusion of novel agricultural technologies. For example, the tomato harvester was developed in response to the need to protect against the risk of labor scarcity due to the imminent end of the Bracero program.

When analyzing the different dimensions to which mechanical harvesters conform as part of the universe of technological innovations in agriculture, one must consider the policy implications and the impact on economic agents. For example, from a policy perspective, mechanization technologies might lead to a decreased demand for labor and, due to economies of size, to an increased concentration of firms. Considering the impact on economic agents, mechanical harvesters are adopted because they can potentially increase revenues, reduce labor input costs, and reduce labor input-related risks (Sunding and Zilberman, 2001). However, to ensure adoption, mechanical harvesters must work well and be economically viable. Once a technology has proven feasible, several factors could impact diffusion patterns, including the inherent risks associated with the agriculture activity, investment costs, uncertainties around the innovation's performance and reliability, and appropriateness for a specific agricultural operation.

In general, mechanical harvesters are crucial for perennial crops because, in contrast to other agricultural crops, perennial crops are labor intensive and labor dependent. Critical masses of temporal labor are needed for specific field activities, the main one (among other activities during the production year) being harvest. This persistent intensive use of labor in perennial crop production in the USA is in large part due to the continued abundant supply of labor from Mexico and other Latin American countries throughout the 20th century. The US agricultural specialty crop industry has long depended on migrant labor (Martin, 2009); for example, the immigrations to California by Chinese workers in the 1870s and Japanese in the 1900s, the Mexican immigration to southwestern states in the 1920s, the 1917–1922 temporary program targeting unskilled Mexican farmworkers (later known informally as the first Bracero program), the 1930s Dust Bowl migration from Oklahoma and Arkansas to western states and the World War II Mexican Bracero Program. Since the end of the Bracero Program in 1963, undocumented migrant labor pools have mostly supplemented agricultural labor in the USA (Martin, 2009). In fact, in 2010, Mexico accounted for 75% of hired farm workers in the USA (Calvin and Martin, 2010).

However, the supply of migrant labor is unpredictable. Given the US economic recession in 2008, fewer migrant workers have been available to harvest fruit and vegetable crops, and increased spending on border enforcement has raised the cost of migration for potential workers (Taylor *et al.*, 2012; Charlton and Taylor, 2016). Indeed, the number of unauthorized Mexican immigrants estimated to be living in the USA showed a decrease of 12.9% in 2012 since its peak of 12.2 million in 2007 (Passel *et al.*, 2012). In addition, Latin America's economic growth and productivity in both farm and non-farm sectors have generally been accelerating relative to such growth in the USA. In Mexico, fertility rates began to drop around 1980 in response to numerous cultural and economic factors and that demographic shift is now showing up in the working-age population. Moreover, imminent changes in Mexico's labor policies are triggering the slowdown of agricultural migrant labor to the USA; for example, the nascent farm labor-organizing activity, the incipient increases in wages, and the unprecedented government guarantee of farmworker wages (Charlton and Taylor, 2016). Moreover, as the agricultural labor force ages and older workers exit, the US farm labor supply is likely to tighten further (Zahniser *et al.*, 2012). Even as the supply of farm labor appears to be shrinking, demand for it has remained relatively constant at an annual average of 1 million workers since 2007. However, labor economists argue that so far, by 2012, there were no conclusive signals of a widespread labor shortage (Zahniser *et al.*, 2012).

Given the imminent changes in the supply of agricultural migrant labor to the USA and the lack of a massive adoption of labor-saving technologies in special perennial crops destined for the fresh market, one wonders if the technological challenges are great enough to consider the possibility of alternative adaptation strategies to high labor costs. One strategy could be switching towards less labor-intensive crops where a greater share of domestic demand is met by importing food from countries with lower labor costs (Gallardo and Brady, 2015). Even if technologies adapt rapidly, allowing for widespread mechanization, there would likely be numerous effects on farm structure in perennial crops, such as increased farm size and geographic transitions to different terrain. What has been observed is that there are significant differences in the production systems across labor-intensive crops that will impact the ability to mechanize (Gallardo and Brady, 2015).

2.2 Economic Analysis of Mechanization of Tree Fruits

There is a large body of research in economic analysis of mechanization of perennial crops and in general in agriculture. This section will present the most commonly used economic models that analyze the decision-making process of adoption and the subsequent diffusion of a new technology, namely mechanical harvesters.

Studies of adoption focus on analyzing the individual firm's decision-making process in adopting the new technology. Diffusion, from an economics perspective, is interpreted as an aggregate adoption, and economic studies focusing on diffusion of new technologies usually measure the market share potential of the new technology (Zilberman *et al.*, 2012). This section includes two subsections: one explaining models analyzing the adoption and the other explaining models analyzing diffusion of agricultural technologies, such as mechanization.

2.2.1 Adoption of mechanization technologies

In studying the adoption of new technologies, the economics and sociology disciplines have been major contributors. Sociologists have usually focused on elaborating the profile of potential adopters and opinion leaders, prospective adopters' perceptions of the new technology, rates of adoption, and information channels used in the decision-making process (Marra *et al.*, 2003). Economists have conceptualized the phenomenon of the adoption of new technologies at the level of the individual firm. The seminal work by Griliches (1957) and Mansfield (1963) revealed the importance of economic variables on the decision to adopt new technologies, and spawned a large literature (Feder *et al.*, 1985; Sunding and Zilberman, 2001). The economic literature on adoption of new technologies, including mechanical harvesters, can be classified according to two criteria. Firstly, there is a distinction between studies that evaluate technologies *ex ante* (before adoption decisions are made) and studies that look at adoption choices *ex post* to try to understand the factors that affect adoption. Secondly, the body of research can also be classified into studies focusing on adoption at the level of the individual firm and studies examining the diffusion of the novel technology within the industry. Studies analyzing firm-level adoption of the technological innovations can vary according to how closely the adoption decision-making process is depicted. We will first analyze alternative *ex ante* models that assess profitability of adoption and then we will discuss *ex post* models that explain adoption patterns. The simplest approach is the net present value (NPV); other more sophisticated approaches, such as the option-value model and its modifications, include elements of risk, uncertainty, reversibility of the decision to adopt, and optimal timing for adoption. Our analysis of *ex post* studies will distinguish between two main approaches of analysis: (i) cross-sectional studies, often identifying the characteristics of decision makers or firms that influence the adoption of the new technology; and (ii) temporal studies, usually focused on the determinants of the timing of adoption. The following subsections offer a review of the approaches used in each category.

2.2.1.1 Net present value (NPV)

A common perspective to assess investment in new technologies is to assess the net profit from the technology compared with the status quo, recognizing that benefits and costs occur at different periods of time. The NPV analysis is the most basic approach to evaluate an investment decision. Because profits and expenditures occur at different time periods, all financial activities must be discounted to the present using a discount factor. If the NPV is greater than zero, then the investment is profitable and worth considering. The NPV rule is a building block in decision theory where a decision maker uses the market interest rate to discount income in different periods of time and invest in those with the highest NPV, considering capital availability. A related concept is internal rate of return (IRR). For every investment, the IRR is the interest rate that results in a zero NPV. IRRs are used to compare different investments.

Sometimes the size of an investment is a decision variable, and it is important to be able to estimate the value of an incremental unit of capital and its associated costs (Dixit and Pyndick, 1994). One approach for such analysis is to assess the impact of changes in investment size on the NPV. There are two other approaches. Jorgenson (1963) compares the 'per period' value of the marginal product of an extra capital unit and the equivalent 'per period' marginal cost of capital ('rental cost'), computed from the purchase price, interest and depreciation rates, and applicable taxes. The firm will stop adoption or investment at the point where 'per period' marginal product equals the 'per period' rental cost. The other approach was developed by Tobin (1969), who compared the capitalized value of the marginal investment in the new technology with its purchase cost. The marginal value of the investment can be estimated if the technology can be sold or ownership traded in a secondary market. Otherwise the marginal value of the investment is calculated by imputing the expected present value of the profits that the technology would yield. As long as the marginal value of the technology is greater than its purchase cost, adoption is economically sound (Dixit and Pyndick, 1994).

The application of the NPV approach needs to be adjusted to accommodate the constraints faced by a decision maker. For example, choices may be different if the decisions are reversible or irreversible, and depend on the degree of uncertainty faced by the decision maker. When choices are irreversible and made under uncertainty, more elaborate decision rules, that allow for waiting and learning, need to be established (Dixit and Pyndick, 1994).

2.2.1.2 Introducing risk and uncertainty to technology adoption models

Investments frequently involve risky choices, including risk of productivity, reliability of products, market risks, etc. Some of the key events that affect investments (e.g. prices of produce harvested by a new machine) change over time. There are two approaches to analyzing risky investment choices. Some approaches designed for adoption choice at a given moment of time (e.g. whether or not to buy a mechanical harvester) emphasize the risk aversion of the decision maker (Marra *et al.*, 2003). Other approaches recognize that decision makers may change their mind over time, and then develop decision rules that establish critical values of key random variables that trigger adoption. Both approaches were designed to explain underinvestment in technologies that are worthwhile using traditional NPV.

The models that introduce risk considerations at a given moment in time rely on the expected utility framework (Just and Zilberman, 1983, 1988; Koundouri *et al.*, 2006). This approach balances expected profits and risk frequently measured by variance of profit of alternatives. Individuals vary in risk aversion and make both discrete (whether or not to adopt) and continuous (how much to adopt) choices. This literature suggests that the criterion for adoption of a new technology is that the NPV of the expected gain is greater than the NPV of the cost of bearing risk (this cost can be represented by the variance multiplied by a coefficient that measures the risk aversion of the decision maker). This approach suggests that more risk-averse individuals are less likely to adopt a specific technology and, when adopting, will make a smaller investment.

The seminal work of Arrow and Fisher (1974) changed the way we analyze investment by recognizing that investment decisions are not only yes/no decisions, but also involve determination of timing. The uncertainty about the future value of an investment and its irreversible costs provides an explanation of why investors delay executing some investment that is profitable in the future. Sunk costs are examples of irreversible outcomes, and there are many others (Marra *et al.*, 2003). The average NPV may be positive, but there may be probability of losses. If there is the option of delaying the decision to invest that allows avoiding unfavorable outcomes, then there is value to delaying, which is called the option-value. For example, a grower may be offered a harvesting technology shown to be profitable today, but he may realize that, because of learning by doing, an improved harvester may be available next year. So this farmer may decide to wait. McWilliams and Zilberman (1996) demonstrated that the gradual adoption of computers in agriculture was motivated by the recognition that prices of computers were decreasing over time.

Dixit and Pindyck (1994) offer a comprehensive review of option-value modeling. The model develops a stochastic dynamic framework that analyzes investment decisions in the presence of uncertainty, irreversibility, and the flexibility to postpone the investment. There is a large body of research in agricultural economics using the option-value model (Chavas, 1994; Purvis *et al.*, 1995; Zhao, 2001; Isik *et al.*, 2001, 2003; Carey and Zilberman, 2002; Baerenklau and Knapp, 2007; Livingston *et al.*, 2015). The early models of option-value considered only one source of uncertainty, such as water price, output price, yield, etc., but more recent models develop techniques that incorporate multiple sources; for example, Torani

(2014) developed a framework in which the decision to invest in a new technology follows a threshold decision rule under the effect of two stochastic processes.

Both the risk aversion and option-value approaches suggest that investments should not necessarily be taken when they are, on average, profitable but should rather take into account the cost of risk or the value of delay. Thus understanding the uncertainties about new technologies is crucial in making sound decisions. The adoption decision of mechanized harvesting involves many types of uncertainty, including technology performance, future price of labor, future price of output, energy costs, and availability of cheaper technologies in the future. The decision maker must give weight to risk aversion as well as improved information and conditions in the future that may justify a delayed decision.

2.2.2 Diffusion of new technologies: heterogeneity and patterns for adoption

Diffusion analyzes the level of penetration of a novel technology in a specific market. The literature on diffusion aims to explain the adoption behavior of large populations. There are different ways of measuring diffusion: one is to measure as the share of producers adopting a technology and the alternative is the share of the area of production utilizing the novel technology (Feder *et al.*, 1985; Sunding and Zilberman, 2001; Jaffe *et al.*, 2002; Foster and Rosenzweig, 2010). The stylized fact behind much of the diffusion literature is that diffusion follows an S-shaped function of time. Rogers (1976) distinguished between three groups: early adopters; followers during a period of takeoff with higher rates of adoption; and laggards during the saturation stage. Sometimes this progress is followed by a period of decline where the technology is replaced by another new one. The literature on diffusion consists of studies that aim to understand the individual adoption process and the resulting diffusion process, as well as statistical studies to identify the socio-economic, structural or demographic variables affecting the decision to invest in a new technology.

There are two types of approaches to explain diffusion of new technologies. One approach is the imitation model introduced by Rogers (1976), which argues that the decision to adopt is driven in large part by observing others; implementation of innovation is thus driven by marketing intensively to likely early adopters and then diffusion is driven by word-of-mouth. An alternative approach is the threshold model developed by David (1975) and further expanded by Stoneman and Ireland (1983) and Feder *et al.* (1985). According to this approach, decision makers consider profit and risk when evaluating new technologies. The new technology tends to improve temporal profits, but entails fixed costs. Due to heterogeneity among decision makers as well as dynamic processes of learning and improvement in technologies, the time of adoption varies among decision makers resulting in an S-shaped diffusion curve.

Threshold models vary in their analysis of adoption by individual decision makers. Several use static models that emphasize risk considerations (e.g. Jensen, 1982), while few use dynamic models. In McWilliams and Zilberman (1996) the timing of adoption is determined by the trade-off between the gain from early adoption versus the decreased fixed cost of the technology by delay. The literature tends to suggest that consideration of dynamic processes that result in changes in key variables have a higher explanatory power compared with static approaches (Feder *et al.*, 1985; Karshenas and Stoneman, 1993; Sunding and Zilberman, 2001; Koundouri *et al.*, 2006).

One of the major sources of heterogeneity that affects timing of adoption is the size of a farm, especially with respect to farm equipment such as mechanical harvesters. Larger farms tend to be early adopters; however, the work by Olmstead and Rhode (2001) suggested that one mechanism to overcome the scale barrier of adoption is by introducing custom services that rent equipment to farmers. Lu *et al.* (2016) suggested that during every period some large-scale farmers adopt a new technology and then rent the use of it to smaller-scale farmers. Over time, the rate of ownership tends to increase as price declines. Another source of heterogeneity may be differences in ability and resource quality. For example, Caswell and Zilberman (1986) showed that adoption of modern irrigation technologies is more likely to occur when water-holding capacity is low. Their analysis suggested that mechanization may occur in locations where labor availability is unreliable.

One branch of the literature that reconciles the threshold and imitation models of diffusion is through focus on multi-stage process models. These models emphasize that adoption involves multiple stages – awareness, assessment, decision, purchase, use, and re-evaluation (Kalish, 1985; Zilberman *et al.*, 2012). Some of these models underscore the importance of learning and judgment associated with adoption (Rogers, 2003). Other studies include learning as a risk-reducing strategy (Chatterjee and Eliashberg, 1990). There are studies focusing on sequential learning across countries (Ganesh *et al.*, 1997) and studies on the importance of referrals of previous adopters (Schmitt *et al.*, 2011) and the effect of social networks on the adoption of new agricultural technologies (Goldenberg *et al.*, 2007).

The literature emphasizes that the adopters of new technologies are concerned about uncertainty revolving around the properties and performance of the new technology and how these uncertainties affect various aspects of agricultural performance. This is precisely the case for mechanical harvesters in the present. To diminish risks associated with performance and malfunctioning, insurance and system back-ups could be an alternative. Mechanisms to protect the risks associated with the performance of a novel technology include warranties, responsive technical support, educational hands-on demonstrations, arrangements such as money-back guarantees, experimentation with new technology, and short-term renting (Sunding and Zilberman, 2001).

2.3 Empirical Findings

In the USA, five tree fruit industries are using some form of mechanical harvesting: Florida oranges for processing, Michigan tart cherries, California olives for canning, California tree nuts, and Washington apples. The degree of commercial utilization of mechanical harvesting varies from complete mechanization (Michigan tart cherries) to harvest aid platforms (Washington apples).

To our knowledge, the majority of published economic analyses related to mechanical harvesting in tree fruits are centered on the Florida citrus industry (Searcy *et al.*, 2007, 2012; Iwai *et al.*, 2009a, b; Blanco and Roka, 2009; Moseley *et al.*, 2012). There is one published study on mechanical harvest for tart cherries (Wright *et al.*, 2006), one on sweet cherries (Seavert and Whiting, 2011) and one on olives (Klonsky *et al.*, 2012). Most of these studies use the NPV concept to estimate the economic potential of mechanical harvesting compared with hand harvesting. Mechanical harvesting is more likely to be adopted when its NPV is greater than the NPV of hand harvesting. The studies using methodologies other than NPV are Wright *et al.* (2006), Iwai *et al.* (2009b) and Moseley *et al.* (2012).

Searcy *et al.* (2012) analyzed the impact of the adoption of mechanical harvesting on Florida orange growers and processors. They simulated two scenarios, with 5% and 95% of the processing orange volume being mechanically harvested. They found that mechanical harvesting enables a higher collection of juice for processing (measured in pound solids) and enables more opportunities for optimal harvest windows. Going from 5% to 95% of the total orange processing volume being mechanically harvested, growers gain US$227/acre. However, processors face higher costs, losing US$23/acre. This suggests that growers would need to subsidize the processor to make both parties better off. Researchers concluded that 'a systems approach' is needed and that mechanical harvesting adoption is not just a farm management decision.

The reasons for the orange juice processors' losses in a scenario where 95% of fruit is mechanically harvested are explained by Searcy *et al.* (2007). Growers are paid on the basis of the pounds solids (sugar content) delivered to the processor. Because this quality indicator is constantly changing when fruit is on the tree and quickly deteriorates after harvest, managing the allotments of fruit to be delivered to a processor is crucial. These allotments are based on the fruit quality and plant processing and storage capacity. Allotments are typically aligned with a hand-harvesting system. Mechanical harvesting introduces changes in the timing and volume of fruit going to the processing plant. Re-planning the whole process involving growers and processors is needed so that all members of the supply chain benefit from mechanical harvesting.

Moseley *et al.* (2012) investigated whether mechanical harvesting had an immediate or lagged effect on fruit yields and tree health measured by declining annual yields. They collected data from four citrus operations in southwest Florida over 10 years (1999–2008). They found no statistically significant yield differences (per acre) for blocks that were mechanically versus hand harvested. Also, there were no cumulative effects of mechanical harvesting on the yields. The authors concluded that the long-term economic sustainability of the Florida citrus industry rests on the premise that tree health must be restored and average production should fluctuate between 400 to 500 boxes per acre.

Iwai *et al.* (2009a) estimated the NPV for the hand and mechanical harvesting of a 'Hamlin' orange operation in southwest Florida. Researchers used a discounted cash flow (DCF) approach. Researches forecast future free cash flow (FCF) values by using a stochastic process for times series. NPV is estimated for three different scenarios: (i) hand harvesting; (ii) mechanical harvesting with harvest recovery rate of 98%, harvest cost reduction of US$0.25 per box; and (iii) mechanical harvesting with harvest recovery rate of 90%, harvest cost reduction of US$0.51 per box. Results for the NPV are US$8314, US$8344 and US$7507, respectively, for the three above-mentioned scenarios. The differences in NPV when using mechanical harvest and when not is 0.36%. This minimal difference might explain the low adoption rate of mechanical harvesting systems in Florida (7.5% of all the state's orange acreage in 2006–2007).

Iwai *et al.* (2009b) used a real options approach (ROA) to investigate the likelihood of adoption of mechanical harvesting by southwest Florida 'Hamlin' orange citrus operations. ROA is an application of a financial option model that takes into consideration the decision maker's option to invest or to wait. In this model, the new technology is adopted when its NPV exceeds the NPV of the current technology by the margin of the option value of the investment. Researchers estimated that the NPV of hand-labor harvesting is at US$11,056/acre compared with mechanical harvesting at US$7012/acre. This suggests that mechanical harvesting must return an additional US$4044/acre in the 2007–2008 season for growers to invest in this new technology. In addition, researchers estimated that an increase in current mechanical harvesting FCF of US$463/acre (FCF growth rate of 4.05%) would be enough to reach the threshold level for mechanical harvesting to be adopted.

Blanco and Roka (2009) conducted a cost/benefit analysis of an abscission agent registration for use by the citrus industry. The abscission compound 5-chloro-3-methyl-4-nitro-1H-pyrazole (CMNP) could enhance the efficiency of current mechanical harvesting equipment and lead to its full economic advantage. The abscission compound enables mechanical harvesting equipment to operate during the late harvest season of 'Valencia' oranges without imposing a yield loss on next season's crop. In addition, harvest cost savings are realized by extending the time of operation of a mechanical system. Abscission expenditures include registration costs, research, and development expenditures. Results show that to obtain a positive NPV for the use of the abscission agent, the minimum usage acreage should be 15,000 acres and the difference between mechanical and hand-harvesting costs should be US$0.10 per box.

Wright *et al.* (2006) used a threshold farm size analysis to measure the ability of Polish tart cherry growers to move from hand to mechanical harvesting. They compared this with an economic valuation used to estimate the point in the life cycle of an established Michigan tart cherry orchard when it becomes feasible to replant to adopt an overhead mechanical harvester. To adopt, Polish tart cherry operations need to be within the 23–53 acres size. For Michigan tart cherry growers to adopt, average yields obtained by using an overhead mechanical harvester should be at least 9654 lb/acre (10.82 t/ha). Michigan growers would likely recuperate their initial investment near the end of the 25-year lifespan. Benefit would augment if harvest begins at year 3 (with a positive NPV for overhead mechanical harvesters at US$4557).

Klonsky *et al.* (2012) conducted an NPV analysis to estimate the profitability of using a mechanical harvester with California black ripe table olives. The assumptions for the orchard in the study were that 90% of the block was 'Manzanillo' olive cultivar and 10% 'Sevillano'. The orchard density was at 202 trees per acre. There were 10 contiguous acres of olives, and the lifespan of the orchard was 25 years. They found that the net returns per acre for mechanical harvesting at an efficiency of 80% (US$414) were US$65 dollars higher than net returns per acre for hand harvesting (US$349).

Seavert and Whiting (2011) compared the NPV of a competitive orchard system (COS) when using hand and mechanical harvesting. A COS targets fruit that would contribute to an adequate rate of return on investment, compensates the grower for the opportunity costs and borrowed capital to establish an orchard, and covers all financial risks involved with the investment. Also a COS provides sufficient cash flow to the growers for replacing orchard blocks in a timely fashion and affording technologies that increase efficiencies and decrease inputs (labor, chemicals, traceability, etc.). The NPV of a COS using hand harvesting is at US$59,499/acre (US$148,748/ha) compared with the NPV when using mechanical harvesting at US$93,592/acre (US$233,981/ha). When using a mechanical harvester the break-even price of sweet cherries to establish an orchard is reduced from US$0.99 to US$0.82/lb (US$2.2 to US$1.83/kilo).

In summary, from the literature reviewed, mechanical harvesting systems used in perennial tree fruit crops are demonstrated to be profitable in the long run compared with hand harvesting under strong assumptions of efficiencies and economies of scale. Note that mechanical harvesting is more feasible for industries producing crops for the processing market rather than for the fresh market. Finally, a total systems approach is the most suited when dealing with mechanization adoption, as this decision could affect current organizational schemes along the supply chain.

2.4 Summary and Conclusion

In this chapter we provide a comprehensive discussion of the adoption and diffusion of mechanization technologies. We argue that the labor-associated risks and the high dependence of perennial crops on labor are the major trigger to develop new technologies. However, these technologies are risky and expensive. Thus, the adoption of these technologies is partial. Analysis of the diffusion of mechanization technologies can benefit from the insights of the extensive economic literature used to analyze the adoption and diffusion of new agricultural technologies. The literature emphasizes the importance of elements such as uncertainty, irreversibility, learning and option to postpone an investment as major explanations of why we witness lags in adoption of technologies that seem to be profitable. For mechanization technologies, one obstacle for adoption is the high initial costs, while another is lack of access to credit. In addition, mechanization entails field workers with a different set of skills compared with current workers, implying high labor transaction costs. The fixed cost constraints may be overcome by rental arrangements, which also serve for learning and may explain low initial sales of machines. Heterogeneity in terms of farmer age and human capital may also explain lag of adoption, because older farmers tend to have higher cognitive costs and operate with a shorter planning horizon. However, external pressures might change the current mechanization panorama. Diminishing farm labor supply may act as a common stressor for perennial crop producers around the world (Charlton and Taylor, 2016). Perennial crop producers face limited alternative options such as to shift their business to less intensive crops or invest in technologies and management practices that would diminish labor dependence.

Acknowledgment

The authors wish to thank Benjamin Gordon (University of California at Berkeley) for his insightful comments to improve the content of this chapter.

References

Arrow, K.J. and Fisher, A.C. (1974) Environmental preservation, uncertainty, and irreversibility. *Quarterly Journal of Economics* 88, 312–319.

Baerenklau, K. and Knapp, K.C. (2007) Dynamics of agricultural technology adoption: age structure, reversibility, and uncertainty. *American Journal of Agricultural Economics* 89(1), 190–201.

Blanco, G. and Roka, F.M. (2009) *Cost/Benefit Analysis of Abscission Registration for Citrus Mechanical Harvesting.* Paper presented at the Southern Agricultural Economics Association (SAEA) Annual Meeting, Atlanta, Georgia, January 31–February 3.

Calvin, L. and Martin, P. (2010) Labor-Intensive U.S. fruit and vegetable industry competes in a global market. *Amber Waves* 8(4), 24–32.

Carey, J.M. and Zilberman, D. (2002) A model of investment under uncertainty: modern irrigation technology and emerging markets in water. *American Journal of Agricultural Economics* 84(1), 171–183.

Carlisle-Cummins, I. (2015) *From Ketchup to California Cuisine: How the Mechanical Tomato Harvester Prompted Today's Food Movement.* Cal Ag Roots, California Institute for Rural Studies. Available at: http://civileats.com/2015/07/15/from-ketchup-to-california-cuisine-how-the-mechanical-tomato-harvester-prompted-todays-food-movement/ (accessed 17 June 2016).

Caswell, M.F. and Zilberman, D. (1986) The effects of well depth and land quality on the choice of irrigation technology. *American Journal of Agricultural Economics* 68(4), 798–811.

Charlton, D. and Taylor, J.E. (2016) A declining farm workforce: analysis of panel data from rural Mexico. *American Journal of Agricultural Economics* 98(4), 1158–1180.

Chatterjee, R. and Eliashberg, J. (1990) The innovation diffusion process in a heterogeneous population: a micro-modeling approach. *Management Science* 36, 1057–1079.

Chavas, J.P. (1994) Production and investment decision under sunk cost and temporal uncertainty. *American Journal of Agricultural Economics* 76(1), 114–127.

David, P.A. (1975) *Technical Choice, Innovation and Economic Growth.* Cambridge University Press, Cambridge, UK.

Dixit, A.K. and Pindyck, R.S. (1994) *Investment Under Uncertainty.* Princeton University Press, Princeton, New Jersey.

Feder, G., Just, R.E. and Zilberman, D. (1985) Adoption of agricultural innovation in developing countries: a survey. *Economic Development and Cultural Change* 33, 285–298.

Foster, A. and Rosenzweig, M. (2010) *Barriers to Farm Profitability in India: Mechanization, Scale, and Credit Markets.* Presented at the World Bank-UC Berkeley Conference on Agriculture for Development, Berkeley, California.

Gallardo, R.K. and Brady, M.P. (2015) Adoption of labor-enhancing technologies by specialty crop producers: the case of the Washington apple industry. *Agricultural Finance Review* 75(4), 514–532.

Ganesh, J., Kumar, V. and Subramaniam, V. (1997) Learning effect in multinational diffusion of consumer durables: an exploratory investigation. *Journal of the Academy of Marketing Science* 25, 214–228.

Goldenberg, J., Libai, B., Moldovan, S. and Muller, E. (2007) The NPV of bad news. *International Journal of Research in Marketing* 24, 186–200.

Griliches, Z. (1957) Hybrid corn: an exploration in the economics of technological change. *Econometrica* 25(4), 501–522.

Huffman, W.E. (2012) The status of labor-saving mechanication in US fruit and vegetable harvesting. *Choices* 27(2), 1–7.

Isik, M., Khanna, M. and Winter-Nelson, A. (2001) Sequential investment in site-specific crop management under output price uncertainty. *Journal of Agricultural and Resource Economics* 26(1), 212–229.

Isik, M., Coble, K.H., Hudson, D. and House, L.O. (2003) A model of entry–exit decisions and capacity choice under demand uncertainty. *Agricultural Economics* 28(3), 215–224.

Iwai, N., Emerson, R.D. and Roka, F.M. (2009a) *Labor Cost and Value of Citrus Operations with Alternative Technology: Enterprise DCF Approach.* Paper presented at the Southern Agricultural Economics Association (SAEA) Annual Meeting, Atlanta, Georgia, January 31–February 3.

Iwai, N., Emerson, R.D. and Roka, F.M. (2009b) *Harvest Cost and Value of Citrus Operations with Alternative Technology: Real Options Approach.* Paper presented at the AAEA and ACCI Joint Annual Meeting, Milwaukee, Wisconsin, July 26–29. Agricultural and Applied Economics Association, Milwaukee, Wisconsin.

Jaffe, A.B., Newell, R.G. and Stavins, R.N. (2002) Environmental policy and technological change. *Environmental and Resource Economics* 22(1–2), 41–70.

Jensen, R. (1982) Adoption and diffusion of an innovation under uncertain probability. *Journal of Economic Theory* 27, 182–192.

Jorgenson, D. (1963) Capital theory and investment behavior. *American Economic Review* 53, 247–259.

Just, R.E. and Zilberman, D. (1983) Stochastic structure, farm size and technology adoption in developing agriculture. *Oxford Economic Papers* 35(2), 307–328.

Just, R.E. and Zilberman, D. (1988) The effects of agricultural development policies on income distribution and technological change in agriculture. *Journal of Development Economics* 28, 193–216.

Kalish, S. (1985) A new product adoption model with price, advertising, and uncertainty. *Management Science* 31(12), 1569–1585.

Karshenas, M. and Stoneman, P. (1993) Rank, stock, order, and epidemic effects in the diffusion of new process technologies. *Rand Journal of Economics* 24, 503–528.

Klonsky, K., Livingston, P., DeMoura, R., Krueger, W.H., Rosa, U.A. *et al.* (2012) *Economics of Mechanically Harvesting California Black Ripe Table Olives.* Poster presented at the International Symposium on Mechanical Harvesting and Handling Systems of Fruits and Nuts. Lake Alfred, Florida, April 2–4. International Society for Horticultural Science, Leuven, Belgium.

Koundouri, P., Nauges, C. and Tzouvelekas, V. (2006) Technology adoption under production uncertainty: theory and application to irrigation technology. *American Journal of Agricultural Economics* 88(3), 657–670.

Livingston, M., Roberts, M.J. and Zhang, Y. (2015) Optimal sequential plantings of corn and soybeans under price uncertainty. *American Journal of Agricultural Economics* 97(3), 855–878.

Lu, L., Reardon, T. and Zilberman, D. (2016) Supply chain design and adoption of indivisible technology. *American Journal of Agricultural Economics* 98(5), 1419–1431.

Lusk, J.L. (2016) *Unnaturally Delicious.* St Martin Press, New York.

Mansfield, E. (1963) The speed of response of firms to new techniques. *Quarterly Journal of Economics* 77, 290–311.

Marra, M., Pannell, D.J. and Abadi Ghadim, A. (2003) The economics of risk, uncertainty and learning in the adoption of new agricultural technologies: where are we on the learning curve? *Agricultural Systems* 75(2–3), 215–234.

Martin, P. (2009) *Improving Poverty? Immigration and the Changing Face of Rural America.* Yale University Press, New Haven, Connecticut.

McWilliams, B. and Zilberman, D. (1996) Time of technology adoption and learning by using. *Economics of Innovation and New Technology* 4(2), 139–154.

Moseley, K.R., House, L.A. and Roka, F. (2012) Adoption of mechanical harvesting for sweet orange trees in Florida: addressing grower concerns on long-term impacts. *International Food and Agribusiness Management Review* 15(2), 83–98.

Olmstead, A.L. and Rhode, P.W. (2001) Reshaping the landscape: the impact of diffusion of the tractor in American agriculture, 1910–1960. *Journal of Economic History* 61(3), 663–698.

Passel, J.S., Cohn, D. and Gonzalez-Barrera, A. (2012) *Net Migration from Mexico Falls to Zero – and Perhaps Less.* Pew Hispanic Center, Pew Research Center, Washington, DC.

Purvis, A., Boggess, W.G., Moss, C.B. and Holt, J. (1995) Technology adoption decisions under irreversibility and uncertainty: an 'ex ante' approach. *American Journal of Agricultural Economics* 77(3), 541–551.

Rasmussen, W.D. (1968) Advances in American agriculture: the mechanical tomato harvester as a case study. *Technology and Culture* 9(4), 531–543.

Rogers, E.M. (1976) New product adoption and diffusion. *Journal of Consumer Research* 2, 290–301.

Rogers, E.M. (2003) *Diffusion of Innovations*, 5th edn. Free Press, New York.

Schmitt, P., Skiera, B. and Van den Bulte, C. (2011) Referral programs and customer value. *Journal of Marketing* 75, 46–59.

Searcy, J., Roka, F.M. and Spreen, T. (2007) *Optimal Harvest Time of Florida Valencia Oranges to Maximize Grower Returns.* Paper presented at the Southern Agricultural Economics Association (SAEA) Annual Meeting, Mobile, Alabama, February 4–7.

Searcy, J., Roka, F.M. and Spreen, T. (2012) *The Impact of Mechanical Citrus Harvester Adoption on Florida Orange Juice Growers.* Poster presented at the AAEA Annual Meeting, Seattle, Washington, August 12–14. Agricultural and Applied Economics Association, Milwaukee, Wisconsin.

Seavert, C.F. and Whiting, M.D. (2011) Comparing the economics of mechanical and traditional sweet cherry harvest. *Acta Horticulturae* 903, 725–730.

Stoneman, P. and Ireland, N. (1983) Technological diffusion, expectations, and welfare. *Oxford Economic Papers* 38, 283–304.

Sunding, D. and Zilberman, D. (2001) The agricultural innovation process: research and technology adoption in a changing agricultural sector. *Handbook of Agricultural Economics* 1A(1), 207–261.

Taylor, J.E., Charlton, D. and Yuńez-Naude, A. (2012) The end of farm labor abundance. *Applied Economic Perspectives and Policy* 34(4), 587–598.

Tobin, J. (1969) A general equilibrium approach to monetary theory. *Journal of Money, Credit, and Banking* 1, 15–29.

Torani, K. (2014) Adoption of renewable energy technologies under uncertainty. Doctoral Dissertation, University of California, Berkeley, California.

Wright, R.T., Martinez, L. and Thornsbury, S. (2006) *Technological Leapfrogging as a Source of Competitive Advantage.* Paper presented at the Southern Agricultural Economics Association (SAEA) Annual Meeting, Orlando, Florida, February 4–8.

Zahniser, S., Hertz, T., Rimmer, M.T. and Dixon, P.B. (2012) *The Potential Impact of Changes in Immigration Policy on US Agriculture and the Market for Hired Farm Labor. A Simulation Analysis.* Economic Research Report 135 (May). Economic Research Service, United States Department of Agriculture, Washington, DC.

Zhao, J. (2001) *Information Externalities and Strategic Delay in Technology Adoption and Diffusion.* Working paper. Iowa State University, Ames, Iowa.

Zilberman, D., Zhao, J. and Heiman, A. (2012) Adoption versus adaptation, with emphasis on climate change. *Annual Reviews of Resource Economics* 4, 27–53.

3 Sensing for Stress Detection and High-throughput Phenotyping in Precision Horticulture

Sindhuja Sankaran*, Chongyuan Zhang and Afef Marzougui

Washington State University, Pullman, Washington, USA

3.1 Overview

Sensing is a critical component of automation in tree fruit production, which includes site-specific management of resources (water, nutrients), precision chemical application for disease control, mechanical harvesting, and other crop protection applications to enhance the production efficiency and environmental stewardship. Another application of advanced sensing technologies that has developed in the past decade is crop scouting. The sensing technologies have a potential to assess abiotic (water stress, nutrient deficiencies, etc.) and biotic (bacterial, viral, fungal diseases, insect damage, etc.) stress conditions in different crops, sometimes in asymptomatic stages. The sensor technologies are greatly beneficial in determining the regions within an orchard that have anomalies (unhealthy or poorly performing trees), rather than specific types of disease or stress condition. The identification of such localized regions through remote sensing technologies can assist in the scouting process, thereby enhancing the scouting efficiency and decreasing the associated costs. Similarly, in the past few years, there has been great interest in high-throughput crop phenotyping using sensing technologies to assist genetics and breeding programs towards crop improvement efforts. The controlled studies (focusing on one or few traits at a given time, e.g. disease resistance) and assessment of relative differences between different cultivars enhance the feasibility of using sensing for crop trait/phenotype

* Corresponding author, email: sindhuja.sankaran@wsu.edu

evaluation as a practical solution to promote crop improvement efforts. In this chapter, commonly used sensing techniques with a focus on sensing applications for stress detection and high-throughput phenotyping, with examples, will be discussed.

3.2 Sensor Technologies

Several sensor technologies are used in precision agriculture and high-throughput phenotyping applications (Table 3.1). The sensing principles of common methods are described below. In addition to those mentioned below, there are several sensing techniques such as X-ray imaging, X-ray fluorescence, and magnetic resonance imaging that can help to assess shoot, root, and crop quality traits.

3.2.1 RGB camera

Monitoring and diagnosing crops could be enhanced with imaging sensors (Sankaran et al., 2015). A red, green, blue (RGB) camera or true-color camera detects a signal and delivers it into three color bands: red, green, and blue. The process of converting the light into this model is assured by charge coupled device (CCD) or complementary metal–oxide–semiconductor (CMOS) sensors. Generally, a Bayer color filter array is used to generate one true-color image pixel by merging a blue, a red and two green sensors. However, confounded spectral response, limited dynamic range, and sensitivity to brightness are the main difficulties that are associated with the use of RGB cameras (Nijland *et al.*, 2014). Another type of range imaging with CCD/CMOS sensors is three-dimensional (3D) imaging (Gupta and Ibaraki, 2014). Multiple RGB cameras can be integrated to generate the 3D image to extract plant architectural features.

3.2.2 Multispectral and hyperspectral sensors/cameras

Spectral reflectance sensors or cameras detect spectral signals reflected by objects or a scene from visible light (350–700 nm), near-infrared (700–1400 nm), and short-wave infrared (1400–2500 nm) regions. Spectral reflectance cameras are similar to digital cameras (RGB cameras) but with a specific filter in spectral bands of interest, instead of red, green, or blue filters. Based on the number of spectral bands, they can be generally classified into multispectral sensors (3–10 spectral bands) and hyperspectral sensors (10s to 100s of spectral bands). The cameras extract a few or a stack of images from the visible–near-infrared spectrum, while some sensors (spectroscopes or radiometers) extract the average visible–near-infrared spectral reflectance from the region of interest. These sensors/cameras allow diagnosis of crops and fruit trees, such as water, nutrient and health status.

Table 3.1. Sensing in fruit crop evaluation.

Crop	Sensors	Features/phenotypes extracted	Associated traits/phenotypes	Reference
Apple	Visible–near-infrared spectrometer	Wavelength features	Bruise and different types of fruit damage	Kleynen *et al.* (2003)
Grape	Spectroradiometer, infrared thermometer	Water index, normalized difference vegetation index (NDVI)	Leaf conductance, leaf and air temperature, pre-dawn water potential	Serrano *et al.* (2010)
Apple	RGB camera, hyperspectral camera	RGB color intensities, multivariate spectral features	Fruit color, maturity	Garrido-Novell *et al.* (2012)
Apple	Portable fluorometer	Chlorophyll a fluorescence (F690+F730 and F690/F730)	Sunburn apple peel	Tartachnyk *et al.* (2012)
Citrus	Fluorescence sensor	Yellow, red, and far-red fluorescence	Nutrient deficient and diseased leaves	Sankaran and Ehsani (2012)
Apple	Thermography, chlorophyll fluorescence camera	Initial fluorescence state, fluorescent quantum efficiency, temperature	Apple scab infection rating	Belin *et al.* (2013)
Tomato	Visible–near-infrared spectrometer	Partial least-square regression features	Quality traits (juice soluble solid content, juice pH, color parameters, firmness and water content)	Ecarnot *et al.* (2013)
Date palm	CCD camera	Image features from R, G and B channels	Hardiness of dried fruit	Manickavasagan *et al.* (2014)
Blueberry	Fluorescence imaging, leaf temperature sensor	Several fluorescence features	Water use efficiency	Estrada *et al.* (2015)
Apple	X-ray CT imaging system	Image features (bitter pit count and area)	Bitter pit rating	Si and Sankaran (2016)

Vegetation indices derived from combinations of spectral bands can also help in identifying the status of the crop or tree fruit. Some of the frequently used vegetation indices include normalized difference vegetation index (NDVI), photochemical reflectance index, normalized difference nitrogen index, and water band index (Harris Geospatial Solutions, 2016). Spectral reflectance bands or vegetation indices can also be applied for monitoring vigor, growth, and maturity. However, with an increased number of spectral bands, the cost and complexity of the sensors inevitably increase. In the case of both active sensors (using artificial light source) and passive sensors (using sunlight as a light source), the quality of output images depends heavily on the illumination conditions.

3.2.3 Fluorescent sensors

The general principle of fluorescent sensors lies in assessing vegetation photosynthetic dynamics by studying fluorescence emission of the plant (Gupta and Ibaraki, 2014). This imaging technique can detect biotic and abiotic stress symptoms such as nutrient deficient, water stress, bacterial, and viral diseases (Gorbe and Calatayud, 2012). Fluorescent sensors are employed in an active mode, which involves the presence of an illumination system to image fluorescence (Chaerle *et al.*, 2007). Stimulated by a specific wavelength, the plant emits fluorescent signals from different spectral regions that are captured by the sensor. The type of emitted signals are related to the type of illumination source bands (Gorbe and Calatayud, 2012). The two common types of fluorescence used in crop assessment are blue-green fluorescence at 400–600 nm range and chlorophyll fluorescence at 650–800 nm range (Sankaran *et al.*, 2010a).

3.2.4 Time of flight sensors

The time of flight (ToF) sensors measure the distance between the sensor and objects through the time elapse between emitting and receiving signals, and speed of signal with or without phase detector. In other words, the distance travelled by the signal is equal to the speed of the signal multiplied by the time elapse, where the signal can be electromagnetic (usually 850 nm) or sound waves (Li, 2014). The ToF sensors consist of a signal emitter, signal receiver, sensor, electronic driver, and computation components. Examples of ToF sensors include ToF camera, light detection and ranging (lidar), sound navigation and ranging (sonar), and radio detection and ranging (radar) systems. In the case of the 3D camera, depth is included to extract 3D measurements of the plant. The camera emits an infrared light and measures the time needed for the signal to be received again (Chéné *et al.*, 2012). The presence of a phase detector increases the accuracy of ToF but decreases its effective range. The ToF sensors are active sensing devices that can be used regardless of the lighting conditions. Functionalities of sonar sensors are limited to measuring distance, plant height, and plant spacing due to their simplicity, while most of the ToF sensors are effective in reconstructing plant structure and measuring leaf area. The cost of ToF sensors is relatively low, compared with spectral reflectance sensors and infrared thermal sensors. Although ToF sensors are active sensors, they can still be subjected to noise from sunlight or engines.

3.2.5 Thermal camera

Infrared thermal cameras create a thermal profile of objects or a scene by detecting the near-infrared radiation or heat signal, which cannot be seen with the naked eye. The temperature can be presented as a pseudo color image showing a thermal profile. The structure of infrared thermal cameras is similar to that of digital cameras but with special components to capture the thermal infrared signal, and maybe equipped with or without internal cooling component. Infrared thermal cameras are applied to monitor the water stress of crops or fruit trees and then help scientists or farmers to choose the best irrigation scheme and drought-resistant varieties. Even with state-of-the-art thermal cameras, blurring edges due to the spontaneous heat exchange between objects, or between objects and air is observed in the images. Blurring edges can result in complicated automated image processing. An RGB image integrated with a thermal image can be more robust for image processing. In field conditions, dynamic changes in environmental conditions (sunlight, cloud cover, wind, etc.) can also introduce errors.

3.3 Sensing for Stress Detection

3.3.1 Case study I: Water stress detection in grapevine

Water is the key element that influences the growth, yield, and quality of grapes, especially for rain-fed grape production areas. Application of sensing to monitor the water status has been exploited for optimizing the irrigation scheduling, and monitoring the status and performance of grapevines. Möller *et al.* (2007) estimated the water status of irrigated grapes through proximal thermal and visible (RGB) imaging. Sensors were placed 15 m above the grape canopy through a truckcrane for proximal sensing, and artificial wet surface and air temperature (+5°C) were used as T_{wet} and T_{dry}, respectively. It was found that, for the entire growing season, crop water stress index (CWSI) was highly and stably correlated with leaf conductance (g_L) ($R^2 = 0.91$, $P < 0.0001$) using T_{canopy} (from central of canopy), T_{wet}, and T_{dry}. The relationship between CWSI and stem water potential (Ψ_{stem}) lacked stability in terms of intercept and slope.

Baluja *et al.* (2012) monitored the water status of rain-fed grapes using thermal and multispectral images with the help of an unmanned aerial vehicle (UAV). Their results showed that the relationship between aerial temperature and g_L or Ψ_{stem} were significant ($R^2 = 0.68$, $P < 0.01$, and $R^2 = 0.50$, $P < 0.05$, respectively). It was reported that modified simple ratio (MSR), NDVI, and simple ratio index (SRI) were significantly correlated with Ψ_{stem} ($R^2 = 0.66$, 0.68 and 0.64, all $P < 0.05$). The greenness index (GI), green normalized difference vegetation index (GNDVI), MSR, NDVI, SRI, transformed chlorophyll absorption reflectance index (TCARI), and TCARI/OSAVI (ratio between TCARI and optimized soil-adjusted vegetation index) were also significantly correlated with g_L ($R^2 = 0.63$, 0.70, 0.78, 0.75, 0.78, 0.80, 0.84, all $P < 0.05$). These results demonstrate the potential of application of proximal and remote sensing for monitoring water status of grapes and deciding irrigation schedule in grapes. Similarly, Zúñiga *et al.* (2016) found that the water-use efficiency in grapevines under subsurface irrigation could be evaluated using NDVI and temperature data extracted from multispectral and thermal cameras mounted on UAV (Fig. 3.1). Other studies on utilizing sensing to evaluate water status in grapevine can be found in Dobrowski *et al.* (2005), Rodríguez-Pérez *et al.* (2007), and Bellvert *et al.* (2014), among others.

3.3.2 Case study II: Identification of fruit damage in apples

For the fresh market, fruit quality is essential for both farmers and distributors; and defects such as sunburn, frost damage, hail damage, bruising, and scarring will not only reduce the fruit value, but also increase the cost of sorting, packing, and storage. Sensing can assist in rapid and non-invasive detection of defects in fruits such as apples. Tartachnyk *et al.* (2012) investigated the fruit characteristics of apples with different peel colors. Chlorophyll breakdown can be detected using laser-induced chlorophyll fluorescence, and a significant decline of chlorophyll was found in 'Granny Smith' (67%) and sunlit sides of 'Fuji' (48%) and 'Braeburn' (17%). In addition, the sum of fluorescence at 690 nm and 740 nm, along with ratio of fluorescence between 690 nm and 730 nm in sunburn peels of 'Granny Smith', 'Fuji' and 'Braeburn' was reduced by 77%, 61% and 35%, and 29%, 11% and 12%, respectively. Kleynen *et al.* (2003) classified healthy (green and red) apples from defective ones (such as scald, hail damage, limb rub, russet, scab tissue, frost damage, rot, visible flesh damage and bruises) using visible–near-infrared spectroscopy. Apples were correctly classified with 98% accuracy and higher, except for rotted apples (92.9%) and bruised apples (bruised within an hour, 11.3%). Four wavelength features (450, 500, 750 and 800 nm) were identified to be effective for the classification of defective apples. Kleynen *et al.* (2005) developed a multispectral vision system based on these four bands and succeeded in classification of defective apples. Such techniques are being used for disorder/disease identification in several horticultural crops (Nicolai *et al.*, 2006; Gorbe and Calatayud, 2012; Jarolmasjed *et al.*, 2016; Wu *et al.*, 2016).

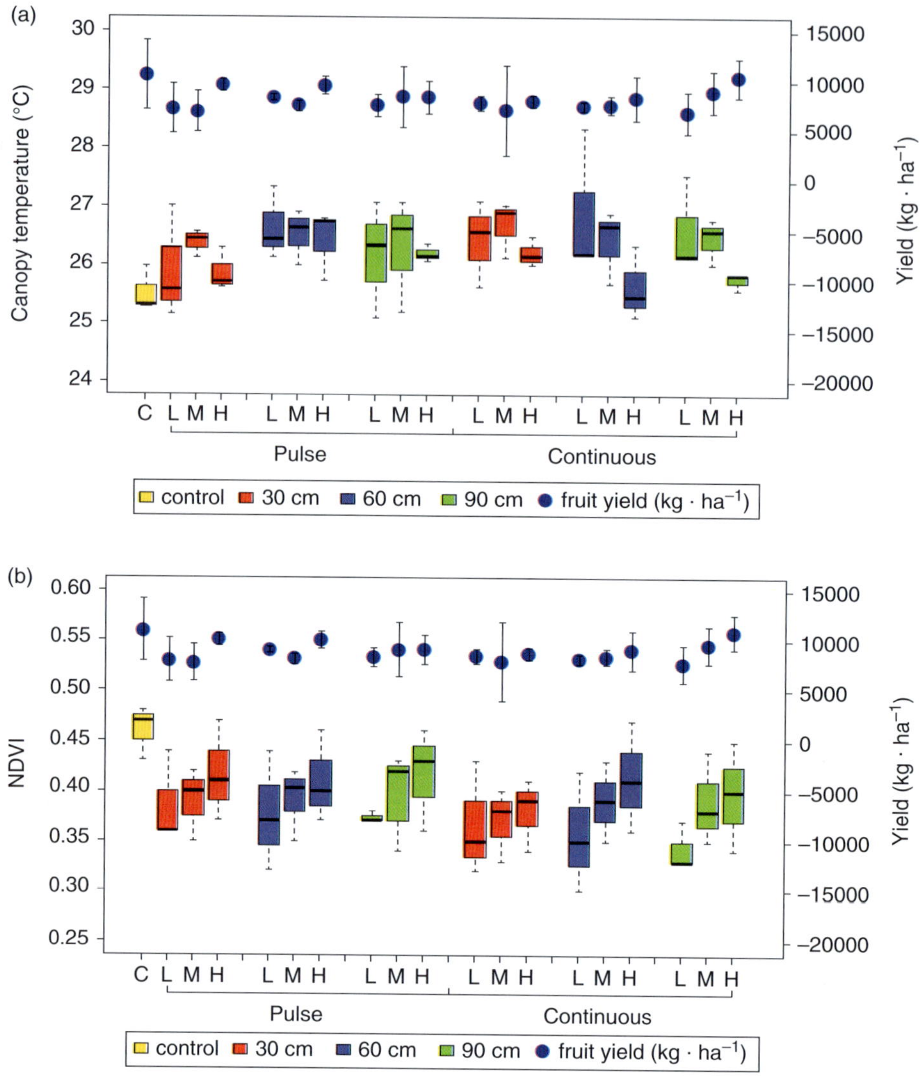

Fig. 3.1. Boxplot of grapevine (a) temperature and (b) NDVI at 37 days before harvesting, along with fruit production data at harvest from reference trees. L, M and H refer to irrigation rates of 15%, 30% and 60%; C refers to control plots receiving 100% irrigation based on evapotranspiration rates. Two irrigation mechanisms, namely the pulse and continuous irrigation mechanisms, were utilized. In the boxplot, the upper and lower whisker represent the maximum and minimum pit ratio values, upper and lower box borders represent the 75th and 25th percentile values, respectively, and the horizontal dark line indicates the median. The sub-surface irrigation treatments were at depths of 30 cm, 60 cm and 90 cm (Zúñiga *et al.*, 2016).

3.3.3 Case study III: Disease detection in citrus

Optical sensing techniques such as visible–near-infrared spectroscopy, mid-infrared spectroscopy and fluorescence spectroscopy, in addition to volatile biomarker detection, offers rapid sensing of plant diseases. These methods offer unique benefits that can greatly aid in citrus disease detection. Several studies are available that indicate the potential of a few proximal and remote sensing techniques for huanglongbing detection in citrus (Table 3.2). Similarly, several platforms have been utilized to evaluate the sensors ranging from hand-held/enclosed devices to aerial imaging. The challenges in huanglongbing detection include variability in chlorosis symptoms between leaves at different growth stages, different levels of infections, and similarity with other conditions (nutrient deficiencies, chemical damage, etc.). In general, using the hand-held sensors or under a controlled environment, the accuracies in disease detection were much higher than with aerial applications. Some of the benefits of ground-based sensing platforms are ease in automating, non-limiting sensor payload, and better control of data acquisition and quality. One of the major limitations is a smaller field-of-view. In this regard, aerial sensing platforms offer rapid data acquisition with better coverage. However, in regard to huanglongbing, given the complexity of symptoms, the identification of a specific type of disease can be challenging.

The suitability of a sensing technique depends on the application for disease detection. For instance, mid-infrared spectroscopy is an optical sensing technique that can be used to acquire the chemical profile of a sample; therefore, the technique can be used for identifying disease in pre-symptomatic stages. However, one of the limitations is the need for sample preparation. A practical application of this technique can be for early-automated detection of huanglongbing during mechanical harvesting. Similarly, multispectral imaging (with specific waveband filters) can be useful for remote sensing applications. Even if a specific type of citrus disease cannot be recognized, the technique can greatly reduce the scouting needs and improve management efficiency by helping to focus on orchard regions where diseased trees can potentially be found. Table 3.3 summarizes some of the benefits and limitations of various sensing techniques. It should be noted that data/image processing and machine learning algorithms are also critical aspects of accurate information about disease detection using sensors.

Table 3.2. Research publications on huanglongbing disease detection in citrus using sensor technologies (not inclusive).

Sensing techniques	References
Visible–near-infrared spectroscopy	Mishra *et al.*, 2011; Sankaran *et al.*, 2011
Mid-infrared spectroscopy	Hawkins *et al.*, 2010; Sankaran *et al.*, 2010b; Cardinali *et al.*, 2012
Fluorescence spectroscopy and imaging	Sankaran and Ehsani, 2012, 2013; Wetterich *et al.*, 2013, 2015
Multispectral and hyperspectral imaging	Kumar *et al.*, 2012; Garcia-Ruiz *et al.*, 2013; Pourreza *et al.*, 2014, 2015a, b
Thermal imaging	Sankaran *et al.*, 2013

Table 3.3. Benefits and limitations of sensor techniques for disease detection in citrus.

Technique	Benefits	Limitations
Visible–near-infrared spectroscopy	Non-invasive, rapid	Sensitive to sunlight changes
Mid-infrared spectroscopy	Chemical signature allows detection in early/pre-symptomatic stages	Requires sample preparation
Fluorescence spectroscopy	Non-invasive, good for specific and overall plant stress	Cannot be used remotely (active sensor), expensive
Hyperspectral/ thermal imaging	Can locate specific areas with stress	Less practical for large orchard, thermal is sensitive to environmental conditions

3.4 Sensing for High-throughput Phenotyping

3.4.1 Case study I: Crop architecture evaluation

Estimation of architectural crop traits/phenotypes such as tree height, tree crown diameter, leaf area and curvature, and leaf orientation are essential to estimate the performance of the crops. Often these architecture traits are associated with light use efficiency and yield potential in a crop. Several studies have investigated 3D crop architecture towards phenotyping in controlled environment and field conditions using a range of sensors in horticultural crops (Paulus *et al.*, 2013; Rose *et al.*, 2015; Schöler and Steinhage, 2015; Nguyen *et al.*, 2016a, b, c).

Chéné *et al.* (2012) used a low-cost depth camera (Microsoft Kinect RGB-depth camera) and segmented depth images of plants (rosebush, yucca, and apple) through a customized algorithm in a controlled environment. The results demonstrated that the leaf curvature and leaf orientation (3–6 degrees difference between manual and camera measurements) could be fairly estimated with the depth camera and the associated algorithm. In addition, leaves with apple scab can be detected using a combination of thermal camera and depth camera that creates a binary mask for each leaf on the diseased apple branch. Such phenotyping studies can also be developed for estimating tree architecture in field conditions.

Díaz-Varela *et al.* (2015) reconstructed the olive tree canopy to estimate tree height and crown diameter with images acquired using a modified RGB camera and UAV. The orthomosaic and digital surface model was applied to reconstruct the image, while using object-based classification to extract tree features. The sensing and associated algorithm were evaluated with 29 genotypes (about 150 trees in total) from one field location, while 30 trees (three varieties) were in another field location. The difference between measured and estimated (image-based) mean tree height and mean crown diameter was within 0.06–0.12 m and 0.15–0.20 m, respectively. A high correlation coefficient was achieved

(R^2 = 0.58; P < 0.0001) between estimated and measured crown diameter in one of the field locations with a larger number of genotypes. The study also found that the accuracies of remote sensing-based tree parameter estimation were higher at hedgerow level (continuous canopy), than those at individual tree level (discontinuous canopy). Other techniques have been described (Moorthy *et al.*, 2011; Moriondo *et al.*, 2016). The sensing techniques for reconstructing 3D tree architecture can also assist in fruit assessment (size, shape, distribution, etc.). In addition, developing sensing techniques for fruit-plant architecture evaluation can also assist in tree management and fruit production towards precision agriculture (Costes *et al.*, 2006).

3.4.2 Case study II: Disease rating

Herzog *et al.* (2015) studied the application of an I-sensor developed to measure the impedance of grape berry cuticle, which allows the assessment of grapevines resistance to *Botrytis cinerea*. This fast phenotyping method could be used to identify quantitative trait locus (QTL) regions towards developing disease-resistant varieties in fruit breeding programs. QTL analysis in breeding programs refers to a statistical method that associates phenotypic data (trait measurements) with genotypic data to identify/explain the genetic basis of a complex variation (Miles and Wayne, 2008). The portable sensor was employed to determine the thickness and permeability by measuring the impedance Z of both the cuticle (C) and cuticle with epicuticular waxes (CW) at a lower frequency of 2 kHz and upper frequency of 30 kHz. These measurements were used to acquire the relative impedance Z_{rel}, which was significantly correlated with *B. cinerea* resistance. The Pearson correlation coefficients between Z_{rel} with that of CW and C were –0.67 and –0.60, respectively (P < 0.0001) (Herzog *et al.*, 2015), when 40 grapevine genotypes were compared for their susceptibility to *B. cinerea*. When the grapevine bunch was grouped based on compactness, a maximum correlation of –0.82 (P < 0.0001) was found between Z_{rel} and C. In addition, regression models were evaluated to relate the I-sensor measurements and bunch compactness in relation to the susceptibility of grape berry against the fungal disease. Although the impedance of the cuticle that was a novel phenotypic trait was found to be significantly useful for QTL analysis within F_1 progeny (crossing of GF.GA-47-42 × ‘Villard Blanc’), it was reported that the application could be enhanced by finding new genetic markers. Thus, considering further environmental factors (such as weather conditions) and increasing the number of genotypes is needed to validate these findings.

Conrad and Bonello (2016) reviewed simple, portable sensors for rapid assessment of trees for disease resistance. The authors investigated the combination of infrared (IR) and Raman spectroscopy to determine and correlate chemical fingerprinting and chemometrics analysis for identifying disease-resistant forest trees. For instance, vibrational spectroscopy techniques using Fourier-transform infrared (FTIR) spectroscopy, near-infrared (NIR) spectroscopy, and Raman spectroscopy have the potential to detect chemical changes in trees by measuring either the level of IR attenuation or the amount of energy exchanged after an external excitation of molecules. After that, chemical fingerprint data were analyzed with chemometric approaches to detect the metabolic variation of infected and non-infected trees. It was reported that the efficacy of such rapid phenotyping techniques needed to be validated with a larger sample diversity and assessing their predictive performance in evaluating tree resistance.

3.4.3 Case study III: Water stress response in apples

Studies on phenotyping apple trees for water stress response have been evaluated in France (Virlet *et al.*, 2012, 2014; Lopez *et al.*, 2015; Gómez-Candón *et al.*, 2016). Virlet *et al.* (2014) assessed 122 apple hybrids to phenotype for water stress using Moran's water deficit index (WDI) as a high-throughput phenotyping approach. The 122 hybrids were derived from a 'Starkrimson' × 'Granny Smith' parent cross and grafted on an M.9 rootstock. The two treatments comprised water stress and well-watered, and remote sensing data was acquired with two digital cameras and a thermal infrared camera mounted on an UAV (data collection in 2011). As ground reference data, stem water potential data was collected with a pressure chamber using standard procedures. In addition to WDI, NDVI and T_s–T_a (surface temperature–air temperature) data were computed. A strong relationship between the stem water potential and the temperature features were achieved (R^2 = 0.79–0.80). The potential of the technique was demonstrated in this work.

In a continuing study, Gómez-Candón *et al.* (2016) tested similar sensors in addition to a modified digital camera with a NIR band to test the water stress treatments in 520 trees (122 apple genotypes with 30 genotypes/subset, data collected in 2013). In addition to multiple days, the data was collected multiple times in a single day to record the diurnal variation in canopy temperature (08:00 h to 16:00 h at 2 h intervals). The results indicated that, irrespective of the time of data collection, the difference in canopy temperature (T_s–T_a) could be found. Interestingly, genotype effect could be observed at 12:00 h and 14:00 h, but the effect was absent at the beginning of the day. In future, more studies of a similar type can be anticipated.

3.5 Summary and Future Directions

The applications of sensors for stress detection in precision agriculture have advanced greatly in recent years, mainly due to the advancement in sensor technologies and lower sensor costs. Although remote sensing applications have been investigated for several decades, the application for decision making in fruit orchards is still limited. One of the major reasons could be the lack of research studies indicating the practical use of the sensor technology over the long term. Literature is available providing proof-of-concept studies, but continuous research over a longer period is needed to establish sensing as a component of orchard management. Moreover, economic analysis of utilizing sensor technology in precision orchard management will be greatly beneficial to establish the sensor technology for stress detection. It is anticipated that, with the launch of nanosatellites, such applications in stress detection will continue to improve. Another potential area that needs to improve along with the technology is data analytics. Similarly, sensing for high-throughput phenotyping is very limited. Preliminary research is starting to appear in the literature, but it will be a long way before these technologies can be adopted in conventional breeding programs.

References

Baluja, J., Diago, M.P., Balda, P., Zorer, R., Meggio, F., Morales, F. and Tardaguila, J. (2012) Assessment of vineyard water status variability by thermal and multispectral imagery using an unmanned aerial vehicle (UAV). *Irrigation Science* 30(6), 511–522.

Belin, É., Rousseau, D., Boureau, T. and Caffier, V. (2013) Thermography versus chlorophyll fluorescence imaging for detection and quantification of apple scab. *Computers and Electronics in Agriculture* 90, 159–163.

Bellvert, J., Zarco-Tejada, P.J., Girona, J. and Fereres, E. (2014) Mapping crop water stress index in a 'Pinot-noir' vineyard: comparing ground measurements with thermal remote sensing imagery from an unmanned aerial vehicle. *Precision Agriculture* 15(4), 361–376.

Cardinali, M.C. do B., Boas, P.R.V., Milori, D.M.B.P., Ferreira, E.J., e Silva, M.F., Machado, M.A. and Bellete, B.S. (2012) Infrared spectroscopy: a potential tool in huanglongbing and citrus variegated chlorosis diagnosis. *Talanta* 91, 1–6.

Chaerle, L., Leinonen, I., Jones, H.G. and Van Der Straeten, D. (2007) Monitoring and screening plant populations with combined thermal and chlorophyll fluorescence imaging. *Journal of Experimental Botany* 58(4), 773–784.

Chéné, Y., Rousseau, D., Lucidarme, P., Bertheloot, J., Caffier, V. *et al.* (2012) On the use of depth camera for 3D phenotyping of entire plants. *Computers and Electronics in Agriculture* 82, 122–127.

Conrad, A.O. and Bonello, P. (2016) Application of infrared and Raman spectroscopy for the identification of disease resistant trees. *Frontiers in Plant Science* 6, 1152.

Costes, E., Lauri, P.E. and Regnard, J.L. (2006) Analyzing fruit tree architecture: implications for tree management and fruit production. *Horticultural Reviews* 32, 1–61.

Díaz-Varela, R.A., de la Rosa, R., León, L. and Zarco-Tejada, P.J. (2015) High-resolution airborne UAV imagery to assess Olive tree crown parameters using 3D photo reconstruction: application in breeding trials. *Remote Sensing* 7(4), 4213–4232.

Dobrowski, S.Z., Pushnik, J.C., Zarco-Tejada, P.J. and Ustin, S.L. (2005) Simple reflectance indices track heat and water stress-induced changes in steady-state chlorophyll fluorescence at the canopy scale. *Remote Sensing of Environment* 97(3), 403–414.

Ecarnot, M., Bączyk, P., Tessarotto, L. and Chervin, C. (2013) Rapid phenotyping of the tomato fruit model, Micro-Tom, with a portable VIS–NIR spectrometer. *Plant Physiology and Biochemistry* 70, 159–163.

Estrada, F., Escobar, A., Romero-Bravo, S., González-Talice, J., Poblete-Echeverría, C., Caligari, P.D. and Lobos, G.A. (2015) Fluorescence phenotyping in blueberry breeding for genotype selection under drought conditions, with or without heat stress. *Scientia Horticulturae* 181, 147–161.

Garcia-Ruiz, F., Sankaran, S., Maja, J.M., Lee, W.S., Rasmussen, J. and Ehsani, R. (2013) Comparison of two aerial imaging platforms for identification of Huanglongbing-infected citrus trees. *Computers and Electronics in Agriculture* 91, 106–115.

Garrido-Novell, C., Pérez-Marin, D., Amigo, J.M., Fernández-Novales, J., Guerrero, J.E. and Garrido-Varo, A. (2012) Grading and color evolution of apples using RGB and hyperspectral imaging vision cameras. *Journal of Food Engineering* 113(2), 281–288.

Gómez-Candón, D., Virlet, N., Labbé, S., Jolivot, A. and Regnard, J.L. (2016) Field phenotyping of water stress at tree scale by UAV-sensed imagery: new insights for thermal acquisition and calibration. *Precision Agriculture* 17(6), 786–800.

Gorbe, E. and Calatayud, A. (2012) Applications of chlorophyll fluorescence imaging technique in horticultural research: a review. *Scientia Horticulturae* 138, 24–35.

Gupta, S.D. and Ibaraki, Y. (eds) (2014) *Plant Image Analysis: Fundamentals and Applications.* CRC Press, Boca Raton, Florida.

Harris Geospatial Solutions (2016) Harris Geospatial Vegetation Indices Background. Available at: http://www.harrisgeospatial.com/docs/BackgroundVegetationIndices.html (accessed 30 September 2016).

Hawkins, S.A., Park, B., Poole, G.H., Gottwald, T.R., Windham, W.R., Albano, J. and Lawrence, K.C. (2010) Comparison of FTIR spectra between huanglongbing (citrus greening) and other citrus maladies. *Journal of Agricultural and Food Chemistry* 58(10), 6007–6010.

Herzog, K., Wind, R. and Töpfer, R. (2015) Impedance of the grape berry cuticle as a novel phenotypic trait to estimate resistance to *Botrytis cinerea. Sensors* 15(6), 12498–12512.

Jarolmasjed, S., Espinoza, C.Z., Sankaran, S. and Khot, L.R. (2016) Postharvest bitter pit detection and progression evaluation in 'Honeycrisp' apples using computed tomography images. *Postharvest Biology and Technology* 118, 35–42.

Kleynen, O., Leemans, V. and Destain, M.F. (2003) Selection of the most efficient wavelength bands for 'Jonagold' apple sorting. *Postharvest Biology and Technology* 30(3), 221–232.

Kleynen, O., Leemans, V. and Destain, M.F. (2005) Development of a multi-spectral vision system for the detection of defects on apples. *Journal of Food Engineering* 69(1), 41–49.

Kumar, A., Lee, W.S., Ehsani, R.J., Albrigo, L.G., Yang, C. and Mangan, R.L. (2012) Citrus greening disease detection using aerial hyperspectral and multispectral imaging techniques. *Journal of Applied Remote Sensing* 6(1), 063542–1.

Li, L. (2014) *Time-of-Flight Camera – an Introduction.* Technical White Paper SLOA 190. Texas Instruments, Dallas, Texas.

Lopez, G., Pallas, B., Martinez, S., Lauri, P.E., Regnard, J.L., Durel, C.E. and Costes, E. (2015) High-throughput phenotyping of an apple core-collection for the identification of genotypes with low sensitivity to drought. In: *VIII International Symposium on Irrigation of Horticultural Crops*, 8–11 June 2015, Lleida, Spain.

Manickavasagan, A., Al-Mezeini, N.K. and Al-Shekaili, H.N. (2014) RGB color imaging technique for grading of dates. *Scientia Horticulturae* 175, 87–94.

Miles, C. and Wayne, M. (2008) Quantitative trait locus (QTL) analysis. *Nature Education* 1(1), 208.

Mishra, A., Karimi, D., Ehsani, R. and Albrigo, L.G. (2011) Evaluation of an active optical sensor for detection of Huanglongbing (HLB) disease. *Biosystems Engineering* 110(3), 302–309.

Möller, M., Alchanatis, V., Cohen, Y., Meron, M., Tsipris, J. *et al.* (2007) Use of thermal and visible imagery for estimating crop water status of irrigated grapevine. *Journal of Experimental Botany* 58(4), 827–838.

Moorthy, I., Miller, J.R., Berni, J.A.J., Zarco-Tejada, P., Hu, B. and Chen, J. (2011) Field characterization of olive (*Olea europaea* L.) tree crown architecture using terrestrial laser scanning data. *Agricultural and Forest Meteorology* 151(2), 204–214.

Moriondo, M., Leolini, L., Staglianò, N., Argenti, G., Trombi, G. *et al.* (2016) Use of digital images to disclose canopy architecture in olive tree. *Scientia Horticulturae* 209, 1–13.

Nguyen, T.T., Slaughter, D.C., Maloof, J.N. and Sinha, N. (2016a) Plant phenotyping using multi-view stereo vision with structured lights. In: *Proceedings SPIE Commercial+ Scientific Sensing and Imaging.* SPIE Conference (9866) on Autonomous Air and Ground Sensing Systems for Agricultural Optimization and Phenotyping, Baltimore, Maryland, 17 May 2016. International Society for Optics and Photonics. doi: 10.1117/12.2229515.

Nguyen, T.T., Slaughter, D.C., Townsley, B.T., Carriedo, L., Maloof, J.N. and Sinha, N. (2016b) In-field plant phenotyping using multi-view reconstruction: an investigation in eggplant. In: *Proceedings of the 13th International Conference on Precision Agriculture*, July 31–August 3, 2016, St Louis, Missouri (unpaginated, online). International Society of Precision Agriculture, Monticello, Illinois.

Nguyen, T.T., Slaughter, D.C., Townsley, B., Carriedo, L., Julin, N.N. and Sinha, N. (2016c) Comparison of structure-from-motion and stereo vision techniques for full in-field 3D reconstruction and phenotyping of plants: an investigation in sunflower. In: *2016 ASABE Annual International Meeting*, Orlando, Florida, 17–20 July. American Society of Agricultural and Biological Engineers, St Joseph, Michigan, p. 1.

Nicolaï, B.M., Lötze, E., Peirs, A., Scheerlinck, N. and Theron, K.I. (2006) Non-destructive measurement of bitter pit in apple fruit using NIR hyperspectral imaging. *Postharvest Biology and Technology* 40(1), 1–6.

Nijland, W., De Jong, R., De Jong, S.M., Wulder, M.A., Bater, C.W. and Coops, N.C. (2014) Monitoring plant condition and phenology using infrared sensitive consumer grade digital cameras. *Agricultural and Forest Meteorology* 184, 98–106.

Paulus, S., Dupuis, J., Mahlein, A.K. and Kuhlmann, H. (2013) Surface feature based classification of plant organs from 3D laserscanned point clouds for plant phenotyping. *BMC Bioinformatics* 14(1), 1.

Pourreza, A., Lee, W.S., Raveh, E., Ehsani, R. and Etxeberria, E. (2014) Citrus huanglongbing detection using narrow-band imaging and polarized illumination. *Transactions of the ASABE* 57(1), 259–272.

Pourreza, A., Lee, W.S., Etxeberria, E. and Banerjee, A. (2015a) An evaluation of a vision-based sensor performance in huanglongbing disease identification. *Biosystems Engineering* 130, 13–22.

Pourreza, A., Lee, W.S., Ehsani, R., Schueller, J.K. and Raveh, E. (2015b) An optimum method for real-time in-field detection of huanglongbing disease using a vision sensor. *Computers and Electronics in Agriculture* 110, 221–232.

Rodríguez-Pérez, J.R., Riaño, D., Carlisle, E., Ustin, S. and Smart, D.R. (2007) Evaluation of hyperspectral reflectance indexes to detect grapevine water status in vineyards. *American Journal of Enology and Viticulture* 58(3), 302–317.

Rose, J.C., Paulus, S. and Kuhlmann, H. (2015) Accuracy analysis of a multi-view stereo approach for phenotyping of tomato plants at the organ level. *Sensors* 15(5), 9651–9665.

Sankaran, S. and Ehsani, R. (2012) Detection of huanglongbing disease in citrus using fluorescence spectroscopy. *Transactions of the ASABE* 55(1), 313–320.

Sankaran, S., and Ehsani, R. (2013) Detection of huanglongbing-infected citrus leaves using statistical models with a fluorescence sensor. *Applied Spectroscopy* 67(4), 463–469.

Sankaran, S., Mishra, A., Ehsani, R. and Davis, C. (2010a) A review of advanced techniques for detecting plant diseases. *Computers and Electronics in Agriculture* 72(1), 1–13.

Sankaran, S., Ehsani, R. and Etxeberria, E. (2010b) Mid-infrared spectroscopy for detection of huanglongbing (greening) in citrus leaves. *Talanta* 83(2), 574–581.

Sankaran, S., Mishra, A., Maja, J.M. and Ehsani, R. (2011) Visible-near infrared spectroscopy for detection of huanglongbing in citrus orchards. *Computers and Electronics in Agriculture* 77(2), 127–134.

Sankaran, S., Maja, J.M., Buchanon, S. and Ehsani, R. (2013) Huanglongbing (citrus greening) detection using visible, near infrared and thermal imaging techniques. *Sensors* 13(2), 2117–2130.

Sankaran, S., Khot, L.R., Espinoza, C.Z., Jarolmasjed, S., Sathuvalli, V.R. *et al.* (2015) Low-altitude, high-resolution aerial imaging systems for row and field crop phenotyping: a review. *European Journal of Agronomy* 70, 112–123.

Schöler, F. and Steinhage, V. (2015) Automated 3D reconstruction of grape cluster architecture from sensor data for efficient phenotyping. *Computers and Electronics in Agriculture* 114, 163–177.

Serrano, L., González-Flor, C. and Gorchs, G. (2010) Assessing vineyard water status using the reflectance based water index. *Agriculture, Ecosystems and Environment* 139(4), 490–499.

Si, Y. and Sankaran, S. (2016) Computed tomography imaging-based bitter pit evaluation in apples. *Biosystems Engineering* 151, 9–16.

Tartachnyk, I., Kuckenberg, J., Yuri, J.A. and Noga, G. (2012) Identifying fruit characteristics for non-invasive detection of sunburn in apple. *Scientia Horticulturae* 134, 108–113.

Virlet, N., Lebourgeois, V., Martinez, S., Labbe, S., Costes, E. and Regnard, J.L. (2012) Phenotyping the response of an apple tree hybrid population to soil water constraint under field conditions: new insights brought by high resolution imaging. In: *II International Symposium on Horticulture in Europe*, July 1–5, Angers, France, pp. 879–886.

Virlet, N., Lebourgeois, V., Martinez, S., Costes, E., Labbé, S. and Regnard, J.L. (2014) Stress indicators based on airborne thermal imagery for field phenotyping a heterogeneous tree population for response to water constraints. *Journal of Experimental Botany* 65(18), 5429–5442.

Wetterich, C.B., Kumar, R., Sankaran, S., Belasque Junior, J., Ehsani, R. and Marcassa, L.G. (2013) A comparative study on application of computer vision and fluorescence imaging spectroscopy for detection of huanglongbing citrus disease in the USA and Brazil. *Journal of Spectroscopy* 2013, 841738.

Wetterich, C.B., Neves, R.F., Junior, J.B. and Marcassa, L.G. (2015) Detection of citrus canker and huanglongbing using fluorescence imaging spectroscopy (FIS) technique and two different computational methods: support vector machine and neural network. In: *Frontiers in Optics*. Annual Meeting of the OSA. Optical Society of America, Washington, DC, p. FW5E–4.

Wu, L., He, J., Liu, G., Wang, S. and He, X. (2016) Detection of common defects on jujube using Vis-NIR and NIR hyperspectral imaging. *Postharvest Biology and Technology* 112, 134–142.

Zúñiga, C.E., Khot, L.R., Jacoby, P. and Sankaran, S. (2016) Remote sensing based water-use efficiency evaluation in sub-surface irrigated wine grape vines. In: *Proceedings SPIE Commercial+ Scientific Sensing and Imaging*. SPIE Conference (9866) on Autonomous Air and Ground Sensing Systems for Agricultural Optimization and Phenotyping, 17–21 April, Baltimore, Maryland. International Society for Optics and Photonics. doi: 10.1117/12.2228791.

4 Light Interception and Canopy Sensing for Tree Fruit Canopy Management

Francisco Rojo[1]*, Jingjin Zhang[2], Shrinivasa Upadhyaya[3] and Qin Zhang[4]

[1]*Escuela de Agronomía, Pontificia Universidad Católica de Valparaíso, Casilla 4-D, Quillota, Chile;* [2]*Shanghai Jiao Tong University, Shanghai, China;* [3]*University of California Davis, Davis, California, USA;* [4]*Washington State University, Prosser, Washington, USA*

4.1 Introduction

Plants need light to perform photosynthesis, a process by which they transform solar energy into chemical energy that is stored in different organs as a carbohydrate. This important substance can then be used by the plants to maintain all the physiological processes that are required for its survival or its reproduction. It is for this reason that solar radiation is a key factor to consider when we want to evaluate plant productivity (i.e. plant biomass or yield, depending on the particular goal that people have in mind). The effect that light has in a crop's productivity has been long exploited in agriculture; trellis systems have been analyzed in terms of their capacity to intercept light, or yield predictions have been used to target the management of crops to achieve potential yield at each location within a field or orchard. However, it was not until the past couple of decades, with the advance of the sensing and data acquisition systems, that many of these ideas were used to develop practical tools that could be used by growers to improve precision in their agricultural practices. This chapter provides a review of the principles, state-of-the-art technology and applications of light interception sensing systems and modeling for use in production agriculture. Case studies will be presented where proximally sensed photosynthetically active radiation (PAR) interception data measured using a mobile platform were used to assist in tree fruit canopy management and estimation of potential yield.

* Corresponding author, email: francisco.rojo@pucv.cl

During photosynthesis, the carbohydrate and oxygen molecules are produced from carbon dioxide and water in a redox chemical reaction. The energy required for this process is provided by the sun (Taiz and Zeiger, 2002). Photosynthesis takes place in a specialized part of a cell called the chloroplast that is located primarily in the cells of the leaves. In the inner membrane of the chloroplasts there are pigments called chlorophylls, which are responsible for absorbing light. Chlorophylls are sensitive predominantly in the blue and red parts of the spectrum (Niklas and Spatz, 2012). In practice, it is usually assumed that plants absorb solar radiation in the range of 400–700 nm, commonly known as the PAR range (Monteith, 1972; Jones, 1992).

Measurements of PAR interception have been widely used to estimate biomass and yield. Many researchers have found a linear relationship between biomass production and the amount of PAR intercepted by plants (Monteith, 1965, 1972; Cannell *et al.*, 1988; Dewar *et al.*, 1998; Loomis and Amthor, 1999; Rosati and Dejong, 2003). The slope of this line is often called radiation use efficiency (RUE) and has been found to be a function of nutrient and plant water status, as well as environmental factors such as temperature (Medlyn, 1998). Daily average values of PAR interception (Robinson and Lakso, 1991), as well as midday PAR interception data (Lampinen *et al.*, 2009), have proven to be useful predictors of actual yield. Theoretical potential yield lines, based on midday PAR interception data, were provided by Lampinen *et al.* (2009) for almond and walnut crops. Their results are shown in Fig. 4.1, where each data point corresponds to data taken in a different orchard. Accurate information about tree canopy architecture and PAR absorption is useful to estimate potential yield. This information can also be very valuable to implement canopy management, and enhance quality and quantity of yield (Lampinen *et al.*, 2006).

4.2 Principles and Technologies

Solar radiation has a direct influence on the production of biomass; therefore, information on how light gets distributed on the earth's surface can help us to understand how much production can be achieved in a given region. Biomass production will also depend on many other factors as well, such as air temperature, relative humidity, water availability, plant disease or cultural practices. However, solar radiation alone has been used in theoretically estimating potential yield that can be achieved when there are no other limitations present. This potential yield can be used as a guide by growers on their management practices, but more elaborate models that include those limitations are needed if the goal is to obtain actual yield. These mathematical models can describe the distribution of solar radiation on the earth's surface, but local conditions (e.g. soil, slope, orientation) can generate the occurrence of spatial variability leading to canopies with different dimensions, densities and light absorption capacities. Canopy sensing systems can help to guide precision

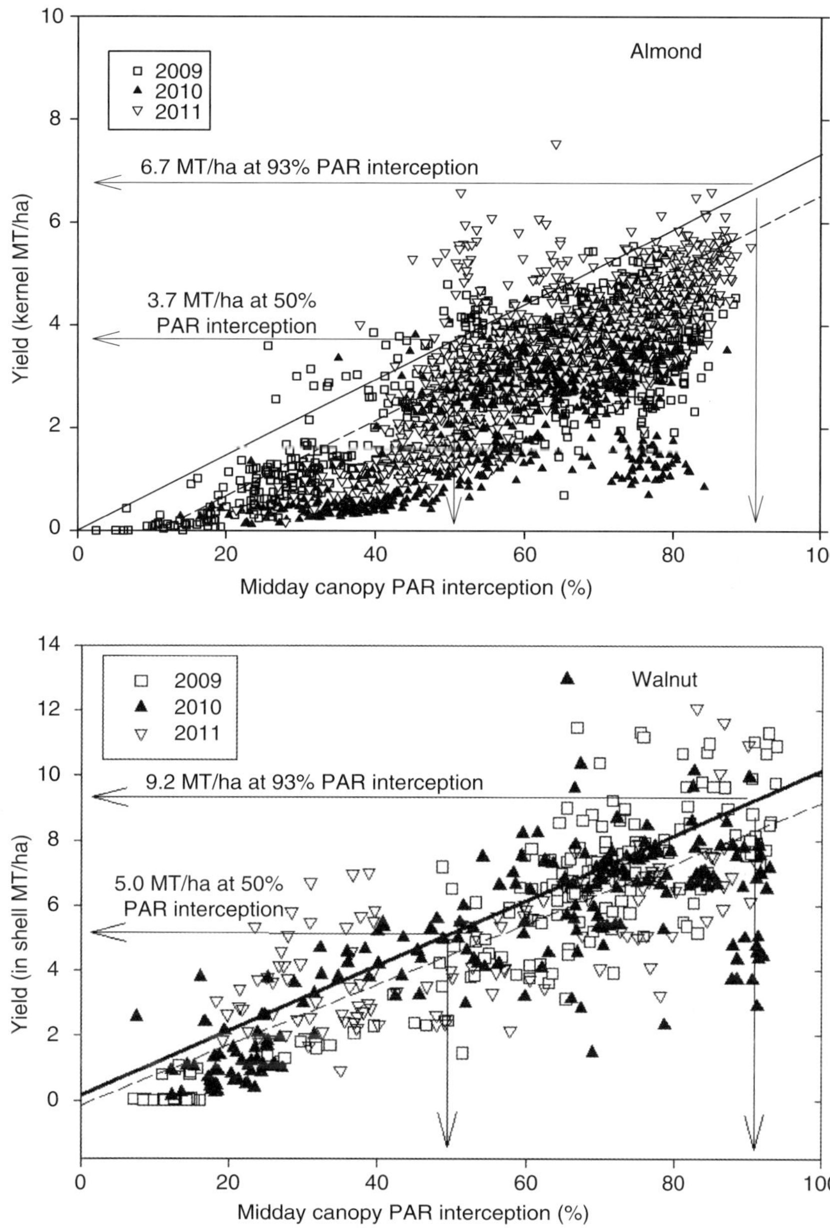

Fig. 4.1. Plot of actual yield versus the midday PAR interception (from Lampinen *et al.*, 2014).

agriculture practices to address this variability and increase production while reducing environmental impact. The following sections will cover modeling and sensing techniques for the estimation of both solar radiation and PAR interception.

4.2.1 Solar radiation and tree productivity

Several models with different levels of complexity have been proposed to estimate incident solar radiation on the surface of the earth. The solar radiation model described below, developed by Hottel (1976), is well suited for practical engineering applications and assumed that the beam component of the solar radiation under optimal sky conditions (G_{bn}) could be described as the product of atmospheric transmittance and extraterrestrial radiation. The model takes into account the zenith angle, Julian day, altitude of the observer and four climate types (Duffie and Beckman, 2013). This model can be represented as:

$$G_{bn} = \tau_b G_{on} = \left[a_0 + a_1 exp\left(-\frac{a_2}{cos\theta}\right)\right]\left[G_{sc}\left(1+0.033cos\left(\frac{360n}{365}\right)\right)\right] \quad (4.1)$$

In equation 4.1, the first term in the square brackets is the atmospheric transmittance for direct radiation (τ_b) and the second term in the square brackets is the extraterrestrial radiation (i.e. solar radiation outside the earth's atmosphere) measured on a plane perpendicular to the solar radiation (G_{on}). G_{bn} is the beam irradiance normal to the sun at the earth's surface, G_{sc} is the solar constant (i.e. 1367 W/m^2), n is the Julian day, θ is the zenith angle, and a_0, a_1 and a_2 are empirical coefficients.

The beam component normal to the earth's surface in a particular location can be computed from the beam normal to the sun (i.e. $G_b = G_{bn}\cos\theta = \tau_b G_{on}\cos\theta$, and the diffuse component can be obtained by replacing the transmittance coefficient τ_b by τ_d (i.e. $G_d = \tau_d G_{on}\cos\theta$). Liu and Jordan (1960) found the transmittance coefficient for diffuse radiation to be linearly related to the atmospheric transmittance (i.e. $\tau_d = 0.271 - 0.294\tau_b$). Then, the total radiation incident on a horizontal surface can be obtained by the sum of beam (G_b) and diffuse radiation (G_d) components as follows:

$$G = G_b + G_d = (\tau_b + \tau_d)G_{on}cos\theta \quad (4.2)$$

Frequently it is assumed that half of the total solar radiation is in the PAR range (Monteith, 1972). Using the earth–sun geometric relationship, the zenith angle for any given time and geographic location can be derived (Threlkeld, 1970). The cosine of the zenith angle can be shown to be (Duffie and Beckman, 2013):

$$cos\theta = cos\phi\ cos\delta\ cos\omega + sin\phi sin\delta \quad (4.3)$$

where ϕ, δ and ω are latitude, declination angle and hour angle, respectively. The latitude is the angular displacement to the north or south of the equator, which is usually known for a given location. The declination angle is the angular position of the sun at solar noon with respect to the equator and can be calculated by the following empirical formula (Duffie and Beckman, 2013):

$$\delta = 23.45 \sin\left(360 \frac{284+n}{365}\right) \tag{4.4}$$

The hour angle is the angular displacement of the sun from the local meridian due to rotation of the earth. The hour angle can be computed using the following expression:

$$\omega = 15(solar\ time - 12) \tag{4.5}$$

where the *solar time* is the apparent angular motion of the sun across the sky and where solar noon is the time the sun crosses the meridian of the observer (Duffie and Beckman, 2013). Solar time is calculated as follows:

$$Solar\ time = standard\ time + \left[4\left(L_{st} - L_{loc}\right) + EoT\right]/60 \tag{4.6}$$

where the *standard time* is the time used by each time zone, which corresponds to the daylight savings time (i.e. clock's time) ± 1 hour, depending of the day of the year. L_{st} is the standard longitude for the local time zone and L_{loc} is the longitude of the observer, both expressed in minutes. *EoT* is a correction factor that accounts for the perturbations in the rotation of the earth that affect the time the sun crosses the meridian of the observer (Duffie and Beckman, 2013). *EoT* can be computed as follows:

$$EoT = 229.2\begin{pmatrix} 0.000075 + 0.001868 cos\left((n-1)\frac{360}{365}\right) \\ -0.032077 sin\left((n-1)\frac{360}{365}\right) \\ -0.014615 cos\left(2(n-1)\frac{360}{365}\right) \\ -0.04089 sin\left(2(n-1)\frac{360}{365}\right) \end{pmatrix} \tag{4.7}$$

Two angles are required to describe the position of the sun at any given time and geographic location; these are the zenith (equation 4.3) and the azimuth angles. The latter can be computed by the following equation (Duffie and Beckman, 2013):

$$\gamma = sign(\omega)\left|\cos^{-1}\left(\frac{\cos\theta\ \sin\phi - \sin\delta}{\sin\theta\ \cos\phi}\right)\right| \tag{4.8}$$

where the *sign* function is equal to +1 if ω is positive and equal to –1 if it is negative.

4.2.2 Sensing canopy light interception

Solar radiation at the surface of the earth can also be measured directly by sensors. This section will describe the technologies available to measure solar radiation and approaches commonly used to obtain the amount of PAR intercepted by crops.

Total solar radiation (i.e. diffuse plus direct components) can be measured by a pyranometer, which in most cases is based on thermopiles or photovoltaic detectors. The thermopile detector, usually covered by a glass hemisphere, absorbs radiative energy which elevates its temperature, producing a thermoelectric voltage that is a function of the solar irradiance that the sensor receives (Paw U, 2000). In some cases, a thermistor can also be used as the detector but this option is less commonly found (Figliola and Beasley, 2011). Another common type of pyranometer is based on photovoltaic detectors (e.g. silicon photocells). The principle of this detector is based on the interaction of a flux of photons with the electrons located in a semi-conductor material that generate an electric current that can be amplified and measured (Figliola and Beasley, 2011). Pyranometers based on photovoltaic detectors are less expensive, but also less precise than the ones using thermopiles (Duffie and Beckman, 2013). However, they are characterized by a fast response time and are spectrally selective (Figliola and Beasley, 2011). A filter diffuser plate is usually added to the silicon photocell, making it sensitive only to specific wavelengths such as the PAR range (Paw U, 2000). The use of the silicon photocell is limited to clear sky conditions (Paw U, 2000).

PAR intercepted by a canopy can be obtained by computing the difference between two measurements of PAR, one at the top and the other at the bottom of the canopy. In practice, the measurement taken at the top of the canopy can be replaced by full sun measurements taken outside the crop row. PAR intercepted by trees has also been obtained by methods based on hemispherical photography (Robinson and Lakso, 1991; Pearcy *et al.*, 2005). Diurnal behavior of the PAR interception has been studied as well, where a typical midday depression has been reported (Pearcy *et al.*, 2005; Guillen-Climent *et al.*, 2012; Zhang *ct al.*, 2015).

In most cases PAR measurements are performed by hand-held devices, which is very time intensive (Grossman and DeJong, 1998). This is an important limitation when large fields need to be sampled over a small time window (Zhang *et al.*, 2015). A photodiode-based scanner was developed by Giuliani *et al.* (2000) to measure whole-canopy light interception. This system was used to digitize the shadows of a single tree at different times throughout the day; from this, tree-crown information was extracted. A mobile lightbar system was developed by Lampinen *et al.* (2012) to obtain PAR interception data from complete rows of trees. They employed light interception data to estimate crop yield over large plots rather than individual trees or blocks of a few trees (Zarate-Valdez *et al.*, 2012). Zhang *et al.* (2015) developed a similar mobile platform to collect PAR under tree canopies, where data were collected at a constant spatial resolution (i.e. sampling rate was allowed to vary to compensate for speed changes). In their approach the non-midday shadows of sweet cherry rows were digitized and transformed into midday shadows.

A knowledge of the solar radiation intercepted by trees during the whole season is needed for more detailed studies where models provide information not only about solar radiation, but also on how much of that radiation is being intercepted by the crops, are needed. This important topic will be covered in the next section.

4.2.3 Modeling canopy light interception

Light attenuation through a non-scattering media is usually described as a negative exponential function, commonly known as the Beer–Lambert law (equation 4.9), which states that the logarithm of the light absorbed by a medium is directly proportional to the path length (Fuwa and Valle, 1963). This expression is commonly used in spectroscopy to estimate the absorption coefficient of different materials (Sassaroli and Fantini, 2004) and is given by:

$$I_h = I_0 e^{-k'h} \tag{4.9}$$

where I_0 is the light intensity falling on the outer layer of the material, I_h is the light intensity after it is transmitted a distance h through the material and k' is the light extinction coefficient (also referred to as the absorption coefficient).

For crop canopies that are closed and have their leaves randomly distributed, light distribution is usually obtained by an equation similar to the Beer–Lambert law, where the path length h is replaced by the cumulative leaf area index (Monteith, 1965; Boote and Pickering, 1994; Annandale *et al.*, 2004; Duursma and Mäkelä, 2007). The leaf area index (LAI) is defined as the area of leaves situated above one square meter of ground (Thornley and France, 2007). Light distribution within the canopy can be expressed in terms of radiation transmitted or intercepted:

$$\textit{Transmitted radiation, } I = I_0 e^{-kL} \tag{4.10}$$

$$\textit{Intercepted radiation, } Q = I_0\left(1 - e^{-kL}\right) \tag{4.11}$$

where Q, I and I_0 are in terms of total radiation or PAR, and k is the light extinction coefficient of the canopy, which is defined as the area of the shadow cast by a canopy on a horizontal surface divided by the summation of the area of all the individual leaves in that canopy. An expression was derived by Campbell and Norman (1998) for extinction coefficient based on the assumption that leaf orientation distribution can be approximated by the surface of an ellipsoid, which is shown below:

$$k = \frac{\sqrt{x^2 + tan\theta^2}}{x + 1.744(x + 1.182)^{-0.733}} \tag{4.12}$$

where x is the ratio of vertical to horizontal projection of the ellipsoid that represents the canopy elements and θ is the zenith angle.

Some authors have derived expressions which take into account that direct and diffuse radiations have different contributions over sunlit and shaded leaves (Bosc, 2000; Thornley and France, 2007). Solar radiation can also be reflected by soil and leaves, which can affect the energy balance of the canopy. Soil reflectance can be important at low values of LAI but it can be neglected when LAI is large. Reflectance of leaves in the PAR range has been estimated to be 0.056 when leaf absorptivity is 0.8 (Campbell and Norman, 1998).

The one-dimensional description of the light intercepted by a tree that was presented in equation 4.11 is well suited for crop canopies that can be considered as a 'homogeneous block' (Thornley and France, 2007), as is the case for field crops. In such models, the total light interception by a field can be computed by multiplying equation 4.11 by the area of interest. However, additional information is required when canopies are better represented by individual units, which are usually simplified by assuming simple shapes (Bosc, 2000). Examples of these types of plants are fruit trees or woodland trees. Some examples of canopy shapes assumed in developing models are ellipsoids (Mann *et al.*, 1979; Wang and Jarvis, 1990; Annandale *et al.*, 2004; Duursma and Mäkelä, 2007), semi-ellipsoids (Thornley and France, 2007), spheres (West and Welles, 1992), cones (Duursma and Makela, 2007), rectangles (Oyarzun *et al.*, 2007) and rhomboids (Connor and Gómez-del-Campo, 2013). Canopies have also been described as arrays of multiple ellipsoidal units (Norman and Welles, 1983). Most of the models of light interception deal with single trees. However, models that include the effect of neighboring trees have been developed as well (Makela and Hari, 1986; Chen *et al.*, 1994; Duursma and Makela, 2007). Three-dimensional (3D) foliage distribution has also been studied using partial differential equations (Beyer *et al.*, 2014).

Another approach that can been used is based on the idea that PAR interception information can be extracted from the shadows of the trees: the interception problem can be transformed from a 3D domain (i.e. canopies) to a two-dimensional (2D) domain (i.e. shadows), where the calculations of path length and shadow overlap can be performed. Finally, the total PAR intercepted by a tree can be calculated by considering the area and intensity of the shadow at any given time, which can be estimated using the light attenuation concept presented earlier in this section, and the dynamic behavior of their shadows. This approach will be discussed in one of the case study sections.

4.3 Applications

In this section, several agricultural applications will be presented for light interception measurement systems. Leaf area index and its correlation with PAR will be discussed, as well as its importance in canopy management and crop production. The relationship between PAR interception and canopy volume will also be discussed, which directly relates to biomass production. Finally, we shall provide examples of how canopy PAR interception can be used to estimate real or potential yield. The following approaches will be discussed: (i) analysis based on daily PAR interception data based on simple regression models; and (ii) models of photosynthesis which use PAR interception data integrated over the entire season. Advantages and disadvantages of each approach will be discussed.

4.3.1 Canopy management

Manual measurement of tree geometric characteristics, such as volume, is a very labor-intensive and time-consuming process (Van der Zande *et al.*, 2006). Non-destructive remote sensing techniques, such as light detection and ranging (lidar), present an interesting alternative to obtain the same type of information in a convenient manner. Lidar technology allows the characterization of objects in space, which is achieved by measuring the time that multiple light pulses take to reach the object and return to the detector (Van der Zande *et al.*, 2006). This information is then summarized in a 3D point cloud, where each point in the cloud represents the position where a laser beam made contact with an object. The geometric characteristics of a canopy are directly related to the tree productivity and can be used to estimate yield or water consumption (Lee and Ehsani, 2009). They can also be used to implement site-specific spray applications (Walklate *et al.*, 2002). In addition, lidar technology can be used to quantify canopy growth and study resource allocation by measuring changes in the canopy structure over time (Rosell and Sanz, 2012).

Even though the ground-based lidar system is a faster and more convenient alternative to acquire canopy information than manual measurements, it still requires a slow ground speed to generate a point cloud density adequate for the proper extraction of the canopy features. Lightbar scans, on the other hand, allow faster data acquisition which is more suitable for practical agricultural applications. Equation 4.10 also suggests that an optical volume of the canopy can be computed by multiplying the path length of the light inside the canopy and the total area projected for that canopy, and that the optical volume should be proportional to the actual volume, where k is the proportionality constant. The recent development of lightbar scan systems has made it possible to obtain the volume of a canopy faster than was possible with conventional lidar systems. The results of comparing optical volume using a lightbar system and real volume measured by the lidar system can be seen in Fig. 4.2.

Tree volume has been found to relate to yield and fruit weight (Pascual *et al.*, 2011), leaf area index (Arno *et al.*, 2013) and leaf area density (Sanz *et al.*, 2013), where leaf area density is half of the total foliage area within a given volume (Beland *et al.*, 2014). Investigators have used this technique to carry out real-time foliar surface estimation for spray applications (Walklate *et al.*, 2002; Palacín *et al.*, 2007; Rosell *et al.*, 2009a, b; Arno *et al.*, 2013; Sanz *et al.*, 2013; Beland *et al.*, 2014). Additionally, lightbar scans and canopy volume estimations can be used to study how pruning affects light interception and fruit production, and to evaluate the effect of different cut distances of mechanized systems.

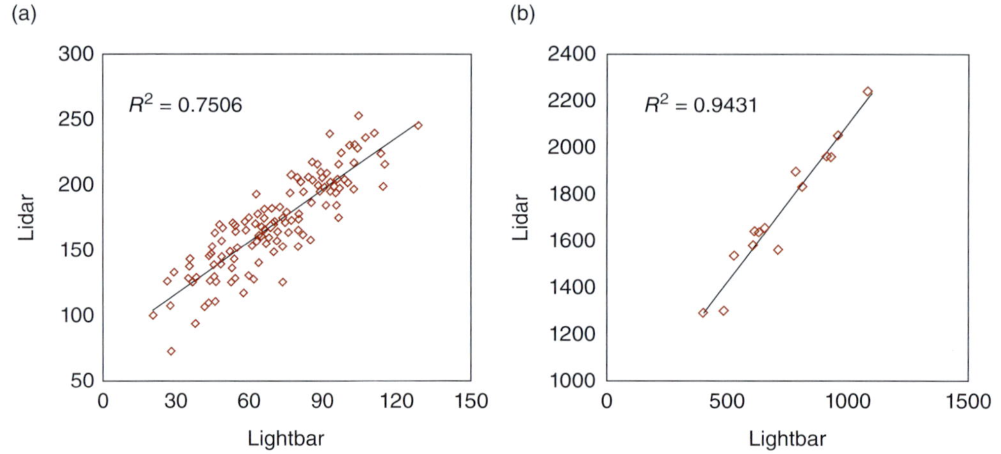

Fig. 4.2. Plot of canopy volume measured by lidar and lightbar for almond trees: (a) blocks of five trees; and (b) whole rows of trees.

4.3.2 Yield estimation

Accurate information about tree canopy architecture and PAR absorption is useful to estimate potential yield. This information can also be very valuable to implement canopy management, and enhance quality and quantity of yield (Lampinen *et al.*, 2006). Potential yield is a theoretical concept that indicates the yield that a genotype can produce under optimal management in the absence of biotic or abiotic stresses (Evans and Fisher, 1999; Acevedo *et al.*, 2002). In reality, the potential yield can rarely be achieved, but it can give an idea of how good the growing conditions were as well as how the crop was managed. Moreover, potential yield can assist in making management decisions (e.g. fertilization), especially if this is known early in the season. On the other hand, actual yield depends on all the limitations that are present in the field which affect the physiological processes of plants, such as water availability, disease or nutrition.

Yield estimation can be computed using simple relations such as the one mentioned in section 4.1 where actual yield was plotted against midday PAR interception. However, the use of these relationships to estimate plant productivity has been questioned, due to the variable nature and the difficulties with measurement (Loomis and Amthor, 1999). More sophisticated models of photosynthesis suitable for a much wider range of conditions often require a knowledge of PAR interception over time (Duursma and Mäkelä, 2007). Thornley and Johnson (1990) derived a simpler expression, where leaf photosynthesis was described by a quadratic function as follows:

$$\vartheta\left(\frac{dP_L}{dt}\right)^2 - (\alpha I + P_{max})\frac{dP_L}{dt} + \alpha I P_{max} = 0 \tag{4.13}$$

In equation 4.13, P_{max} (g/m^2/s) is the light saturated photosynthesis, α is the quantum efficiency (g/J of CO_2), I (W/m^2) is the downward light flux density upon the leaf surface and ϑ is a coefficient related to the curvature of the light response curve (range from 0 to 1). P_{max} and α depend on environmental variables such as temperature, CO_2 concentration, and nitrogen supply (Johnson *et al.*, 2010). The solution for dP_L/dt is known as the non-rectangular hyperbola model and is given by:

$$\frac{dP_L}{dt} = \frac{\alpha I + P_{max} - \sqrt{\left[\left(\alpha I + P_{max}\right)^2 - 4\theta\alpha I P_{max}\right]}}{2\vartheta} \qquad (4.14)$$

This equation depends on three parameters, which can be obtained by fitting the data obtained by a gas exchange system. For the particular case of $\theta = 0$, the quadratic component vanishes (equation 4.15), which results in the rectangular hyperbola model shown below:

$$\frac{dP_L}{dt} = \frac{\alpha I P_{max}}{\alpha I + P_{max}} \qquad (4.15)$$

Non-rectangular hyperbola models have been found to better fit the light-response curve data (Thornley and Johnson, 1990), as they contain an additional parameter to account for higher-order effects. However, this also increases the complexity of the model. Net photosynthesis is the difference between how much carbohydrate is accumulated by the plant through photosynthesis and how much is expended by the plant for growth and maintenance respiration. The amount of net photosynthesis will determine the resources available for growth (i.e. yield, biomass, and storage). Dark respiration rate (dR_d/dt) can be subtracted from the gross photosynthesis models to obtain an expression for net photosynthesis as follows:

$$\frac{\partial}{\partial L}\left(\frac{\partial P_n^L}{\partial t}\right) = \frac{\alpha I P_{max}}{\alpha I + P_{max}} - \frac{dR_d}{dt} \qquad (4.16)$$

In equation 4.16, partial derivatives have been included (left side) to highlight the fact that the expression represents a leaf located in a multi-layer canopy system, where dP_n^L/dt is the rate of net photosynthesis in the leaf (g/m^2). P_{max} has been found to be proportional to the external CO_2 concentration, thus equation 4.16 can also be expressed in terms of CO_2 concentrations (Acock *et al.*, 1978).

Light interception by crop canopy and leaf photosynthesis concepts can also be combined to develop models of canopy photosynthesis (Thornley and Johnson, 1990). Equation 4.16 can be integrated over the entire light penetration path through the canopy to obtain an expression for photosynthesis per unit ground area as a function of the light interception (Charles-Edwards *et al.*, 1986; Thornley and Johnson, 1990; Upadhyaya and Koller, 2005):

$$\frac{dP_n}{dt} = \frac{\alpha P_{m0} Q}{\alpha I_0 k + P_{m0}} - \frac{dR_c}{dt} \tag{4.17}$$

where P_{m0} (g/m^2/s) is the light-saturated photosynthesis of an upper leaf, $Q = I_0(1\text{-}e^{-kLAI})$ (W/m^2) is the light intercepted, dR_c/dt is the rate of canopy dark respiration per unit ground area (g/m^2/s), and dP_n/dt is the rate of canopy net photosynthesis per unit ground area (g/m^2/s).

Only a fraction of the carbon assimilated by photosynthesis is used to provide energy through respiration and the rest is used to create plant biomass. Some crop models have represented plant biomass as the sum of substrate and structural compartments (Dewar *et al.*, 1998; Thornley, 2011). A third compartment has also been used to represent the carbon that is stored (Seginer, 2003). However, the latter is not always considered, due to the lack of quantitative information, especially in trees (Allen *et al.*, 2008). Potential yield can be estimated using a conservation of mass approach where net assimilation evaluated during the whole season can be equated to the sum of the masses of the different structures (e.g. fruit, leaves, root, etc.).

4.4 Case Studies

In this section, three case studies will be presented to describe a continuous lightbar scan system, mapping procedures used for digital shadows in an orchard, and an example of an empirical model of canopy PAR interception.

4.4.1 Case 1: Systems to continuously measure light interception of orchard crops

4.4.1.1 System description

The mobile measuring system introduced in this section is rather a research prototype for assessing canopy light interception (Zhang *et al.*, 2015). This system consists of two main subsystems: a mobile platform and a data collection system. The main function of the mobile platform is to carry and move the data collection system in field for light interception measurement. In this case, an electric 'Gator' (E-Gator, Deere & Company) is selected as the mobile platform and the lightbar unit is attached to it in the front (Fig. 4.3).

The key elements of this mobile measuring system are:

- an integrated ceptometer unit from Decagon Devices (item 1 in Fig. 4.3), model AccuPAR LP-80, referred to as the lightbar unit hereafter;
- a factory-calibrated quantum sensor from LI-COR (item 2 in Fig. 4.3), model LI-190 SZ;

- an encoder from Koyo Electronics Industries (item 5 in Fig. 4.3), model TRD-S; and
- a self-designed signal input–output interface box, consisting of signal conditioning circuits and a computerized data acquisition module from National Instruments (item 3 in Fig. 4.3).

The lightbars, quantum sensor and the encoder provided inputs to the system through a signal conditioning circuit, which also converted sensor current outputs to voltage signals. A computerized data acquisition module was used to acquire the sensed data.

The lightbar unit was constructed using a total of 80 independent photosensors, spaced at 0.01 m and grouped into eight ten-photosensor units to output one cumulative reading from each unit in every reading cycle, which resulted in eight data points per sampling cycle. The quantum sensor was used to measure the photosynthetic photon flux density (PPFD) (μmol/s/m^2) under open sky conditions to provide a shadow-free reference radiation measurement. The encoder was used to measure platform moving speed for regulating the measurement system sampling rate.

4.4.1.2 Spatial resolution control

The spatial resolution of the PAR measurement system was defined as the physical distance between adjacent recorded data in the platform moving

Fig. 4.3. The mobile measurement system: (1) a lightbar; (2) a quantum sensor; (3) a signal input–output interface box, consisting of signal conditioning circuits and an NI DAQ module; (4) the E-Gator; and (5) an encoder, mounted on the rear wheel axle of the E-Gator.

direction (along the tree row, defined as the *R*-axis hereafter) or perpendicular to the moving direction (across the inter-row spacing, defined as the *C*-axis hereafter). In this case, the 0.01 m × 0.10 m physical dimension of each sensing unit resulted in a fixed spatial resolution of 0.01 m/sample in the *R*-axis and 0.10 m/sample in the *C*-axis for current lightbar arrangement as shown in Fig. 4.3. As the *C*-axis is perpendicular to the platform moving direction, the controllable spatial resolution discussed hereafter refers only to the one in the *R*-axis. Restricted by the dimensions of the sensing unit, the minimal achievable spatial resolution in R-axis would be 0.01 m/sample.

During field measurements, due to unpredictable field conditions such as uneven ground, the mobile platform was normally driven at a variable speed with frequent stops, which resulted in a variation in spatial resolution. To measure canopy PAR interception precisely, a sampling rate control scheme was developed to maintain a constant spatial resolution, by dynamically synchronizing the sampling rate of data acquisition system to the platform moving speed. Figure 4.4 is a flowchart of the control scheme.

Based on this design, the *R*-axis spatial resolution (*R*, m/sample) is determined by the sampling rate (*S*, sample/s) and the platform traveling speed (*V*, m/s) using the following equation:

$$R = \frac{V}{S} \tag{4.18}$$

where *V* is the platform moving speed and can be calculated based on the measured frequency reading using the following equation:

$$V = \frac{F \times L}{PPR} \tag{4.19}$$

where *L* (m) is E-Gator wheel perimeter, *PPR* is pulse per revolution, a measurement parameter of the encoder, and *F* (Hz) is the frequency reading output from the encoder. In this case, *PPR* is chosen to be 200, and *L* is 2.0 m, resulting in spatial resolution 0.01 m/sample.

4.4.2 Case 2: Mapping PAR interception

4.4.2.1 Geometrical transformation of the shadow to represent PAR interception

Due to the effect of the sun position on tree canopy projection, the recorded canopy projection position must be corrected to the ideal position as if the sun were directly above the canopy to make the map accurately represent the distribution of PAR interception. Here, geometrical transformations of the projection for two types of canopy architectures are presented: Y-trellis and upright fruiting offshoot (UFO) canopy architecture. Figure 4.5 shows the field views and sketches of these two architectures.

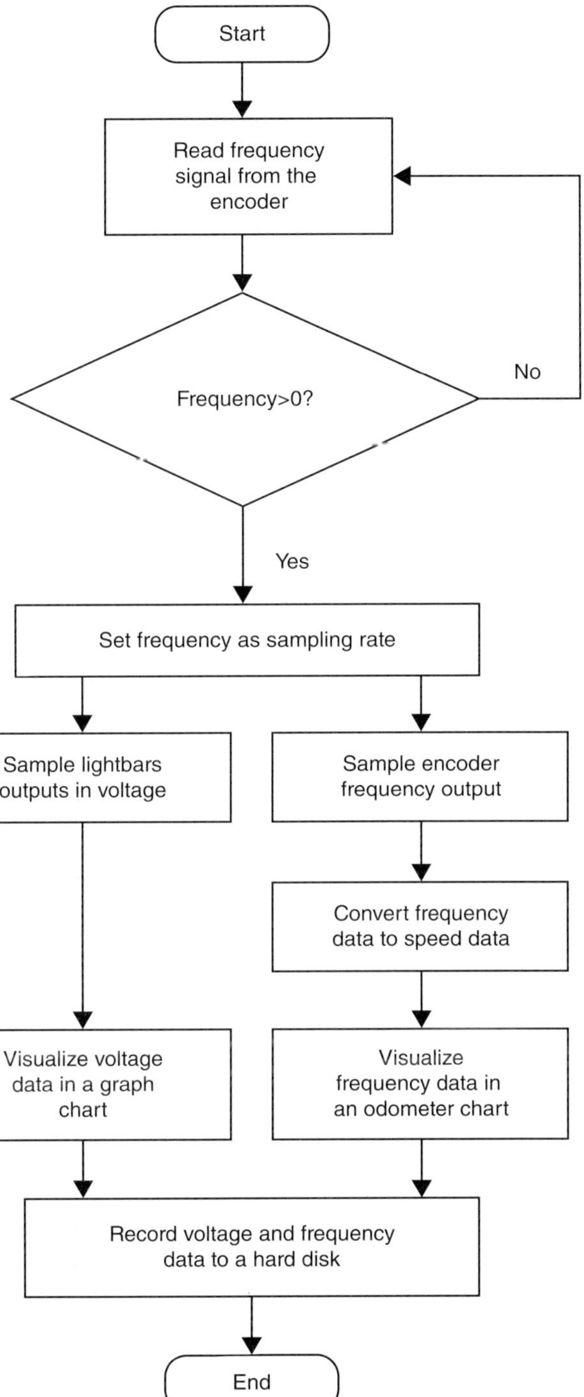

Fig. 4.4. Flowchart of the sampling rate control scheme.

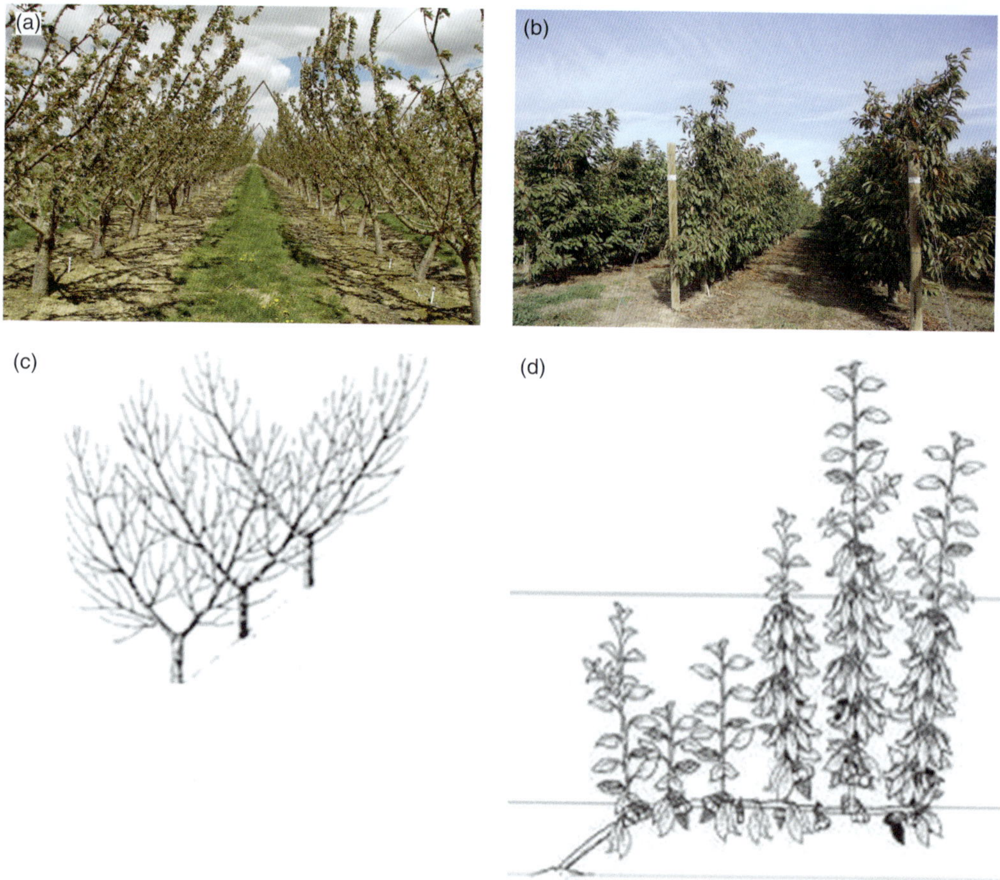

Fig. 4.5. Illustration of two canopy architectures: (a) Y-trellis, field view; (b) UFO, field view; (c) Y-trellis, sketch; (d) UFO, sketch.

4.4.2.2 Y-trellis canopy architecture

Figure 4.6 illustrates the relationships of a simplified 'Y'-shaped canopy architecture with the inter-row direction defined as the *x*-axis (i.e. *C*-axis) and intra-row direction defined as the *y*-axis (i.e. *R*-axis). A 3D coordinate system (*x*-, *y*-, and *z*-axis) is utilized by taking the trunk position on the ground as the origin of the coordinate system. The geometric relationship between measured projection (P_m) in the raw data matrix, its corresponding actual position on the canopy (P_a), and the corrected projection (P_c) is determined by introducing the azimuth angle (α, in clockwise direction from the north), zenith angle (β, in clockwise direction from the vertical), canopy elevation angle (γ), and the height of the branch point (h) into the system. When transformed correctly, the *x* and *y* coordinates of P_c would be the same as the ones of P_a.

The geometric relationship between the coordinates of P_m and P_a is determined using the following coordinates transformation equations:

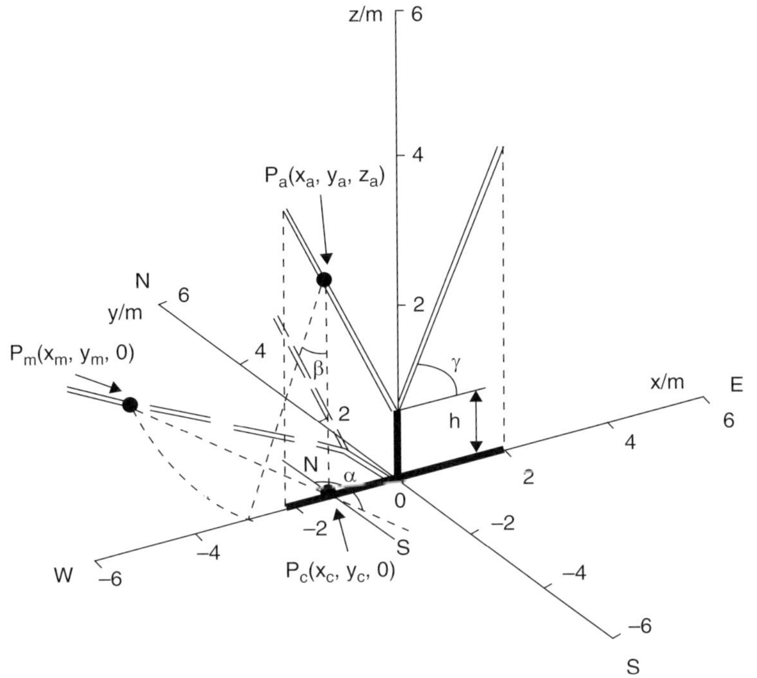

Fig. 4.6. Sketch of the Y-trellis canopy and its projection before and after correction. The double-straight-line structure represents the tree in the field, the double-dash line represents the measured projection of the tree canopy on the ground, and the bold solid line lying on *x*-axis shows the corrected projection as if the sun were right above the tree canopy. P_a, actual position on canopy; P_c, corrected projection; P_m, measured projection; α, azimuth angle; β, zenith angle; γ, canopy elevation angle; *h*, height of branch point (see Section 4.4.2.2). E, east; S, south; W, west; N, north.

$$\begin{cases} x_m = x_a - z_a \times \tan\beta \times \sin(180^\circ - \alpha) \\ y_m = y_a + z_a \times \tan\beta \times \cos(180^\circ - \alpha) \\ z_a = |x_a| \times \tan\gamma + h \end{cases} \tag{4.20}$$

Solving the equation set results in a set of equations for correction of projection coordinates as follows:

$$\begin{cases} x_c = x_a = \begin{cases} \dfrac{x_m + h \times \tan\beta \times \sin(180^\circ - \alpha)}{1 - \tan\gamma \times \tan\beta \times \sin(180^\circ - \alpha)} \text{if } x_m \geq -h \times \tan\beta \times \sin(180^\circ - \alpha) \\ \dfrac{x_m + h \times \tan\beta \times \sin(180^\circ - \alpha)}{1 - \tan\gamma \times \tan\beta \times \sin(180^\circ - \alpha)} \text{if } x_m < -h \times \tan\beta \times \sin(180^\circ - \alpha) \end{cases} \\ y_c = y_a = y_m - (|x_a| \times \tan\gamma + h) \times \tan\beta \times \cos(180^\circ - \alpha) \end{cases} \tag{4.21}$$

In this case, the spatial resolutions of the $m \times n$ raw data matrix are 0.10 and 0.01 m/sample respectively along *C*-axis and *R*-axis. In performing the coordinate correction, two $m \times n$ P_m position matrices, namely a 'measured data x-coordinate matrix' (X_m) and a 'measured data y-coordinate matrix' (Y_m) are defined to indicate the x- and y-coordinates of each measured data. By defining the middle column data position on the first row of the raw data matrix as the origin of the measured data coordinates, all elements in these two position matrices are determined using the following equations:

$$\begin{cases} x_{m(i,j)} = -0.05 \times (n-1) + (j-1) \times 0.10, \ for\, j = 1, 2, \ldots, n \\ y_{m(i,j)} = 0.01 \times (i-1), for\, i = 1,2, \ldots, m \end{cases} \tag{4.22}$$

where $x_{m(i,j)}$ is the element at the ith row and j^{th} column of X_m, and $y_{m(i,j)}$ is the element at the ith row and j^{th} column of Y_m.

The P_c position is defined by two $m \times n$ matrices indicating the corrected x- and y-coordinates of each measured data, referred as 'corrected data x-coordinate matrix' (X_c^*) and 'corrected data y-coordinate matrix' (Y_c^*) using the following equations:

$$\begin{cases} x_{c\ (i,j)}^{*} = j \\ y_{c\ (i,j)}^{*} = \left[y_{c(i,j)} / 0.01 \right], \ for\, i = 1,2, \ldots, m, j = 1,2, \ldots, n \end{cases} \tag{4.23}$$

where m and n are the number of rows and columns respectively of P_c, $x_{c(i,j)}^*$ is the element at the ith row and j^{th} column of X_c^*, $y_{c\ (i,j)}^*$ is the element at the ith row and j^{th} column of Y_c^*, and $y_{c(i,j)}$ is calculated using equations 4.21 and 4.22. The spatial resolution of X_c^* along the columns is noted as Δx_c, and the spatial resolution of Y_c^* along the columns is noted as Δy_c (in this case, $\Delta y_c = 0.01$ m). The scale coefficient (SC) is defined as:

$$SC = [\Delta x_c / \Delta y_c] \tag{4.24}$$

Corrected data matrix (M) is defined using following equations:

$$\begin{cases} m_{(a,b,c)} = FIPAR_{(i,j)} \\ a = y_{c\ (i,j)}^{*} \\ b = \left(x_{c\ (i,j)}^{*} - 1 \right) \cdot SC + 1 \\ c = x_{c\ (i,j)}^{*} \cdot SC, for\, i = 1,2, \ldots, m, j = 1,2, \ldots, n \end{cases} \tag{4.25}$$

where $FIPAR$ is the matrix of canopy light interception data, m is the number of rows of $FIPAR$, $FIPAR_{(i,j)}$ is the element at the ith row and j^{th} column of FIPAR matrix, $m_{(a,b,c)}$ is the elements from the bth column to c^{th} column at the ath row of corrected data matrix (M), and SC is the scale coefficient. M is used to generate canopy map with projection correction (i.e. corrected canopy map).

Figure 4.7 shows the canopy mapping before and after geometrical transformation at three different times: approximately 1 h before midday; midday; and approximately 1 h after midday. The three sets at the top

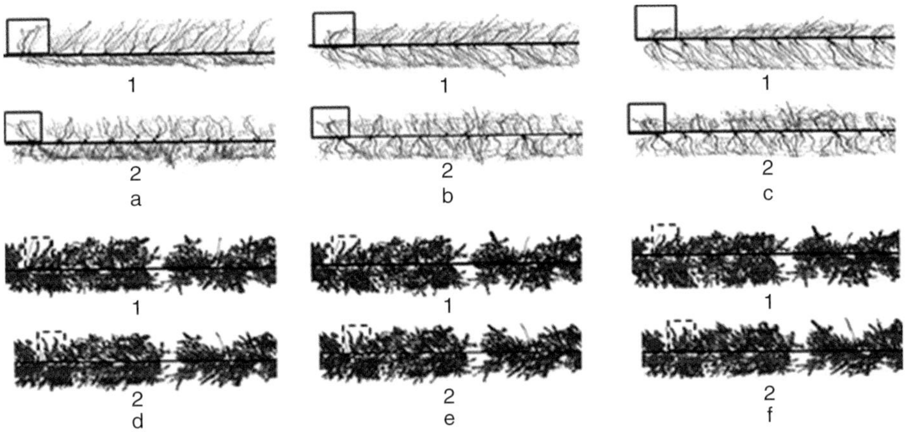

Fig. 4.7. Measured and corrected canopy maps at different times at dormant and full canopy stages: (a) at dormant stage, approximately 1 h before midday; (b) at dormant stage, midday; (c) at dormant stage, approximately 1 h after midday; (d) at full canopy stage, approximately 1 h before midday; (e) at full canopy stage, midday; and (f) at full canopy stage, approximately 1 h after midday. In each figure, '1' marks the measured canopy maps, '2' marks the corrected canopy map, and the horizontal solid line indicates the trunk-aligned line.

represent canopy dormant stage, and the three sets at the bottom represent full canopy stage. The distortion is clearly shown in the figure numbered (1) of each set before geometrical transformation, compared with the figure numbered (2) after the geometrical transformation.

4.4.2.3 UFO canopy architecture

Figure 4.8 illustrates the relationships of a simplified UFO canopy architecture between tree actual position and projection position, with x-axis representing inter-row direction, y-axis representing intra-row direction, and z-axis representing the vertical upward direction.

The geometrical relationship between an actual position (P_a) and the corresponding projection position (P_p) is shown in Fig. 4.8 by taking into account the sun azimuth angle (α) and zenith angle (β). This relationship can be described by using the following set of equations:

$$\begin{cases} x_p = z_a \times \tan\beta \times \sin(\alpha - 180) \\ y_p = y_a + z_a \times \tan\beta \times \cos(\alpha - 180) \end{cases} \tag{4.26}$$

The influence of the sun position (α and β) causes distortion on canopy projection. To calculate measured light interception for individual blocks from each PAR data set, it is critical to correct canopy projection as if the sunbeam were on the direction perpendicular to the row orientation (i.e. α = 90 degrees or 270 degrees).

The *x*- and *y*-coordinates of the original PAR projection are defined by an original *x*-coordinate matrix (X_p) and an original *y*-coordinate matrix (Y_p) using the following equations:

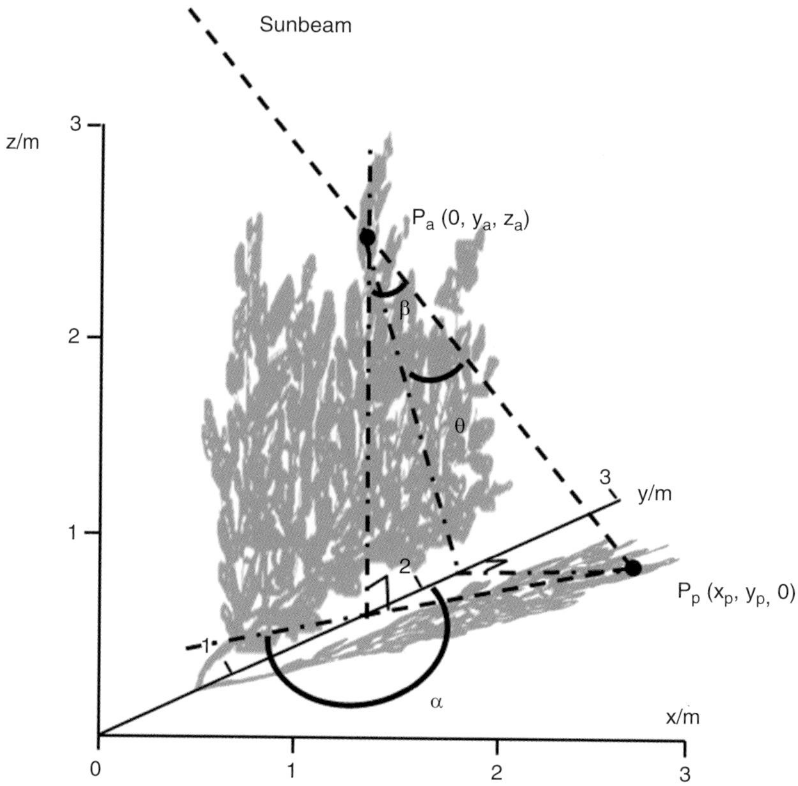

Fig. 4.8. An illustration of a UFO tree showing the relationship of actual position and projection position. The gray object on the *y-z* plane represents the tree in the field, and the gray object on the *x-y* plane represents the projection of the tree canopy on the ground; α = the sun azimuth angle, β = the sun zenith angle.

$$\begin{cases} x_{p(i,j)} = 0.1j \\ y_{p(i,j)} = 0.01(i-1), for\, i == 1, 2, \ldots, m,\ j = 1, 2, \ldots, n \end{cases} \tag{4.27}$$

where $x_{p(i,j)}$ is the element at the ith row and j^{th} column of X_p, and $y_{p(i,j)}$ is the element at the ith row and j^{th} column of Y_p. The corrected *x*- and *y*-coordinates of canopy projection are obtained by the following equations:

$$\begin{cases} x_{c(i,j)} = x_{p(i,j)} \\ y_{c(i,j)} = y_{p(i,j)} - \dfrac{x_{p(i,j)} \cos(180-\alpha)}{\sin|180-\alpha|}, \text{for } i = 1, 2, \ldots, m,\ j = 1, 2, \ldots, n \end{cases} \tag{4.28}$$

where $x_{c(i,j)}$ is the element at the ith row and j^{th} column of corrected *x*-coordinate matrix (X_c), and $y_{c(i,j)}$ is the element at the ith row and j^{th} column of corrected *y*-coordinate matrix (Y_c). Therefore, the PAR data set after projection correction (M_c) is obtained as:

$$\begin{cases} m_{c(i,v)} = m_{(i,j)} \\ k = \left[100 \times y_{c(i,j)}\right] \\ v = x_{c(i,j)} / 0.1, \text{ for } i = 1, 2, \ldots, m, \ j = 1, 2, \ldots, n \end{cases} \tag{4.29}$$

where $m_{(i,j)}$ is the element at the ith row and j^{th} column of the original PAR data set (M), $m_{c(k,v)}$ is the element at the kth column and v^{th} row of corrected PAR data set (M_c).

Figure 4.9 shows the canopy mapping before and after geometrical transformation at approximately 2 h before midday as a representative. The distortion is clearly shown in Fig. 4.9(a) before geometrical transformation, compared with the one in Fig. 4.9(b) after the geometrical transformation.

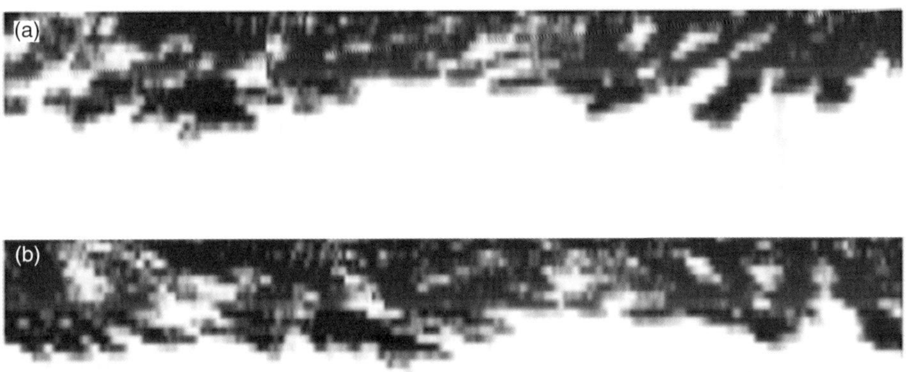

Fig. 4.9. (a) Measured and (b) corrected canopy maps of UFO canopy, approximately 2 h before midday.

4.4.3 Case 3: Modeling canopy PAR interception for estimating potential yield

A canopy PAR interception system retrofitted on to a Kawasaki Mule was used to measure diurnal PAR interception during the 2012 growing season in almond and walnut orchards. The PAR interception system consisted of PAR sensors distributed along a lightbar system, which was divided into 16 measurement units of 40 cm each. A rotary encoder and a differential GPS were also included in the system. All data were recorded using a Campbell Scientific CR3000 data acquisition system at a rate of 10 Hz. The information collected by this system was used to create a map of PAR intercepted, which was obtained using the following formula:

$$U_t = \sum_{i=1}^{n} \left(I_{FS} - I_i\right) \Delta A \tag{4.30}$$

where I_{FS} is PAR incident at the top of the tree (i.e. full sun), I_i is PAR transmitted to the bottom of the tree by the ith pixel in the shadow, ΔA is the pixel area and U_t is the total amount of PAR intercepted by a block of five trees. Also, the area of the block was computed by multiplying the number of pixels by the pixel area.

PAR intercepted by trees is needed to calculate net photosynthesis using equation 4.16. Our goal was to use one or a few measurements from the lightbar system to estimate overall PAR intercepted for the whole season. Diurnal and solar noon scans were collected at Nickels Soil Laboratory in Arbuckle, California, on six occasions. Figure 4.10 shows a visual representation of diurnal and seasonal data acquired by the lightbar system during the 2012 growing season. Each scan corresponds to 20 blocks of almonds and 16 blocks of walnuts, where each block correspond to five trees. Six blocks each of almonds and walnuts were left for model validation. The orchard spacing for the almond orchard was 4.3 m and 6.2 m between trees and rows, respectively. The orchard spacing for the walnut orchard was 4.6 m and 6.8 m between trees and rows, respectively. The row orientations of both orchards were north–south.

Diurnal PAR interception can be described in terms of area and intensity of the shadows, which depend on the sun position and canopy density. Zenith angle (θ) was found to be useful to describe the diurnal behavior of PAR intercepted by a block of five trees (U_t), when midday PAR interception (U_n) was known. The expression for U_t in terms of U_n is given by:

$$U_t = U_n(t) * \frac{F_t}{F_n}(\theta) \tag{4.31}$$

where F_t and F_n are the PAR components of the incident radiation (i.e. full sun) times the area covered by the shadow of the block at a given time, t and noon, respectively.

Using equation 4.31 we can estimate PAR intercepted at any time of the day, if we know noon time incident radiation, U_n. However, U_n is not available for each day during the growing season. In order to develop a practical system for crop management, we need to work with just a few sets of data

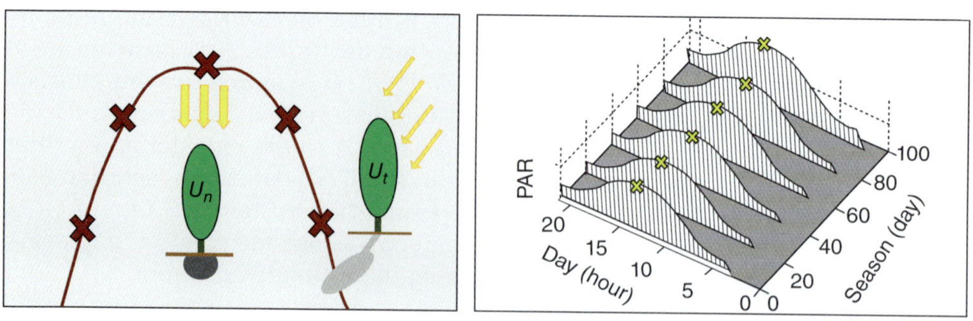

Fig. 4.10. Representation of the diurnal and seasonal lightbar scans.

(ideally one) gathered with the lightbar (i.e. just one midday PAR interception measurement early in the season). Two strategies have been followed to address this problem, based on an assumption on growth of the canopy: (i) there is no growth during the season; and (ii) there is growth during the season which can be described by an empirical growth curve.

The first approach has the advantage that only one measurement is needed, which can be obtained early in the season, but this method is appropriate for mature trees only. In addition, the non-growth approach does not consider the fact that at solar noon shadows change not only due to tree growth, but also due to changes in sun elevation. In the second approach, midday PAR intercepted data for a few days are used to develop an empirical curve, which describe PAR interception throughout the season. Then, midday PAR interception for every day can be estimated using the seasonal growth curve.

Midday PAR interception was found to increase as the season progressed during 2012. No such increases in growth or PAR interception were observed in almond and walnut crops during the 2013 and 2014 growing seasons. A second-order polynomial was used to fit the midday PAR interception data to develop growth curves for almond and walnut trees for the 2012 growing season. Figure 4.11 shows the average growth curves for 20 and 16 blocks of almond and walnut trees, respectively. Midday PAR interception as function of time (i.e. Julian days (*J*)) were described by:

$$PAR = 2.2764J^2 - 772.27J + 165354 \quad (4.32)$$

$$PAR = 6.5153J^2 - 2158.4J + 301820 \quad (4.33)$$

for almond and walnut trees, respectively. The coefficients of determination for the relationship between midday PAR interception and Julian

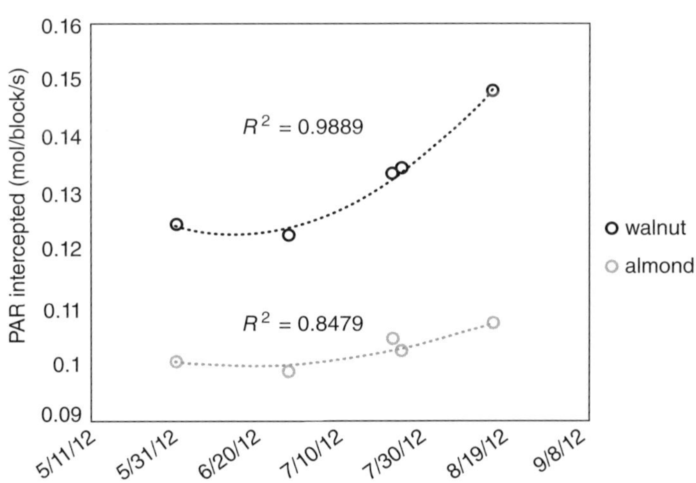

Fig. 4.11. Growth curves for almond and walnut trees for the 2012 growing season.

days were 0.85 and 0.99 for almond and walnut trees, respectively. This figure also shows that walnut trees intercepted higher values of PAR and had a higher growth rate than almond trees.

One of the objectives of this study was to develop a procedure to estimate diurnal and seasonal PAR interception based on one scan (or just a few scans) obtained early in the season. For the case of canopy growth during the season (2012 season), a relative growth curve was generated by dividing the midday PAR intercepted value for a given day by the midday PAR intercepted on a selected reference day. The seasonal values of midday PAR interception for each block were obtained for any given day by multiplying the data obtained on the reference day by its corresponding growth factor, whereas in the no-growth case of the 2013 and 2014 seasons, midday PAR interception for each block was assumed to be the same during the whole season. The value of midday PAR interception computed in this way corresponds to the parameter U_n in equation 4.31.

Diurnal PAR interception can be described in terms of area and intensity of the shadows, which depend on the position of the sun and the canopy architecture. Diurnal data were analyzed to see if it was possible to predict PAR intercepted at different times during the day using a single midday data set. Zenith angle (θ) was found to be useful to describe the diurnal behavior of PAR intercepted by a block of five trees (U_t), when midday PAR interception (U_n) was known. Equation 4.31 shows this relationship, where F_t is the PAR component of the incident radiation (i.e. full sun) times the area covered by the shadow of the block at time t and F_n is the corresponding value for the particular case of solar noon. The ratio F_t/F_n was found to be a function of the zenith angle as shown in Fig. 4.12a (almond trees) and Fig. 4.12b (walnut trees). The following polynomial equations were used to describe this relationship:

$$PAR = -0.0004Zn^2 + 0.0213Zn + 0.7433 \tag{4.34}$$

$$PAR = -0.0003Zn^2 + 0.0175Zn + 0.791 \tag{4.35}$$

for almond and walnut trees, respectively. The coefficients of determination for the relationship between diurnal PAR intercepted and zenith angle were 0.88 and 0.94 for almond and walnut trees, respectively.

PAR interception at any given time during the day can be obtained by multiplying F_t/F_n by PAR interception at noon time (U_n). The ratio F_t/F_n incorporates the temporal component (i.e. diurnal behavior) while U_n incorporates the spatial component (i.e. depends on measured data, which is different for each block). Diurnal and seasonal PAR interception (mol/block/s) were estimated for each block of five trees throughout the 2012 growing season. The results for two blocks (one for almond and one for walnut trees) are shown in Fig. 4.13 in which both diurnal and seasonal effects can be seen for both growth and no growth conditions.

PAR interception was found to be lower in the early morning and late afternoon, and reached peak values slightly before as well as slightly

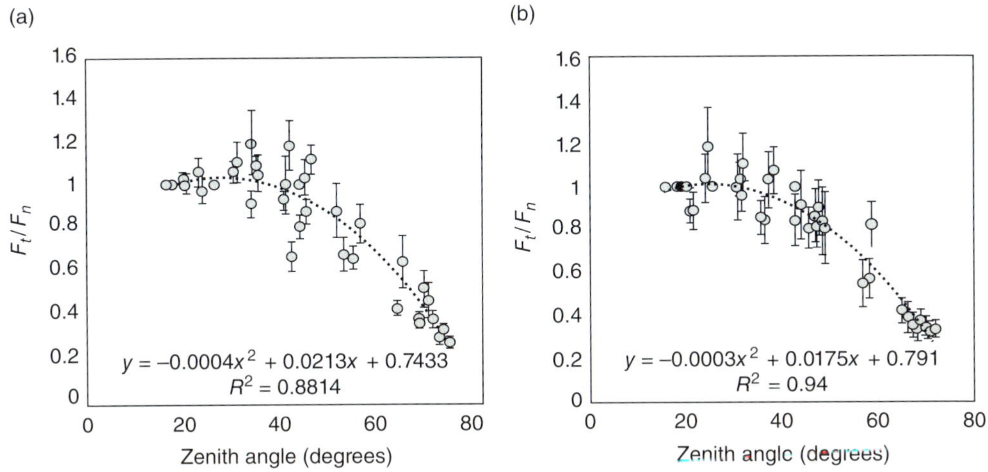

Fig. 4.12. Empirical curve for diurnal full sun data: (a) almond trees; (b) walnut trees.

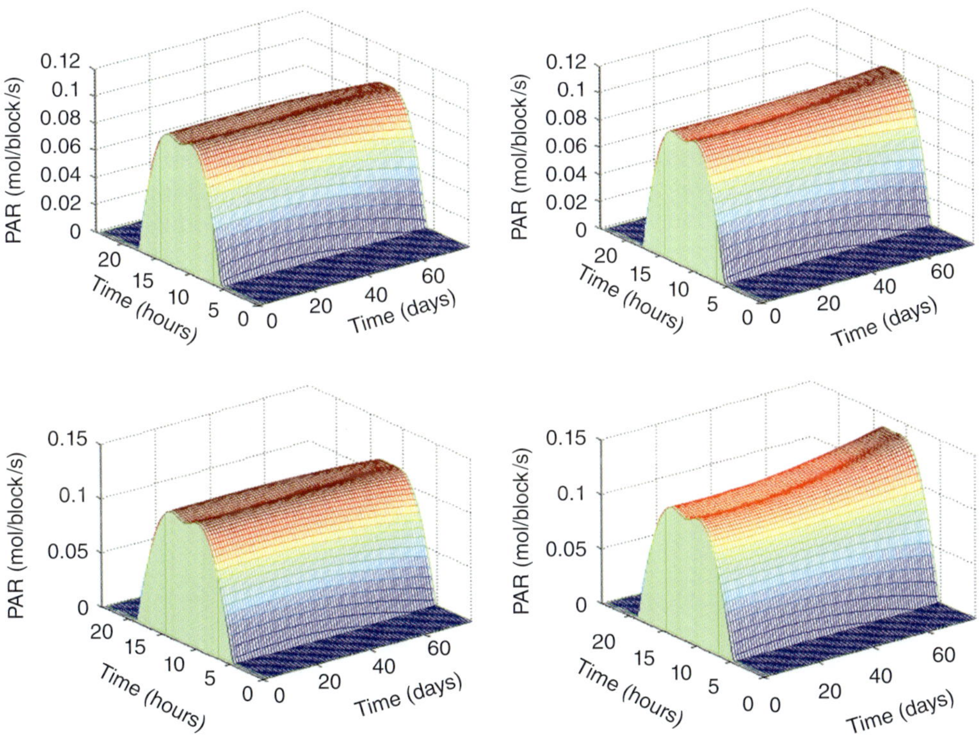

Fig. 4.13. Diurnal and seasonal behavior of PAR intercepted as estimated by the empirical model. Top left: almond no growth (2013 and 2014 seasons). Top right: almond with growth (2012 season). Bottom left: walnut no growth (2013 and 2014 seasons). Bottom right: walnut with growth (2012 season).

after solar noon. It is interesting to see that the PAR absorption dipped at solar noon. This outcome can be explained by two phenomena acting together. First, shadows are larger at sunrise and sunset, and decrease near solar noon, when the sun is almost directly above the canopy. Second, PAR intercepted per surface area was found to be higher at solar noon, decreasing as sun altitude decreased. Since the PAR transmitted by the canopy is related to the integral of the light intensity over the area of the shadow, the smaller area of the shadow led to lower PAR interception even though the intensity was higher at solar noon (Hottel, 1976; Farber and Morrison, 1977).

Figure 4.14 shows the comparison between measured and estimated values of the ratio F_t/F_n used in equation 4.31 (also presented in Fig. 4.12). Note that six blocks were not previously used in developing prediction equations. These equations were:

$$y = 0.9396x + 0.0918 \tag{4.36}$$

$$y = 0.8818x + 0.1611 \tag{4.37}$$

for almond and walnut crops, respectively. Coefficient of determination values (r^2) of 0.89 and 0.93 were found for almond and walnut trees, respectively. After validation, the complete data set was pooled together and new second-order polynomial equations were developed for both crops.

Equation 4.31 was used to estimate PAR intercepted on the same days and times when lightbar scans were obtained. The data corresponding to 28 June 2012 was used as a reference day to estimate midday PAR intercepted for all other days, and a zenith angle was calculated using equation 4.3. The results were compared with the daily PAR intercepted as measured by the lightbar system (Fig. 4.15). Coefficient of determination values (r^2) of 0.87 and 0.86 were found for almond and walnut trees, respectively, indicating that PAR intercepted at any time can be estimated using this

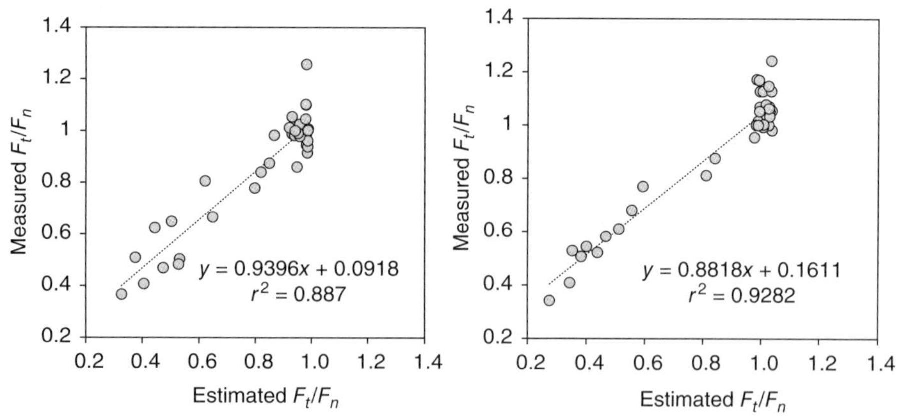

Fig. 4.14. Validation of the ratio F_t/F_n used to account for daily variation of PAR interception for almond (left) and walnut trees (right).

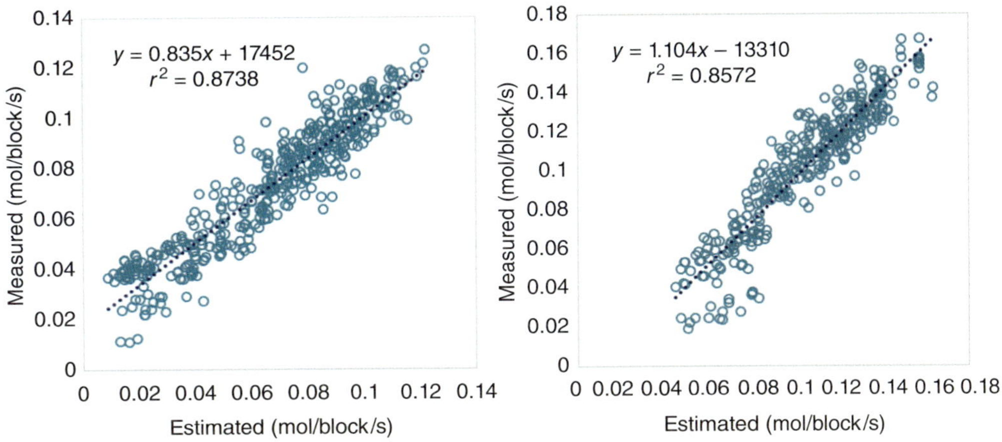

Fig. 4.15. Validation of the empirical model of canopy PAR interception for almond (left) and walnut (right) trees.

prediction technique. The prediction equations for the relationship between estimated and measured values of canopy PAR interception were:

$$y = 0.835x + 17452 \tag{4.38}$$

$$y = 1.104x - 13310 \tag{4.39}$$

for almond and walnut trees, respectively (with units of μmol/block/s). Estimated values of PAR interception can be used to compute the amount of carbon assimilated throughout the season and the potential yield.

4.5 Challenges and Opportunities

This section provides a review of the challenges that PAR interception sensing systems still have to overcome to be fully adopted commercially. The current processing of lightbar data involves several steps that require manual supervision. Fully automated lightbar post-processing software would be very useful, where georeferenced shadow maps can be easily created, together with basic information about the area of the shadows cast by a block of trees and the PAR intercepted by them.

Models of potential yield can be upgraded to incorporate the effect of ambient temperature on physiological processes, by letting the light-saturated photosynthesis parameter and the dark respiration rate change with temperature (equation 4.13), following a Q_{10} type of behavior. In addition, potential yield information should be integrated into a decision support system (DSS) that provides information that is easy to understand so that growers can utilize such a DSS to improve their management practices.

Due to the high importance of the area of the shadow in the overall amount of PAR being intercepted by the trees, a simple technique based on an inexpensive RGB camera can be utilized (Zarate-Valdez *et al.*, 2015);

such a camera can be incorporated into an unmanned aerial vehicle (UAV) or a handheld mobile device. Manual measurements have the inconvenience of being time consuming and labor intensive if several samples have to be taken, whereas a camera mounted on a UAV can take pictures over large areas quickly. However, delineation of individual canopies is still needed to extract features of individual trees (e.g. projected area). This can be performed manually (Getzin *et al.*, 2014) or by automatic procedures (Hirschmugl *et al.*, 2007). Canopy dimension can also be extracted from the shadows of the trees, because shadows depend on the tree's dimensions. Measurements of tree shadows have been made using a ground-based system (Giuliani *et al.*, 2000; Rojo *et al.*, 2014) and a UAV (Sharma *et al.*, 2013).

Multiple linear regression (MLR) could be used to develop a model for PAR interception as a function of the area of the shadow and the cosine of the zenith angle. In addition, digital images (RGB) obtained by UAVs could be used to estimate PAR interception if they are properly calibrated using a lightbar system. Advantages of the use of UAVs are that they can obtain information for a large area faster than ground-based platforms and over a wide window of time (i.e. not necessarily close to solar noon because zenith angle is being considered), which makes them an interesting alternative to estimate PAR interception and potential yield.

4.6 Summary

This chapter has introduced the topic of light interception and canopy sensing for tree fruit canopy management. The chapter explained why canopy light interception is useful information for agriculture and how modern canopy sensing systems can be used to develop decision support systems that can help growers. A brief literature review covered the following topics: (i) solar radiation and tree productivity; (ii) sensing canopy light interception; and (iii) modeling canopy light interception. The chapter also presented several applications of PAR interception measurement systems for assisting fruit tree canopy management, and estimating yield. Relationships between PAR interception and canopy volume have been discussed and examples of how canopy PAR interception can be used to estimate real or potential yield have been provided. In addition, case studies have been presented describing two mobile platforms that measure PAR under the canopies to estimate light interception data. It also described how geometrical transformation of the shadow can be used to represent midday PAR interception, with an empirical model describing the dynamic behavior of the shadow.

References

Acevedo, E., Silva, P. and Silva, H. (2002) Wheat growth and physiology. In: *Bread Wheat Improvement and Production*. Food and Agriculture Organization (FAO), Rome, pp. 39–70.

Acock, B., Charles-Edwards, D.A., Fitter, D.J., Hand, D.W., Ludwig, L.J., Wilson, W.J. and Withers, A.C. (1978) The contribution of leaves from different levels within a tomato crop to canopy net photosynthesis: an experimental examination of two canopy models. *Journal of Experimental Botany* 29(111), 815–827.

Allen, M., Prusinkiewicz, P. and DeJong, T. (2008) Using L-systems for modeling source–sink interactions, architecture and physiology of growing trees: the L-PEACH model. *New Physiology* 166(3), 869–880.

Annandale, J.G., Jovanovic, N.Z., Campbell, G.S., Sautoy, N.D. and Lobit, P. (2004) Two-dimensional solar radiation interception model for hedgerow fruit trees. *Agricultural and Forest Meteorology* 121, 207–225.

Arno, J., Escola, A., Valles, J., Llorens, J., Sanz, R. *et al.* (2013) Leaf area index estimation in vineyards using a ground-based LiDAR scanner. *Precision Agriculture* 14, 290–306.

Beland, M., Widlowski, J. and Fournier, R. (2014) A model for deriving voxel-level tree leaf area density estimates from ground-based LiDAR. *Environmental Modelling & Software* 51, 184–189.

Beyer, R., Letort, V. and Cournède, P.-H. (2014) Modeling tree crown dynamics with 3D partial differential equations. *Frontiers in Plant Science, Plant Biophysics and Modeling*, 5, 329.

Boote, K.J. and Pickering, N.B. (1994) Modeling photosynthesis of row crop canopies. *HortScience* 29(12), 1423–1434.

Bosc, A. (2000) EMILION, a tree functional-structural model: presentation and first application to the analysis of branch carbon balance. *Annals of Forest Science* 57, 555–569.

Campbell, G.S. and Norman, J.M. (1998) *An Introduction to Environmental Biophysics*, 2nd edn. Spring-Verlag, New York.

Cannell, M.G.R., Sheppard, L.J. and Milne, R. (1988) Light use efficiency and woody biomass production of poplar and willow. *Forestry* 61(2), 125–136.

Charles-Edwards, D.A., Doley, D. and Rimmington, G.M. (1986) *Modelling Plant Growth and Development*. Academic Press, Sydney.

Chen, S., Ceulemans, R. and Impens, I. (1994) A fractal-based populus canopy structure model for the calculation of light interception. *Forest Ecology and Management* 67, 97–110.

Connor, D. and Gómez-del-Campo, M. (2013) Simulation of oil productivity and quality of N/S oriented olive hedgerow orchards in response to structure and interception of radiation. *Scientia Horticulturae* 150, 92–99.

Dewar, R.C., Medlyn, B.E. and McMurtrie, R.E. (1998) A mechanistic analysis of light and carbon use efficiencies. *Plant, Cell and Environment* 21, 573–588.

Duffie, J. and Beckman, W. (2013) *Solar Engineering of Thermal Processes*, 4th edn. Wiley & Sons, Inc., Hoboken, New Jersey.

Duursma, R.A. and Mäkelä, A. (2007) Summary models for light interception and light-use efficiency of non-homogeneous canopies. *Tree Physiology* 27, 859–870.

Evans, L. and Fisher, R. (1999) Yield potential: its definition, measurement, and significance. *Crop Science* 39, 1544–1551.

Farber, E.A. and Morrison, C.A. (1977) Clear-day design values. In: Jordan, R.C. and Liu, B.Y.H. (eds) *Applications of Solar Energy for Heating and Cooling of Buildings*. American Society of Heating, Refrigerating and Air-Conditioning Engineers (ASHRAE), New York.

Figliola, R. and Beasley, D. (2011) *Theory and Design for Mechanical Measurements*. John Wiley, New York.

Fuwa, K. and Valle, B. (1963) The physical basis of analytical atomic absorption spectrometry, the pertinence of the Beer-Lambert law. *Analytical Chemistry* 35(8), 942–946.

Getzin, S., Nuske, R.S. and Wiegand, K. (2014) Using unmanned aerial vehicles (UAV) to quantify spatial gap patterns in forests. *Remote Sensing* 6, 6988–7004.

Giuliani, R., Magnanini, E., Fragassa, C. and Nerozzi, F. (2000) Ground monitoring the light-shadow windows of a tree canopy to yield canopy light interception and morphological traits. *Plant, Cell and Environment* 23, 783–796.

Grossman, Y.L. and DeJong, T.M. (1998) Training and pruning system effects on vegetative growth potential, light interception, and cropping efficiency in peach trees. *Journal of the American Society for Horticultural Science* 123, 1058–1064.

Guillen-Climent, M.L., Zarco-Tejada, P.J., Berni, J.A.J., North, P.R.J. and Villalobos, F.J. (2012) Mapping radiation intercepted in row-structured orchards using 3D simulation and high-resolution airborne imagery acquired from a UAV. *Precision Agriculture* 13, 473–500.

Hirschmugl, M., Ofner, M., Raggam, J. and Schardt, M. (2007) Single tree detection in very high resolution remote sensing data. *Remote Sensing of Environment* 110, 533–544.

Hottel, H.C. (1976) A simple model for estimating the transmittance of direct solar radiation through clear atmospheres. *Solar Energy* 18, 129–134.

Johnson, I., Thornley, J., Frantz, J. and Bugbee, B. (2010) A model of canopy photosynthesis incorporating protein distribution through the canopy and its acclimation to light, temperature and CO_2. *Annals of Botany* 106, 735–749.

Jones, H. (1992) *Plant and Microclimate, a Quantitative Approach to Environmental Plant Physiology*, 2nd edn. Cambridge University Press, Cambridge, UK.

Lampinen, B., Edstrom, J., Ramos, D., Metcalf, S., Negrón, C. and Cutter, S. (2006) *Comparison of Growth and Productivity of Pruned and Unpruned Young Howard Walnut Trees as Impacted by Crop Load.* Walnut Research Report 127. California Walnut Board, Folsom, California.

Lampinen, B., Browne, G., Upadhyaya, S., Udompetaikul, V., Slaughter, D. *et al.* (2009) *Development and Testing of a Mobile Platform for Measuring Canopy Light Interception and Water Stress in Almond.* Annual Research Report. Almond Board of California, Modesto, California.

Lampinen, B.D., Udompetaikul, V., Browne, G.T., Metcalf, S.G., Stewart, W.L. *et al.* (2012) A mobile platform for measuring canopy photosynthetically active radiation interception in orchard systems. *HortTechnology* 22(2), 237–244.

Lee, K. and Ehsani, R. (2009) A laser scanner based measurement system for quantification of citrus tree geometric characteristics. *Applied Engineering in Agriculture* 25(5), 777–788.

Liu, B. and Jordan, R. (1960) The interrelationship and characteristic distribution of direct, diffuse and total solar radiation. *Solar Energy* 4(3), 1.

Loomis, R.S. and Amthor, J.S. (1999) Yield potential, plant assimilatory, and metabolic efficiencies. *Crop Science* 39, 1584–1596.

Makela, A. and Hari, P. (1986) Stand growth model based on carbon uptake and allocation in individual trees. *Ecological Modeling* 33, 205–229.

Mann, J.E., Curry, G.L. and Sharpe, P.J.H. (1979) Light interception by isolated plants. *Agricultural Meteorology* 20, 205–214.

Medlyn, B.E. (1998) Physiological basis of the light use efficiency model. *Tree Physiology* 18, 167–176.

Monteith, J.L. (1965) Light distribution and photosynthesis in field crops. *Annals of Botany* 29(1), 17–37.

Monteith, J.L. (1972) Solar radiation and productivity in tropical ecosystems. *Journal of Applied Ecology* 9, 747–766.

Niklas, K. and Spatz, H. (2012) *Plant Physics.* University of Chicago Press, Chicago, Illinois.

Norman, J.M. and Welles, J.M. (1983) Radiative transfer in an array of canopies. *Agronomy Journal* 75, 481–488.

Oyarzun, R., Stockle, C. and Whitting, M. (2007) A simple approach to modeling ratio interception by fruit-tree orchards. *Agricultural and Forest Meteorology* 142, 12–24.
Palacín, J., Palleja, T., Tresanchez, M., Sanz, R. and Llorens, J. (2007) Real-time tree-foliage surface estimation using a ground laser scanner. *IEEE Transactions Instrumentation and Measurement* 56(4), 1337–1383.
Pascual, M., Villar, J.M., Rufat, J., Rosell, J.R., Sanz, R. and Arno, J. (2011) Evaluation of peach tree growth characteristics under different irrigation strategies by LIDAR system: preliminary results. *Acta Horticulturae* 889, 227–232.
Paw U, K.T. (2000) Meteorological instrumentation and observation. *Atmospheric Science* 124, University of California, Davis, California.
Pearcy, R., Muraoka, H. and Valladares, F. (2005) Crown architecture in sun and shade environments: assessing function and trade-offs with a three-dimensional simulation model. *New Physiologist* 166(3), 791–800.
Robinson, T. and Lakso, A. (1991) Bases of yield and production efficiency in apple orchard systems. *Journal of the American Society for Horticultural Science* 116(2), 188–194.
Rojo, F., Dhillon, R., Upadhyaya, S., Jenkins, B., Roach, J., Lampinen, B. and Metcalf, S. (2014) *Modeling canopy light interception for estimating potential yield in almond and walnut.* Paper No. 1456, presented at 12th International Conference of Precision Agriculture, Sacramento, California, 20–23 July 2014. International Society of Precision Agriculture, Monticello, Illinois, p.138.
Rosati, A. and Dejong, T. (2003) Estimating photosynthetic radiation use efficiency using incident light and photosynthesis of individual leaves. *Annals of Botany* 91, 869–877.
Rosell, J.R. and Sanz, R. (2012) A review of methods and applications of the geometric characterization of tree crops in agricultural activities. *Computers and Electronics in Agriculture* 81, 124–141.
Rosell, J.R., Llorens, J., Sanz, R., Arno, J., Ribes-Dasim, M. *et al.* (2009a) Obtaining the three-dimensional structure of tree orchards from remote 2D terrestrial LIDAR scanning. *Agricultural and Forest Meteorology* 149, 1505–1515.
Rosell, J.R., Sanz, R., Llorens, J., Arno, J., Escola, A. *et al.* (2009b) A tractor-mounted scanning LIDAR for the non-destructive measurement of vegetative volume and surface area of tree-row plantations: a comparison with conventional destructive measurements. *Biosystems Engineering* 102, 128–134.
Sanz, R., Rosell, J.R., Llorens, J., Gil, E. and Planas, S. (2013) Relationship between tree row LIDAR-volume and leaf area density for fruit orchards and vineyards obtained with a LIDAR 3D Dynamic Measurement system. *Agricultural and Forest Meteorology* 171(172), 153–162.
Sassaroli, A. and Fantini, S. (2004) Comment on the modified Beer-Lambert law for scattering media. *Physics in Medicine and Biology* 49(14), N225–N257.
Seginer, I. (2003) A dynamic model for nitrogen-stressed lettuce. *Annals of Botany* 91, 623–635.
Sharma, R.C., Kajiwara, K. and Honda, Y. (2013) Automated extraction of canopy shadow fraction using unmanned helicopter-based color vegetation indices. *Trees* 27, 675–684.
Taiz, L. and Zeiger, E. (2002) *Plant Physiology*, 3rd edn. Sinauer Associates Inc, Sunderland, Massachusetts.
Thornley, J. (2011) Plant growth and respiration re-visited: maintenance respiration defined – it is an emergent property of, not a separate process within, the system – and why the respiration: photosynthesis ratio is conservative. *Annals of Botany* 108, 1365–1380.
Thornley, J.H.M. and France, J. (2007) *Mathematical Models in Agriculture, Quantitative Method for the Plant, Animal and Ecological Sciences*, 2nd edn. Cromwell Press, Trowbridge, UK.
Thornley, J.H.M. and Johnson, I. (1990) *Plant and Crop Modeling: a Mathematical Approach to Plant and Crop Physiology.* Clarendon Press, Oxford.

Threlkeld, J. (1970) *Thermal Environmental Engineering*, 2nd edn. Prentice-Hall, Englewood Cliffs, New Jersey.

Upadhyaya, S.K. and Koller, M. (2005) Prediction of processing tomato yield using a crop growth model and remotely sensed aerial images. *Transactions of ASAE* 48(6), 2335–2341.

Van der Zande, D., Hoet, W., Jonckheere, I., van Aardt, J. and Coppin, P. (2006) Influence of measurement set-up of ground-based LiDAR for derivation of tree structure. *Agricultural and Forest Meteorology* 141, 147–160.

Walklate, P.J., Cross, J.V., Richarson, G.M., Murray, R.A. and Baker, D.E. (2002) Comparison of different spray volume deposition models using LIDAR measurements of apple orchards. *Biosystems Engineering* 82(3), 253–267.

Wang, Y.P. and Jarvis, P.G. (1990) Description and validation of an array model MAESTRO. *Agricultural and Forest Meteorology* 51, 257–280.

West, P.W. and Welles, K.F. (1992) Method of application of a model to predict the light environment of individual tree crowns and its use in a eucalypt forest. *Ecological Modelling* 60, 199–231.

Zarate-Valdez, J., Whiting, M., Lampinen, B., Metcalf, S., Ustin, S. and Brown, P. (2012) Prediction of leaf area index in almonds by vegetation indexes. *Computers and Electronics in Agriculture* 85, 24–32.

Zarate-Valdez, J., Metcalf, S., Stewart, W., Ustin, S. and Lampinen, B. (2015) Estimating light interception in tree crops with digital images of canopy shadow. *Precision Agriculture* 16(4), 425–440.

Zhang, J., Whiting, M. and Zhang, Q. (2013) *Sensor-based Canopy Mapping Using Photosynthetically Active Radiation Interception in Y-Trellis Tree Fruit Orchards*. ASABE paper no. 13-1596279. American Society of Agricultural and Biological Engineers, St Joseph, Michigan.

Zhang, J., Whiting, M. and Zhang, Q. (2015) Diurnal pattern in canopy light interception for tree fruit orchard trained to an upright fruiting offshoots (UFO) architecture. *Biosystems Engineering* 129, 1–10.

5 Precision Orchard Systems

MATTHEW WHITING*

Washington State University, Prosser, Washington, USA

5.1 Introduction

Without intervention, most orchard trees will grow to great height and girth, forming a globular or triangular structure. Iteratively, throughout the domestication of tree fruit crops for large-scale production, orchardists and pomologists, in both empirical and theoretical ways, have learned and studied horticultural management strategies to improve both the quantity and quality of fruit produced. In designing and planting a new orchard, growers face increasing pressures to reduce the environmental footprint of production, meet local and global market demands for produce safety, and efficiently and consistently provide a healthy and safe product in a changing climate. Tree fruit orchard systems (i.e. the strategic manipulation of fruiting habit and vegetative growth) are varied, depending on crop, cultivar, rootstock, and location. The decision to plant a new orchard is challenging, because it is difficult to change any key element, cultivar, rootstock, tree spacing, and training system, once the trees are planted (Palmer and Warrington, 2000). The orchardist's goal is deceptively simple: produce high quantities of quality fruit in a manner that is sustainable and profitable. The growth and culture of fruit trees has evolved considerably over centuries since their domestication and cultivation. Interestingly, the modification of canopy architecture has come nearly full circle, from single or groups of trees espaliered in a courtyard garden (Fig. 5.1), to the low-density hedgerow systems of early commercial orchards (Fig. 5.2), back to high-efficiency fruiting walls of ultra high-density modern systems (Fig. 5.3).

* Email: mdwhiting@wsu.edu

Fig. 5.1. Fruit trees espaliered to formal structures of fruiting walls in gardens.

Fig. 5.2. Traditional low-density multiple-leader sweet cherry orchard based on vigorous seedling rootstocks.

Fig. 5.3. Modern fruiting wall orchard system of 'Jazz'/M.9 apple with seven horizontal fruiting tiers.

In modern orchards, the adoption of precision management strategies for vegetative growth and fruiting is routine, necessary for the perennial production of quality produce from mature orchards. Modern orchardists are skilled and knowledgeable in the manipulation of fruiting and vegetative growing points for sustainable and efficient production of quality fruit. Indeed, the evolution of precision orchard systems and the management techniques utilized are largely horticultural, being developed through trial and error, by innovative orchardists. Thus, there is a relative paucity of true scientific literature on the subject. This chapter will therefore focus on the key technological developments that have enabled the adoption of precision orchard systems, and, equally important, enabled the maintenance of mature, compact, productive orchard systems. Issues of light use efficiency and harvest index are fundamental and yet not fully understood (Palmer, 2011). The grower's decision is increasingly difficult with the availability of new scion and rootstock genotypes. The prospect of designing orchards to readily adopt significant automation, mechanization, and/or robotics technologies production is compelling – adoption to date has lagged behind other agricultural industries despite the high labor requirement for tree fruit production. It may be argued that the key factor inhibiting the utilization of automation is the aged and complex, largely random, orchard systems that the majority of fruit are produced from (Fig. 5.2).

5.2 Canopy Architecture and Training Systems

Traditionally, tree fruit orchards (e.g. apple, citrus, cherry, peach, pear) have been low density, based on vigorous or seedling rootstocks, and trained to open-center or globular type architectures. Tree densities were as low as < 200 trees/ha, and very little pruning or canopy management practices were adopted. The large canopies were difficult to harvest, requiring the use of tall ladders. These systems are not well suited to mechanization of orchard management, though mechanical pruning (e.g. restricting height typically) and harvest for processed crops such as tart cherry (*Prunus cerasus* L.) or citrus have been adopted. The unwieldy nature of these complex canopies does not lend them to precise management of growth or fruiting, and the processes of harvest and pruning were especially labor intensive. The first key development enabling the transition to higher efficiency orchard systems was the advent of size-controlling rootstocks. The ability to significantly reduce vegetative vigor led to improvements in canopy light relations, fruit quality, and production efficiency (discussed below). Currently, modern orchard systems are planted at tree densities that often exceed 5000 trees/ha. Robinson (2008) reports that the optimum tree density for US apple orchards is ca. 2500 trees/ha, given the high tree costs; yet this density could be 5000–6000 trees/ha with less expensive trees and higher returns with new premium cultivars. It has been suggested that the optimum tree density for apple spindle systems is 4000 trees/ha in Germany (Weber, 2001). In sweet cherry and peach, tree densities have increased progressively with the availability of size-controlling rootstocks (though the density of trees *per se* is not particularly important).

Early studies of apple orchard productivity and fruit quality concluded that the large trees (6+ m) were unsuitable due to poor light distribution (Heinicke, 1963; Looney, 1968). Subsequent orchard systems designs were variations on a central leader architecture with a pyramidal or conic tree shape and tree densities of 250–750 trees/ha. Trees were planted on moderately size-controlling rootstocks and were trained to have four to five tiers of horizontally trained fruiting wood. Without greater vigor control, however, upper limbs generally became excessively large and shaded lower portions of the tree, resulting in poor quality fruit (Reginato *et al.*, 2008). There followed many adaptations of the central leader architecture with the adoption of rootstocks that offered excellent vigor control. These include the spindle (super spindle, tall spindle), vertical axe, and solaxe (Robinson, 2008; Hrotkó, 2013). These were the first orchard systems that required trellis systems for support, increasing the costs of planting further.

In newly planted orchards, the grower's goals are to fill the inter-tree space rapidly to maximize light interception and orchard precocity. There has been much investigation of orchard systems (i.e. scion and rootstock genotype, tree and row spacing, training system) in both pome and stone fruit, with particular emphasis given to apple. Barritt *et al.* (2008) reported that light interception was greatest in years 7–10 for apple trees trained to the hybrid tree cone system compared with tatura trellised, angled spindle, and vertical double row. In addition, when pruned to a pedestrian height (i.e. 2 m versus 3 m), yield was reduced, likely from reductions in light interception at the lower ratio of tree height to row spacing. Despite greater yield in the first 6 years in peach and nectarine trained to a Y-trellised architecture, the small vase system was recommended due to high initial investment of the Y-trellis and lower annual costs of production in the small vase (Caruso *et al.*, 2015). Orchard system and genotype affect production economics. In a comparison of ten apple orchard systems and tree densities ranging from 1111 to 2460 trees/ha, pruning costs varied among cultivars/system by nearly six-fold (US$0.11–0.66/19 kg box) (Barden, 2002).

5.3 Rootstocks for Vigor Control

The advent and widespread adoption of size-controlling rootstock genotypes has been critical to the adoption of compact orchard systems and the precision canopy management strategies that are commonplace in some tree fruit crops. These rootstocks have been categorically adopted in apple for more than 50 years, and rootstocks offering varying degrees of scion vigor control have become available for sweet cherry in recent decades. The most important series of size-controlling rootstocks for sweet cherry has been from Germany, the ‘Gisela’ selections. There are limited options for growers of pear and plum, but several prospects for peach (Layne, 1987; Reighard, 2002). Rootstocks are utilized for improving precocity, reducing tree size, improving fruit quality, and improving resistance to pests and environmental and edaphic conditions (Rom and Carlson, 1987). Rootstocks affect scion precocity, productivity, and fruit quality (Gao *et al.*, 1992; Hirst and Ferree, 1995) as well as mineral uptake (Fallahi *et al.*, 2002; Amiri *et al.*, 2014). The role of rootstocks in precision orchard systems is fundamental, and there are myriad research papers describing various rootstocks and their influence on canopy vigor, precocity, productivity, and crop value. These will not be reviewed here.

5.4 Light Interception and Productivity

Productivity of orchards is very closely related to total light interception. This has been repeatedly demonstrated for hedgerow systems, and particularly for apple (Jackson and Palmer, 1972; Palmer, 1981; Barrit, 1989; Robinson and Lakso, 1991; Wünsche *et al.*, 1996; Wünsche and Lakso, 2000; Grappadelli and Lakso, 2007). The interception of incident irradiation is dependent upon canopy architecture and leaf area primarily. The necessity of alleyways between rows for machinery generally limits maximum light interception to ca. 70%. Increasing light interception in traditional hedgerow architectures typically reduces fruit quality from excessive intra-canopy shading. This problem is overcome in modern orchard systems by maintaining compact fruiting walls and, in many cases, extending the canopy over the row in a Y- or V-trellised architecture (see below). Barritt (1989) reported strong positive correlation between orchard light interception and leaf area and fruit production. Palmer (1981) similarly revealed a positive correlation between apple orchard light interception and yield, with trees at the more rectangular arrangement of 2.75 m × 0.8 m intercepting less light than trees at a less dense arrangement of 2.0 m × 1.25 m. Wünsche *et al.* (1996) studied the role of light interception by different shoot types in apple and revealed the importance of light interception by spur leaves. Y-trellised systems were higher yielding (ca. 60 tons/ha) than pyramidal shaped trees (ca. 40 tons/ha) and exhibited 20% to 30% more spur canopy light interception, though yields in their research were low by modern Y-trellised standards (e.g. 80–100 t/ha). In addition, Wünsche and Lakso (2000) reported a positive correlation between fruit yield and spur leaf light interception, supporting the hypothesis that fruit yield is better correlated with light interception by spur leaves versus extension shoot leaves. The authors concluded that to maximize orchard yield and fruit quality, open, well illuminated, spur-rich tree canopies will be important. These and other findings underscore the importance of both light interception and light distribution in the production of quality fruit.

Light interception *per se* is not always related to maximum potential orchard yield because there exists significant intra-canopy shading, depending on leaf area index, and canopy architecture. To wit, whole-tree net carbon dioxide (CO_2) exchange rate of mature sweet cherry trees was related negatively to leaf area index (Fig. 5.4). However, the relationship between daily whole-tree net CO_2 assimilation and leaf area index reveals

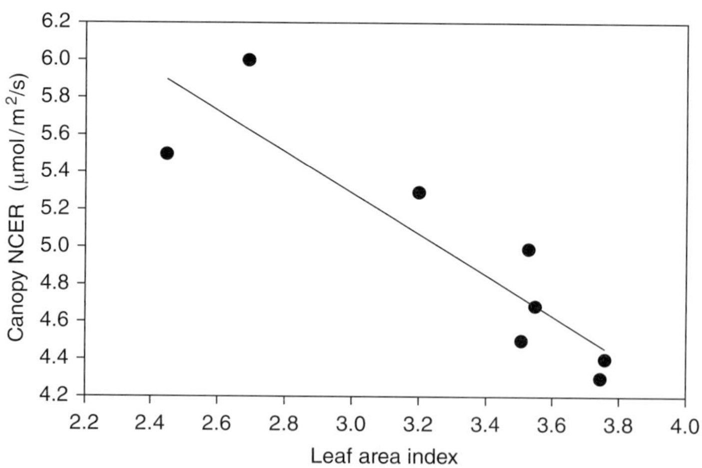

Fig. 5.4. Relationship between whole-canopy net CO_2 exchange rate (NCER) and canopy leaf area index in mature 'Bing' sweet cherry trees trained to a multiple-leader vase architecture.

a different trend; one that suggests there is an optimum canopy leaf area index. As intra-canopy shading increases with greater leaf area index, the rate of net CO_2 assimilation declines linearly within the range evaluated (i.e. ca. 2.4–3.8), yet this is offset, to a point, with increasing capacity for assimilation. Beyond ca. 3.2–3.3 leaf area index, whole-tree CO_2 assimilation declines due to excessive intra-canopy shading. Corelli and Sansavini (1989) demonstrated that light intensity decreases more rapidly from the upper to lower regions of the tree when planted in double- or triple-row systems, and that this reduces leaf CO_2 uptake from intra-canopy shading. These relationships, as well as orchard light interception, are typically modeled (e.g. Jackson and Palmer, 1972, 1979) due to the difficulty of collecting whole-canopy leaf area data and, in particular, recording reliable whole-canopy net CO_2 exchange rates. These relationships have yet to be investigated in modern, planar orchard systems. New research in this area is warranted as orchardists strive to optimize canopy volume (i.e. leaf area index) and the ability of trees to intercept incident irradiation. Recent work has revealed a potential advantage of Y-trellised fruiting wall systems compared with vertical systems due to the greater light interception in the former (Zhang *et al.*, 2015). Diurnal light interception of sweet cherry trained to a Y-trellised upright fruiting offshoots (UFO) architecture was much higher than trees trained to a vertical UFO architecture, particularly in the hours prior to, and immediately after, solar noon. This is clearly due to the ‘footprint’ of these systems – the Y-trellised canopy is trained over the orchard alleyway, intercepting more light at high zenith angles. There is little to no effect of orchard system on light interception at the beginning or end of the day.

Canopy light distribution is equally important, particularly at full canopy. In a study of apple orchard systems (i.e. cultivar/rootstock/architecture), Robinson and Lakso (1991) reported highest light interception in Y-trellised trees compared with slender spindle or central leader. In addition, the Y-trellised trees had the highest light conversion efficiency, determined as the weight of fruit per light interception. The pyramidal shape of central leader and tall spindle tree architectures limits the ability of these systems to intercept light. In contrast, the Y-trellised architectures, in which the canopy grows over the tractor alley, are able to intercept greater light levels, and this often leads to higher yield potential, and greater efficiency of conversion to fruit (Robinson and Lakso, 1989).

5.5 Variability in Fruit Quality

Orchardists adopt precision canopy management strategies to minimize variability among fruit for key quality parameters including size, color, soluble solids, etc. Modern planar orchard systems, and the strategic pruning and growth manipulations that go into them, are designed to optimize canopy light interception and distribution, and thereby improve fruit quality. There are few intensive studies on the variability in fruit quality in orchard systems. Proebsting and Murphey (1987) reported coefficients of variation of 19–24% for firmness and 14% for weight among sweet cherry fruit. In 'Granny Smith' apple trees trained to the 'Solaxe' central leader architecture, a 'fractionator' tree-sampling technique estimated high variability in the distribution among fruit in starch content and fruit mass, whereas fruit soluble solids content and firmness were less variable (Vega *et al.*, 2013). In this study of a non-planar architecture, fruit starch, soluble solids content, and firmness were not related to fruit position in the canopy (i.e. height or aspect). Also in 'Granny Smith', Warrington *et al.* (1996) found that fruit yield was greatest in canopy regions with the highest light transmission values, with the greenest and firmest fruits found in the inner canopy regions. There was little variability in fruit quality comparing directional quadrants, though fruit size was greater on the north side of each canopy, irrespective of training system. Similarly, Volz *et al.* (1995) reported improved 'Royal Gala' fruit size in fruit from the north side of trees, but reported only a weak correlation between canopy factors and fruit quality. In a study of greenhouse-grown nectarine trees trained to a Y-shape, fruit color and soluble solids content were greater in the upper third of the canopy, but the largest fruit were in the middle third of the trees (Kong *et al.*, 2011). In addition, fruit from well exposed regions of the canopy had higher soluble solids content and color than fruit from shaded canopy regions. An intensive study of three consecutive 'Fuji' apple trees trained to a tall spindle architecture revealed significant variability in fruit quality attributes, with similar variability within and among trees. The three-dimensional (3D) position of every fruit was mapped and quality attributes were determined for each fruit separately (Fig. 5.5; Whiting, M., unpublished).

There were no clear relationships between fruit position and fruit quality by directional analyses nor for height in the canopy (Figs 5.6 and 5.7). This is likely due to the excellent light distribution throughout canopies of these young trees that had not yet filled the between-tree, within-row space. Yet, in spite of the grower strategically hand thinning fruitlets to balance crop load, there remained high variability in fruit quality including soluble solids content (ca. 9.0–17.5%), weight (83–390 g), and a more than two-fold range in fruit firmness. Similar work in sweet cherry revealed 1.5–2-fold differences in individual fruit quality attributes among fruit on mature 'Bing' and 'Regina' limbs, though selective fruit thinning is not routinely carried out for this crop. Investigation of position–quality relationships in sweet cherry similarly uncovered no correlation, though fruit soluble solids content was slightly greater in the canopy exterior, on the youngest fruiting wood. These results underscore the importance of better understanding of fruit developmental physiology, the important steps from flower bud initiation to flowering and fruit growth, to maximize fruit quality potential in precision orchard systems.

The benefits of a well illuminated canopy (i.e. both high light interception and distribution) led to the development and commercial production of reflective fabrics and mulches to be laid down in inter-row

Fig. 5.5. Model of three consecutive third-leaf 'Fuji' trees illustrating fruit position, size and color. Size of model fruit is proportional to actual size and color is indicative of percentage of red coloration of fruit skin. Data points were collected with Topcon laser total station; 3D virtual trees and fruit were constructed using Matlab.

alleyways (Fig. 5.8). There are several commercial products available and all provide some benefit through reflecting 'lost' light (i.e. that which was not absorbed by the canopy) back into the canopy. In nectarine, reflective mulches have improved fruit production and fruit weight and the concentrations of phenolic compounds in the exocarp (Andreotti *et al.*, 2009). Similar results have been shown in d'Anjou pears – tree yield was ca. 20% higher from trees grown with reflective fabric compared with trees without. This improvement was attributed to increases in photosynthetically active radiation of ca. 25%, 90% and 30% in the exterior, mid and interior canopy zones, respectively (Einhorn *et al.*, 2011). In sweet cherry, full-season treatment with reflective fabric increased shoot length by ca. 30%, annual trunk cross-sectional increment by 90%, and

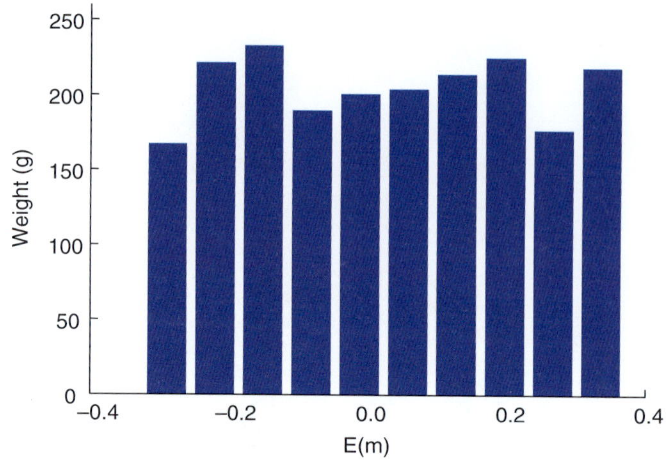

Fig. 5.6. Variability in mean 'Fuji' apple fruit weight (g) based on fruit position (E–W aspect, E) where 0 is the central tree axis. Data collected from commercial 3-year-old trees trained to tall spindle architecture and planted in N–S rows.

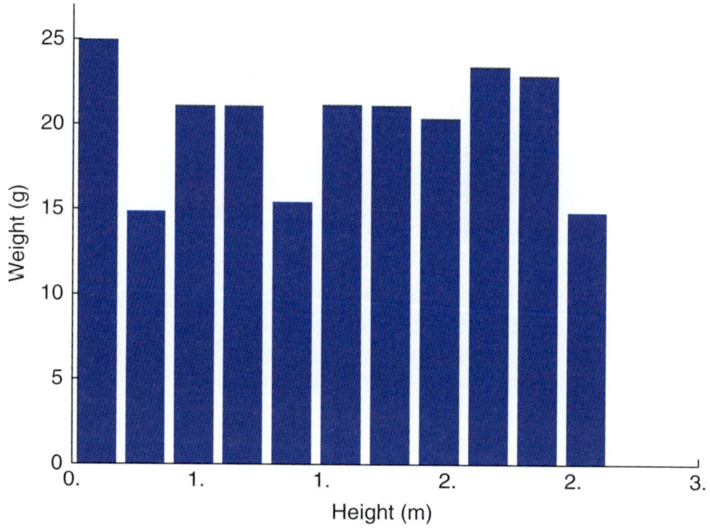

Fig. 5.7. Variability in mean apple fruit weight (g) based on fruit position (canopy height, H) where 0 is the orchard floor.

net CO_2 exchange rate of leaves in the canopy interior by 50%, compared with untreated trees of 'Bing', by improving canopy source–sink relations (Whiting et al., 2008). In addition, 'Bing' fruit maturity was advanced by an estimated 5 days with reflective fabric treatment. In apple, several studies have documented that reflective mulch treatments have hastened fruit maturation, and improved fruit color and market value, compared with untreated trees (Iglesias and Alegre, 2009; Hanrahan *et al.*, 2011; Privé *et al.*, 2011; Overbeck *et al.*, 2013). These reflective ground

Fig. 5.8. Reflective fabric laid down in the alleyway of 'Rainier' sweet cherry trained to the UFO system in Washington State.

covers are particularly useful in moderate-density orchards where total light interception is low and there is significant intra-canopy shading, yet many narrow fruiting wall orchards benefit from their use as well, primarily due to improvements in fruit quality and the potential to increase yield. Indeed, improved source–sink balance from reflective fabrics has increased the carrying capacity in sweet cherry orchards as fruit quality is maintained at higher fruit load.

5.6 Orchard Systems for Harvest Mechanization

The high labor requirements of traditional tree fruit production systems are driving innovation in orchard systems and this is most urgent for harvest, the costs of which often account for more than 50% of annual production costs. In many regions of the world, the cost of skilled harvest labor continues to increase while the availability of this workforce declines. There is an imminent need to improve harvest efficiency of orchard crops, and future orchard systems must consider current and potential harvest technologies. Compact fruiting wall architectures are particularly well suited to utilization of platforms for harvest. A Y-trellised configuration will be important for economical mechanical harvest with minimal fruit damage.

In apple, there have been recent developments towards both a massive harvest approach (i.e. shake-and-catch) (DeKleine and Karkee, 2015) as well as a robotic harvest system. While the shake-and-catch approach may be utilized in 3D architectures, early trials have revealed the necessity for compact planar architectures for robotic harvest, with high visibility of fruit being a requirement (Silwal *et al.*, 2016). Recent trials in apple investigating the relationship between limb length and fruit removal using a shake-and-catch approach found that fruit removal rates were poor when limbs were longer than ca. 15 cm (M. Karkee, personal communication). Dormant pruning of fruiting laterals to less than 15 cm was effective for improving fruit removal. Recent precision pruning trials on 'Jazz' apple in Washington State found that fruit detachment efficiency under shake-and-catch harvest was ca. 96% or 65% when lateral fruiting shoots were pruned to 10 cm or 20 cm, respectively (M. Karkee and M. Whiting, unpublished).

In sweet cherry, an exceptionally labor-intensive crop to harvest due to large tree size and high fruit number per tree, there is recent progress toward mechanizing harvest (Peterson *et al.*, 2003; Larbi *et al.*, 2015). From research on Y-trellised 'Bing' in the USA with a prototype mechanical mass-harvest system, machine harvest costs were US$0.04/kg compared with US$0.55/kg for hand harvest (Seavert and Whiting, 2011). The study concluded that mechanically harvested sweet cherry orchards would be more profitable than traditional orchard systems, but commercial adoption of the harvest system has yet to be realized due to limitations in orchard systems: the harvester requires a Y-trellised architecture that is not widespread in sweet cherry. Further research with a similar harvest system showed potential for improving fruit removal rates with multiple harvest passes in a Y-trellised architecture (He *et al.*, 2015).

The limb angle in Y-trellised fruiting walls is important for efficient mechanical harvest and reducing fruit damage. Experimental Y-trellised sweet cherry orchards at Washington State University have been trained to ca. 60 degrees from horizontal to reduce the drop to the harvester catching surface, and reduce the potential for fruit–branch impact during harvest (Fig. 5.9). In this system, fruit removal and recovery rates are high, and fruit damage is comparable to that from hand harvest (Peterson *et al.*, 2003; Ampatzidis *et al.*, 2012). The relatively flat fruiting walls create a high light environment in the middle of the canopy, and, as a result, often a proliferation of vigorous shoots. More research is needed to better understand the effects of limb angle on sucker shoot growth, and fruit damage during mechanical harvest.

5.6.1 Case study: Upright Fruiting Offshoots system for sweet cherry

Traditionally, sweet cherry has been trained to an open-center, multiple-leader architecture based on vigorous seedling rootstocks and low tree densities (e.g. 200–400 trees/ha) (Whiting *et al.*, 2005). These production systems can take 5–7 years to start fruiting and 10–12 years to achieve full production, and have high labor requirements at maturity due to the large tree size. Adopting precocious rootstocks improved production economics for growers with the ability to harvest fruit in the first 3–5 years after planting (Lang, 2001; Whiting *et al.*, 2005). Further, the introduction of size-controlling rootstocks was fundamental to the transition from

Fig. 5.9. High-density Y-trellised sweet cherry orchard trained to upright fruiting offshoots (UFO) architecture for mechanical harvest.

traditional systems to higher-density architectures that offer improved labor efficiency (Lang, 2000, 2001). The modern training systems are relatively easy to maintain after their establishment, with appropriate training and pruning methods: yields can be higher with early production of good fruit quality; foliar applications are better accomplished due to increased penetration within the canopy; cultural practices are performed efficiently because of minimal need for ladders; and the costs for labor are reduced. The UFO architecture has evolved from precision orchard systems trials at Washington State University, particularly those investigating mechanical harvest efficiency. The secondary and tertiary lateral branching of most other orchard systems develops a complex architecture to which it becomes difficult to apply simple pruning and training rules. In addition, in early mechanical harvest tests the lateral branching in traditional sweet cherry architectures resulted in poor fruit removal and recovery rates, and high fruit damage. The UFO training system for sweet cherry is a trellised high-density system that forms a compact fruiting wall architecture at maturity that may be configured vertically or to a Y-trellis. Trees are planted at a 45 degree angle and trained slightly above the horizontal to form a permanent scaffold from which upright fruiting leaders are trained at about 0.2 m apart (Long *et al.*, 2015; Zhang *et al.*, 2015) (Fig. 5.10). This planar architecture strategically facilitates the adoption of mechanization for tasks such as pruning, harvest and thinning; improves labor efficiency; and simplifies cultural practices (Ampatzidis and Whiting, 2013; Zhang *et al.*, 2015). In a study of hand harvesting in 11 commercial sweet cherry orchards, the highest mean (± SE) harvest rates (0.94 ± 0.02 kg/min

Fig. 5.10. Single row of sweet cherry trained to a vertical UFO architecture.

and 0.78 ± 0.03 kg/min) were recorded in 'Cowiche'/'Gisela5' and 'Tieton'/'Gisela5' orchards trained to the UFO system, respectively. High harvest efficiency in these orchards was likely the result of the planar, simplified architecture and because most fruit were accessible from the ground.

Pruning a mature UFO orchard consists of a two-step process: renewal of one or two of the most vigorous upright leaders per year with a stub cut, and removal of all lateral branches on upright leaders with thinning cuts (Long *et al.*, 2015) (Fig. 5.11). After a brief explanation of these pruning rules, unskilled laborers made > 95% correct cuts when dormant pruning a Y-trellised sweet cherry UFO orchard. The simplification of pruning and training will be fundamental for any future orchard system, particularly as robotic pruning technologies are developed.

5.7 Future Precision Orchard Systems

Current orchard systems are based upon size-controlling rootstocks with superior scion genotypes, bred for large size, excellent coloring, and high consumer appeal. New orchards are being planted at ultra-high densities in order to fill the intra- and inter-row space rapidly, and developing fruiting sites as quickly as possible. Fruiting wall architectures are standard for apple orchards and more realistic for sweet cherry with size-controlling rootstocks and new architectures like the UFO. The key advantage of these narrow architectures is their ability to accommodate mechanization, automation, and precision manipulation of vegetative and flowering/fruiting sites (e.g. manually thinning apple flowers/clusters). There is little doubt that planar systems will be favored in the next decades due

Fig. 5.11. Sweet cherry tree trained to UFO system. Illustrated prior to harvest in the second year (left) and dormant after second growing season (right), indicating the removal of all lateral branches.

to their production efficiencies, excellent light interception, distribution, and high harvest indices. Further, there appears to be great justification for the Y-trellised architecture compared with vertical fruiting walls because of greater light interception and, therefore, yield potential. Light, the most important input to any orchard system, comes at no cost to the grower. This simple fact must be understood and optimized in future orchard systems.

The need to provide sufficient inter-row space to accommodate machinery has limited the distance between rows in orchards; this is typically is ca. 2.5–3.0 m. There is some recent interest in narrow row spacings (e.g. 1.5–2.0 m) of vertical fruiting walls, and adopting over-the-row machinery for all orchard processes. Experimental orchards have been planted in Washington and New Zealand (S. Tustin, personal communication) to examine yield and fruit quality potential from such systems. Experience with existing vertical fruiting walls suggests that ratios of canopy height to row spacing may be ca. 1.25:1 without negative effects on floral initiation or fruit quality in the lower regions of the canopy, provided that the canopy is porous to light. Whether this ratio may be exceeded in narrow-row orchard systems remains to be seen. Radical changes to orchard systems have been proposed for decades, and not all have been successful commercially (Luckwill, 1978).

References

Amiri, M.E., Fallahi, E. and Safi-Songhorabad, M. (2014) Influence of rootstock on mineral uptake and scion growth of 'Golden Delicious' and 'Royal Gala' apples. *Journal of Plant Nutrition* 1, 16–29.

Ampatzidis, Y.G. and Whiting, M.D. (2013) Training system affects sweet cherry harvest efficiency. *HortScience* 48, 547–555.

Ampatzidis, Y., Zhang, Q. and Whiting, M. (2012) Comparing the efficiency of future harvest technologies for sweet cherry. *Acta Horticulturae* 965, 195–198.

Andreotti, C., Ravaglia, D. and Costa, G. (2009) Innovative light management to improve production, sustainability, overall quality, and the phenolics composition of nectarine (*Prunus persica* cv. Stark Red Gold). *Journal of Horticultural Science and Biotechnology* 84, 145–149.

Barden, J.A. (2002) The influence of cultivar and orchard system on pruning time per tree, per hectare, and per unit of yield. *Journal of American Pomological Society* 56, 202–206.

Barritt, B.H. (1989) Influence of orchard system on canopy development, light interception and production of third-year Granny Smith apple trees. *Acta Horticulturae* 243, 121–130.

Barritt, B.H., Konishi, B. and Dilley, M. (2008) Performance of four high density apple orchard systems with 'Fuji' and 'Braeburn'. *Acta Horticulturae* 772, 389–394.

Caruso, T., LoBianco, R., Marra, F.P. and Guarino, F. (2015) Evaluation of small vase and Y-trellis orchard systems for peach and nectarine production in Mediterranean regions. *Acta Horticulturae* 1084, 465–470.

Corelli, L. and Sansavini, S. (1989) Light interception and photosynthesis related to planting density and canopy management in apple. *Acta Horticulturae* 243, 159–174.

DeKleine, M.E. and Karkee, M. (2015) Semi-automated harvesting prototype for shaking fruit tree limbs. *Transactions of the ASABE* 58, 1461–1470.

Einhorn, T.C., Turner, J. and Laraway, D. (2011) Reflective fabric improves intra-canopy light levels and increases yield in a low-density, mature 'd'Anjou' pear orchard. *Acta Horticulturae* 909, 565–570.

Fallahi, E., Colt, W.M., Fallahi, B. and Chun, I.J. (2002) The importance of apple rootstocks for tree growth, yield, fruit quality, leaf nutrition and photosynthesis with an emphasis on 'Fuji'. *Hort Technology* 12, 38–44.

Gao, Y., Motosugi, H. and Sugiura, A. (1992) Rootstocks effects on growth and flowering in young apple trees grown with ammonium and nitrate nitrogen. *Journal of the American Society for Horticultural Science* 117, 446–452.

Grappadelli, L.C. and Lakso, A.N. (2007) Is maximizing orchard light interception always the best choice? *Acta Horticulturae* 732, 507–518.

Hanrahan, I., Schmidt, T.R., Castillo, F. and McFerson, J.R. (2011) Reflective ground covers increase yields of target fruit of apple and pear. *Acta Horticulturae* 903, 1095–1100.

He, L., Zhou, J., Zhang, Q. and Karkee, M. (2015) Evaluation of multipass mechanical harvesting on 'Skeena' sweet cherries trained to Y-trellis. *HortScience* 50, 1178–1182.

Heinicke, D.R. (1963) The micro-climate of fruit trees. II. Foliage and light distribution patterns in apple trees. *Proceedings of the American Society for Horticultural Science* 83, 1–11.

Hirst, P.M. and Ferree, D.C. (1995) Rootstock effects on shoot morphology and spur quality of 'Delicious' apple and relationships with precocity and productivity. *Journal of the American Society for Horticultural Science* 120, 622–634.

Hrotkó, K. (2013) Development in fruit trees production systems. *AgroLife Science Journal* 2, 28–35.

Iglesias, I. and Alegre, S. (2009) The effects of reflective film on fruit color, quality, canopy light distribution, and profitability of 'Mondial Gala' apples. *HortTechnology* 19, 488–498.

Jackson, J.E. and Palmer, J.W. (1972) Interception of light by model hedgerow orchards in relation to latitude, time of year, and hedgerow configuration and orientation. *Journal of Applied Ecology* 9, 341–357.

Jackson, J.E. and Palmer, J.W. (1979) A simple model of light transmission and interception by discontinuous canopies. *Annals of Botany* 44, 381–383.

Kong, Y., Yao, Y., Ma, C. and Li, B. (2011) Effect of canopy position on some fruit quality parameters of greenhouse-grown nectarine. *Acta Horticulturae* 893, 925–930.

Lang, G.A. (2000) Precocious, dwarfing, and productive – how will new cherry rootstocks impact the sweet cherry industry? *HortTechnology* 10, 719–725.
Lang, G.A. (2001) Critical concepts for sweet cherry training systems. *Compact Fruit Tree* 34, 70–73.
Larbi, P.A. Karkee, M., Amatya, S., Zhang, Q. and Whiting, M. (2015) Modification and field evaluation of an experimental mechanical sweet cherry harvester. *Applied Engineering in Agriculture* 31, 387–397.
Layne, R.E.C. (1987) *Peach Rootstocks.* In: Rom, R.C. and Carlson, R.F. (eds) *Rootstocks for Fruit Crops.* Wiley, New York, pp. 185–216.
Long, L., Lang, G., Musacchi, S. and Whiting, M. (2015) *Cherry Training Systems.* Pacific Northwest Extension Publication PNW 667. Oregon State University, Corvallis, Oregon.
Looney, N.E. (1968) Light regimes within standard size apple trees as determined spectrophotometrically. *Proceedings of the American Society for Horticultural Science* 93, 1–6.
Luckwill, L.C. (1978) Meadow orchards and fruit walls. *Acta Horticulturae* 65, 237–244.
Overbeck, V., Schmitz-Eiberger, M.A. and Blanke, M.M. (2013) Reflective mulch enhances ripening and health compounds in apple fruit. *Journal of the Science of Food and Agriculture* 93, 2575–2579.
Palmer, J.W. (1981) Computed effects of spacing on light interception and distribution within hedgerow trees in relation to productivity. *Acta Horticulturae* 114, 80–88.
Palmer, J.W. (2011) Changing concepts of efficiency in orchard systems. *Acta Horticulturae* 903, 41–49.
Palmer, J.W. and Warrington, I.J. (2000) Underlying principles of successful apple planting systems. *Acta Horticulturae* 513, 357–363.
Peterson, D., Whiting, M.D. and Wolford, S. (2003) Fresh market quality sweet cherry harvester. *Applied Engineering Agriculture* 19, 539–543.
Privé, J.P., Russell, L. and Leblanc, A. (2011) Impact of reflective groundcover on growth, flowering, yield, and fruit quality in Gala apples in New Brunswick. *Canadian Journal of Plant Science* 91, 765–772.
Proebsting, E.L. and Murphey, A.S. (1987) Variability of fruit quality characteristics within sweet cherry trees in central Washington. *HortScience* 22, 227–230.
Reginato, G.H., Carrasco, Ó. and Garcia de Cortázar, V. (2008) Evolution of planting and training systems in apple and peach orchards over 25 years of intensive fruit industry development in Chile. *Acta Horticulturae* 772, 431–440.
Reighard, G.L. (2002) Current directions of peach rootstock programs worldwide. *Acta Horticulturae* 592, 421–427.
Robinson, T.L. (2008) The evolution towards more competitive apple orchard systems in the USA. *Acta Horticulturae* 772, 491–500.
Robinson, T.L. and Lakso, A.N. (1989) Light interception, yield and fruit quality of 'Empire' and 'Delicious' apple trees grown in four orchard systems. *Acta Horticulturae* 243, 175–184.
Robinson, T.L. and Lakso, A.N. (1991) Bases of yield and production efficiency in apple orchard systems. *Journal of the American Society for Horticultural Science* 116, 188–194.
Rom, R.C. and Carlson, R.F. (eds) (1987) *Rootstocks for Fruit Crops.* Wiley, New York.
Seavert, C.F. and Whiting, M.D. (2011) Comparing the economics of mechanical and traditional sweet cherry harvest. *Acta Horticulturae* 903, 725–730.
Silwal, A., Karkee, M. and Zhang, Q. (2016) A hierarchical approach to apple identification for robotic harvesting. *Transactions of the ASABE* 59, 1079–1086.
Vega, M.V.M., Wulfson, D., Clemmensen, L.H. and Toldham-Andersen, T.B. (2013) Using multilevel systematic sampling to study apple fruit (*Malus domestica* Borkh.) quality and its variability at the orchard scale. *Scientia Horticulturae* 161, 58–64.
Volz, R.K., Palmer, J.W. and Gibbs, H.M. (1995) Within-tree variability in fruit quality and maturity for 'Royal Gala' apple. *Acta Horticulturae* 379, 67–74.

Warrington, I.J., Stanley, C.J., Tustin, D.S., Hirst, P.M. and Chashmore, W.M. (1996) Light transmission, yield distribution, and fruit quality in six tree canopy forms of 'Granny Smith' apple. *Journal of Tree Fruit Production* 1, 27–54.

Weber, M.S. (2001) Optimizing the tree density in apple orchards on dwarf rootstocks. *Acta Horticulturae* 557, 229–234.

Whiting, M., Lang, G. and Ophardt, D. (2005) Rootstock and training system affect sweet cherry growth, yield, and fruit quality. *HortScience* 40, 582–586.

Whiting, M.D., Rodriguez, C. and Toye, J. (2008) Preliminary testing of a reflective ground cover: sweet cherry growth, yield, and fruit quality. *Acta Horticulturae* 795, 557–560.

Wünsche, J.N. and Lakso, A.N. (2000) The relationship between leaf area and light interception by spur and extension shoot leaves and apple orchard productivity. *HortScience* 35, 1202–1206.

Wünsche, J.N., Lakso, A.N., Robinson, T., Lenz, F. and Denning, S. (1996) The bases of productivity in apple production systems: the role of light interception by different shoot types. *Journal of the American Society for Horticultural Science* 121, 886–893.

Zhang, J., Whiting, M. and Zhang, Q. (2015) Diurnal pattern in canopy light interception for tree fruit orchard trained to an upright fruiting offshoots (UFO) architecture. *Biosystems Engineering* 129, 1–10.

6 Variable Rate Irrigation on Center Pivots

R. Troy Peters*

Washington State University, Prosser, Washington, USA

6.1 Introduction

Variable rate irrigation (VRI), also sometimes referred to as 'precision' or 'site-specific' irrigation, is the ability of an irrigation system to apply different amounts of water to different areas of the field. Center pivot irrigation is the largest single method of irrigation in the USA. This chapter discusses the various VRI options for center pivots, when they might save water, save energy, and create higher crop yields, and when it might be unreasonable to expect these kinds of improvements. Some of the remaining challenges associated with VRI are discussed, and a simple soil-water balance model is used to illustrate water savings estimates for various soils and how VRI might be used to take advantage of significant in-season rainfall events. The benefits of VRI on other irrigation systems such as drip can be readily extrapolated from the discussion centering on center pivots since similar principles apply.

6.2 Variable Speed Irrigation versus Variable Zone Irrigation

Recently center pivot manufacturers and some third-party equipment dealers have been offering VRI as an option or upgrade on their pivots in a couple ways: variable speed irrigation, and variable zone irrigation.

* email: troy_peters@wsu.edu

6.2.1 Variable speed irrigation

Variable speed irrigation does not require additional hardware on the pivot. The overall water flow rate to the pivot remains constant. It simply uses a more sophisticated control panel that will slow down or speed up the pivot to apply more or less water in different areas of the field. Many of the newer pivot control panels already have variable speed capability built into them. After-market solutions for variable speed irrigation from third-party equipment dealers usually mount on the last tower of the pivot and have an integrated global positioning system (GPS) receiver to determine field position. They usually interrupt and modify the movement control signal to the last tower to control the speed of the pivot in different areas of the field. Despite variable speed irrigation's obvious limitations to variations only in pie-shaped wedges (Fig. 6.1), variable speed irrigation is fairly low cost (US$2000–4000) since the only modifications to the pivot are to the pivot's electronic control algorithms. These costs will likely decrease over time.

Some additional useful applications for variable speed technology are as follows.

- On a pivot that cannot go all the way around (a 'wiper') it is possible to vary the speed going into or coming out of the hard stops (ends of the field where the pivot must reverse direction) to avoid running the pivot in overly wet areas, in an attempt to reduce wheel-tracking issues. For example, if the wiper is applying 12.5 mm in a pass (25 mm for every back-and-forth wipe), the pivot might speed up to apply 5 mm of water in the 20 degrees of angle before the hard stop so that the field stays drier. Then, after reversing, it might slow down to apply 20 mm until it reaches the 20 degree mark again, where it speeds up slightly again to return to applying 12.5 mm.
- In areas of the field where infiltration is an issue due to tight soils or steep slopes, it is possible to speed up to wipe back and forth across that area of the field to allow additional time between water applications for water to infiltrate and move deeper into the soil before water is again applied. For example, if there is always runoff on a slope between 20 and 40 degrees and the grower is applying 18 mm of water in a clockwise rotation, the pivot could speed up at 20 degrees to apply 6 mm over the trouble spot, reverse at 40 degrees to apply 6 mm, travel back to 20 degrees where the pivot would again reverse to apply 6 mm (for a total of 18 mm on the trouble spot). The pivot would then slow down at 40 degrees to apply 18 mm again to the rest of the field. The same total amount of water is applied to the trouble spot, but the back-and-forth movement gives more time between water applications for the water to move into the soil in that spot and hopefully increase/reduce the infiltration and runoff.
- Speed up slightly when climbing hills to compensate for tire slippage (Chavez *et al.*, 2010).

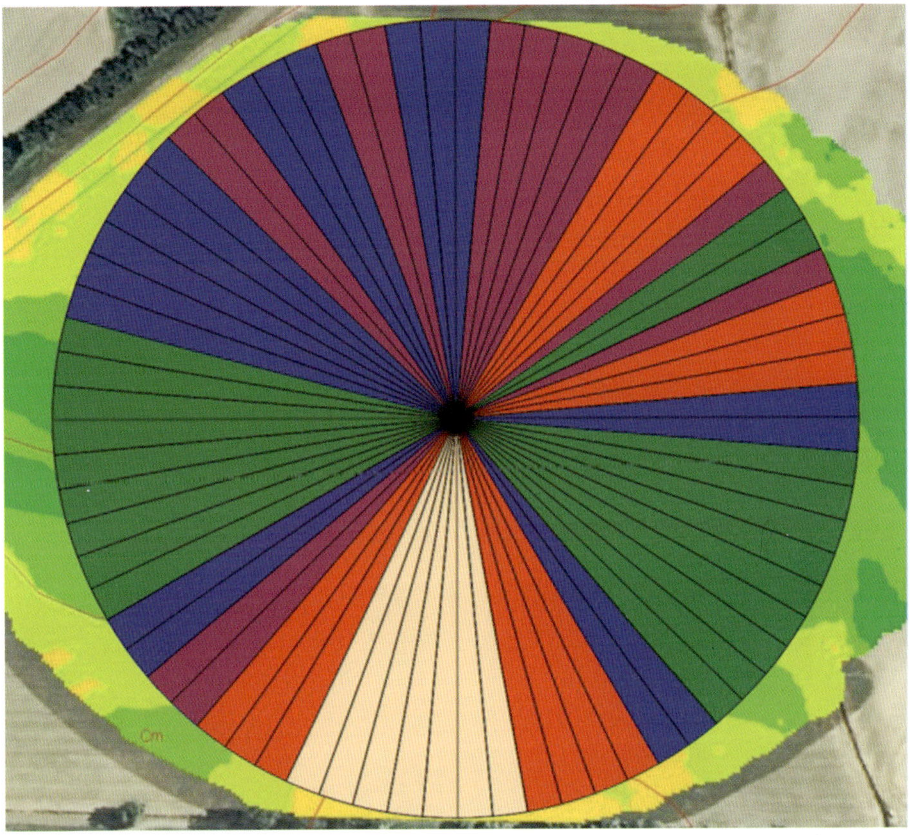

Fig. 6.1. Variable speed irrigation. The pivot varies the travel speed to apply variable amounts of water to defined zones within the field. Colors indicate areas with different amounts of water applied. Image used with permission from pivotirrigation.com.au.

6.2.2 Variable zone irrigation

Variable zone irrigation includes the ability to vary the speed of the center pivot as it moves in a circle *and* varies the application rate along the pivot lateral (Fig. 6.2). Variations in the application rate along the lateral work in conjunction with variations in the pivot speed, creating the ability to apply a wide variety of irrigation depths to different areas of the field. The application rate along the lateral is usually varied by pulsing sprinklers on and off for various amounts of time. In some cases zones of sprinklers are controlled independently, in other cases each sprinkler is controlled independently. Because additional hardware must be mounted on the pivot, as well as more sophisticated control technology, variable zone irrigation is significantly more expensive than variable speed irrigation (US$15,000–25,000) (Milton *et al.*, 2006). These costs will also likely decrease over time. Variable zone irrigation is much better at responding to the spatial variations in the field. Turning sprinklers on and off varies the

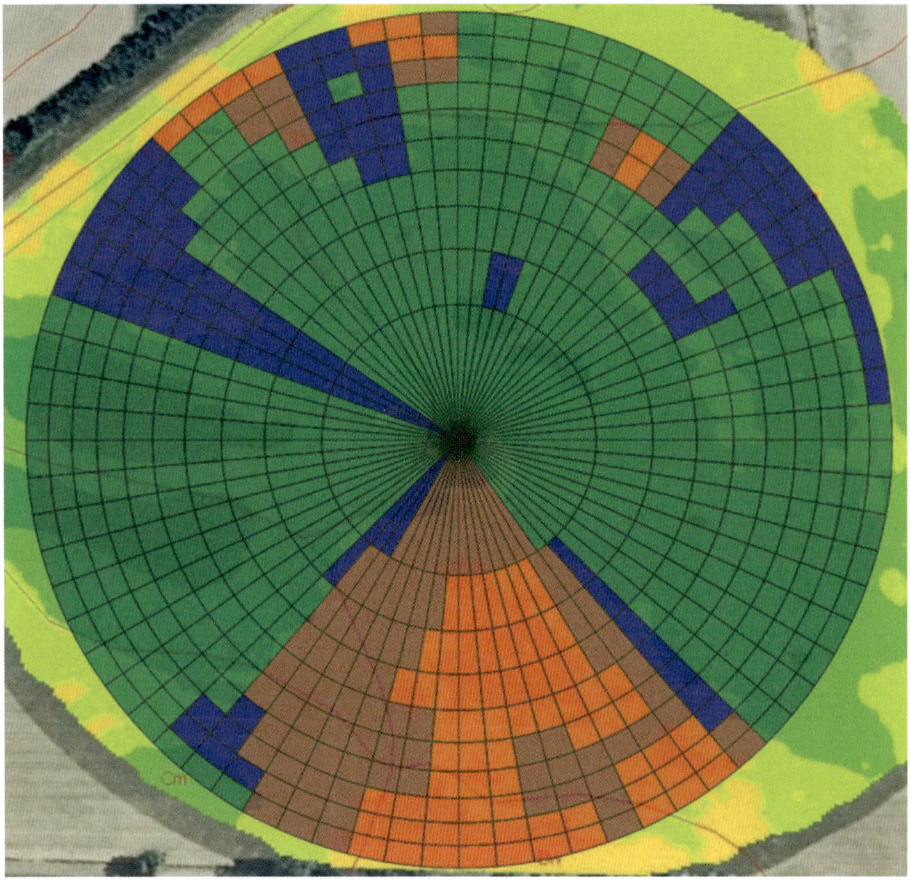

Fig. 6.2. Variable zone irrigation. The pivot varies both travel speed and application rate along the lateral to apply variable amounts of water to defined zones within the field. Colors indicate areas with different amounts of water applied. Image used with permission from pivotirrigation.com.au.

overall flow rate to the pivot and therefore a water delivery system that can absorb these variations is necessary.

6.3 Variable Rate Irrigation in Response to Variable Soils

The water use of healthy crops with access to sufficient water and nutrients will not be significantly dependent on what kind of soil they are grown in. Crops grown in sandy soils will not use significantly more or less water than crops grown in silt or clay soils. So, for example, even in a field with highly variable soils, all areas of the field will be using 6 mm of water every day. Because of this, applying different amounts of water to different areas of the field only makes sense if the crops are getting water from another source besides where the center pivot irrigation system is applying it, or if the crops are using less water in some areas of the field due to disease or pest pressure. More discussion on this follows below.

6.3.1 Variable rate irrigation in response to variations in soil water-holding capacity

While interacting with many growers the following statement is commonly heard: 'I apply more water to the sandier areas of my field during each irrigation.' Crops grown in sandy soils do not need more water. In fact, the soil cannot hold additional water if it is applied to it. If sandy soils are watered more each time, then the additional water will be lost to deep percolation. They need to be watered in smaller amounts but more frequently. Because of this, if the entire field is managed as a whole to prevent water stress and water losses to deep percolation in the sandy areas of the field (smaller amounts, more frequently), then all other areas of the field will also be fine (Figs 6.3 and 6.4).

To demonstrate this concept, some simulations were done using an irrigation scheduling tool (Irrigation Scheduler Mobile; http://weather.wsu.edu.ism; Peters, 2014) to model what the soil water content would look like in a sandy area of the field (Fig. 6.5) and in a silty area of the field (Fig. 6.6) if the whole field were managed for the sand. A similar simulation was

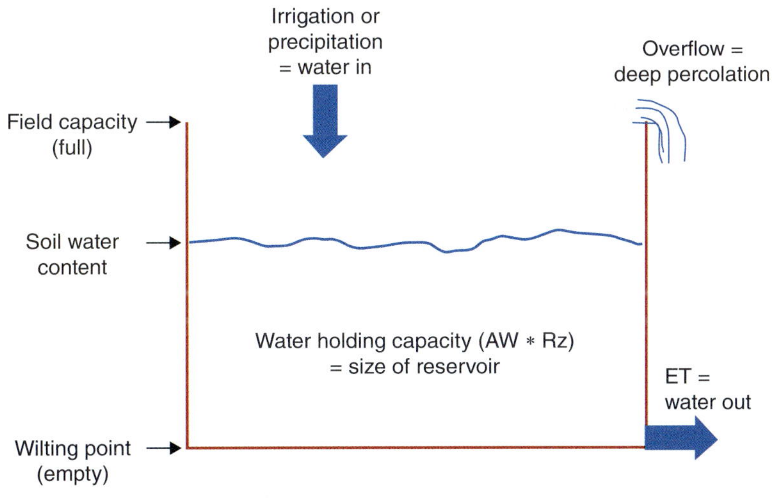

Fig. 6.3. Soil serves as a reservoir for water and nutrients. The size of the reservoir depends on the soil's water-holding capacity (how much water it can hold per meter of root depth) (AW), and the rooting depth of the soil or crop (Rz). Irrigation or precipitation that infiltrates into the soil when there is space in the soil to hold that water is stored for later use by the crop. If more water is applied to the soil than the soil can hold, then that extra water is lost (leached) from the bottom of the root zone (shown as overflow). Crop water use, or evapotranspiration (ET), is largely independent of the soil type.

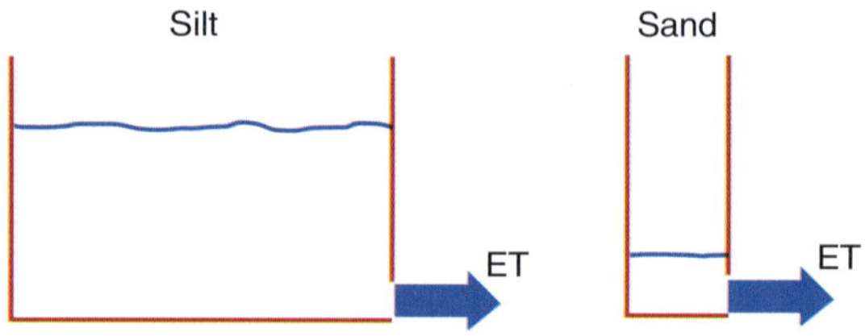

Fig. 6.4. If the same field has areas that are both silt and sand, then if they both started full, after a given amount of time the sandy areas will be getting dry and exhibiting crop water stress, while the silty areas will appear fine. If the entire field is managed for no stress, or no water losses to deep percolation in the sand (overflow in the diagram), then the silty areas will also be fine. If more water is applied to the sand when refilling the soil, that additional water will be lost to deep percolation.

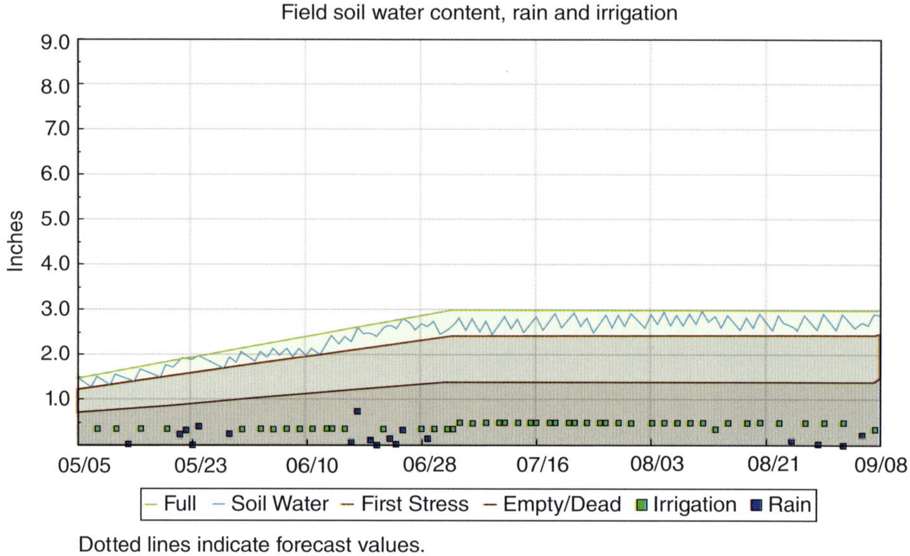

Fig. 6.5. Soil water content over time in relation to the full (field capacity), first stress (management allowable depletion) and empty/dead (permanent wilting point). Chart for a fine sand. In this situation a total of 25 mm of water was lost during the season to deep percolation due to untimely rainfall events.

done of the sandy (Fig. 6.7) and silty (Fig. 6.8) areas of the field if instead the whole field were managed for the silt. It can be seen that, when the entire field is managed for the soils with the lowest water-holding capacities (sandier soils), all other areas of the field are fine. This is not the case, however, if the field were managed for the soil with the larger water-holding capacity (the silt). In that case crops grown in the sand would show water stress.

6.3.1.1 Managing for sandy soils

We simulated the water use in a potato field with areas of both fine sand and silt loam soils with water-holding capacities of 66 mm/m and 192 mm/m

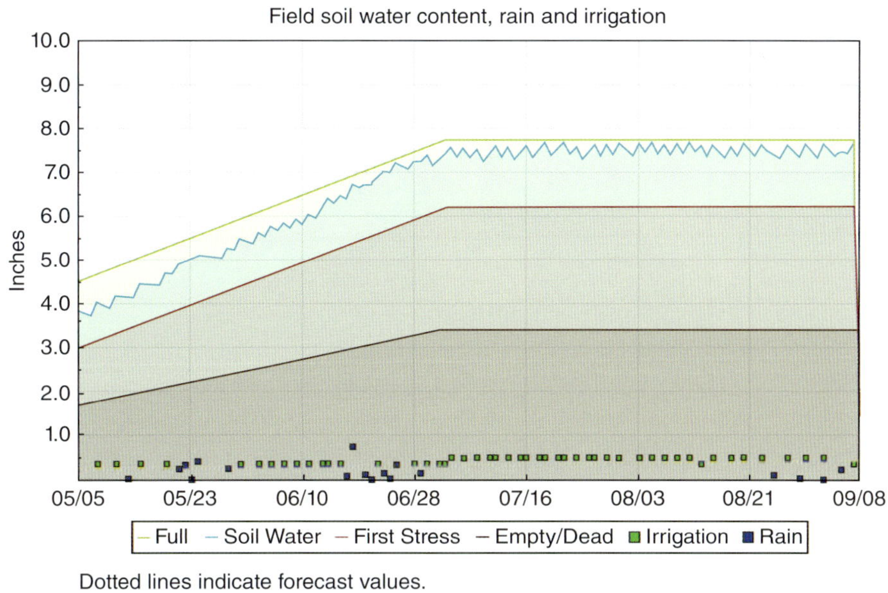

Fig. 6.6. Soil water chart for a silt loam soil managed for the fine sand.

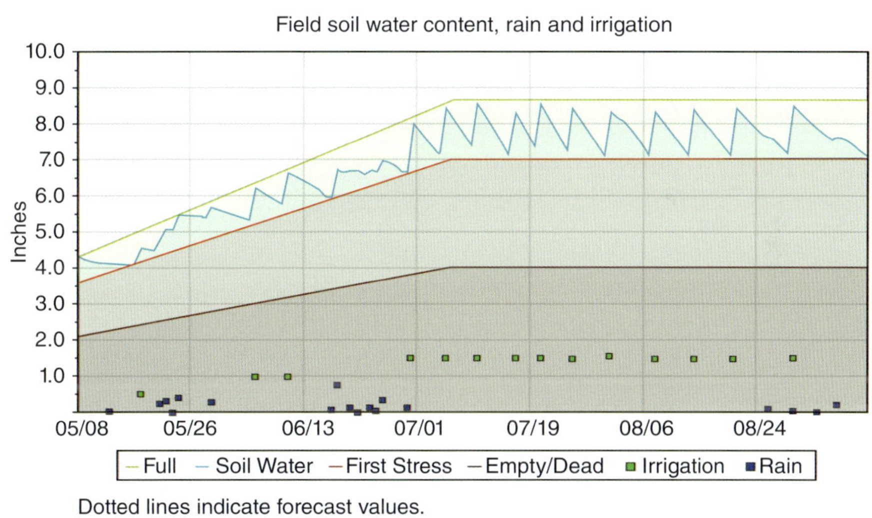

Fig. 6.7. Soil water content chart for irrigation for silt on silt. No water stress. Negligible deep percolation.

(0.8 and 2.3 in/ft), respectively. We did a simulation for a potato crop with a beginning root zone depth of 0.3 m (12 in), and an ending root zone depth of 0.31 m (24 in). The growing root zone causes the graphs to increase during the first part of the season. Figure 6.5 shows the soil water content over time of the sandy soil that was carefully irrigated for sandy soils such that there was no water stress (soil water content remained between the 'Full' and 'First Water Stress' lines) and limited

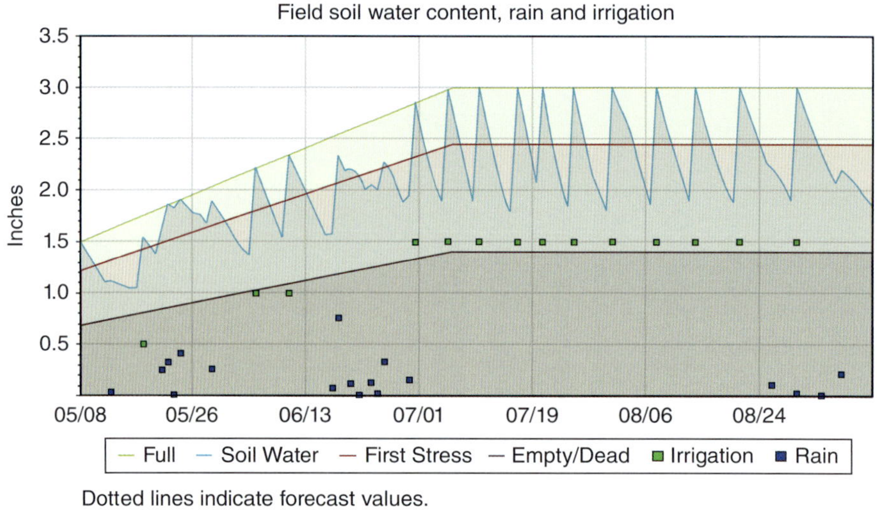

Fig. 6.8. Soil water content chart for irrigation for silt on sand.

water loss to deep percolation (leaching). Because the water-holding capacity of sands is small, frequent irrigations (green squares) of small amounts are required to avoid crop water stress and water losses to deep percolation.

Figure 6.6 shows how the silt loam soil section of this same block would fare in that same field that was managed for the sandy soil. The applied irrigation dates and amounts are the same for both scenarios. Under this management scenario, the following results (Table 6.1, Figs 6.5 and 6.6) are seen.

- There is no water stress in either block.
- The total crop water use (evapotranspiration) (ET) in both the sandy and in the silt soils of the block are exactly the same.
- The total losses to deep percolation are exactly the same, and occur on the same dates.
- There is no stress in either areas of the field and therefore the yields in both areas will be the same.
- At the end of the season the silty soil will have much greater residual water available than the sand.

6.3.1.2 Managing for silty soils

What if the water for the whole field were managed for the deep silt or clay soils? In this case, much more water can be applied at each irrigation event and these events can be much less frequent. The soil water content under this management strategy on the silt soils over time is shown in Fig. 6.7.

Figure 6.8 shows how the sandy areas of the field would fare if the whole field were managed uniformly for the silty soils. While the same

Table 6.1. Comparison of the different sections of the field (sandy or silty soils) when the whole field is managed uniformly for the sand. All water depths are in millimeters.

Scenario	Figure ref.	Season total evapo-transpiration (ET)	Total irrigation	Total rainfall	Deep percolation	Yield loss
Sandy soil. Managed for the sand, by replacing deficits	Fig. 6.5	590	527	78	25	0
Silt loam soil that is managed for the sand (above)	Fig. 6.6	590	527	78	25	0

Table 6.2. Comparison of two different areas of the field if the whole field is managed uniformly for the silt loam soil with deeper irrigations applied less frequently. All water depths are in millimeters.

Scenario	Figure ref.	Season total evapo-transpiration (ET)	Total irrigation	Total rainfall	Deep percolation	Yield loss
Silt soil that is managed to managed to replace the deficits in silt (deeper irrigations)	Fig. 6.7	583	466	78	0	0
Sandy soil that is managed for silt (deeper irrigations)	Fig. 6.8	485	466	78	91	17%

amount of water is applied on both the silt and the sand sections, under this scenario in the sandy areas the results are as follows.

- More water (91 mm) is lost to deep percolation (Table 6.2). More water is applied to the soil than it can hold in the root zone at the time of application.
- There is a 17% yield reduction of the crops in the sandy areas due to water stress.
- The crops would use 98 mm less water (ET) in the sandy areas due to shutting down as a result of water stress.

6.3.2 Variable rate irrigation in response to runoff in some areas

The following is also often heard as a justification for VRI: 'I have runoff on the steeper slopes, and the crop is water stressed in that area of the field so I apply more water to those slopes.' If water is already running off of a slope, applying more water in those areas will not result in additional infiltration in those areas, but instead will result in all of the additional applied water also running off, possibly causing erosion, and that additional runoff water may pond in the low spots of the field, making the overall irrigation and crop uniformity problems in the field worse. If the water is running off, then less water, not more, needs to be applied to slopes in a pass to ensure that the applied water infiltrates into the soil. But to ensure that these areas of the field do not fall behind the rest of the field, this means speeding the pivot up on the entire field as spatial variation would result in these areas falling permanently behind. The 'wiping' method described in the variable speed irrigation section above can help to reduce or eliminate runoff. As an alternative to speeding up the pivot, or as an additional runoff prevention measure, runoff in these steep sloped areas can be mitigated by changing the tillage methods, and possibly the crop row orientation. Modifying the sprinkler system so that it applies water at a slower rate can also help to improve infiltration. This might include using boombacks or draping every other sprinkler around the outside of the truss rods, or using sprinklers with a much larger wetted radius. If the soil is hydrophobic (water balls up and runs off dry soil instead of infiltrating) then using soil surfactants may also help with infiltration.

Because of these factors, in low-rainfall areas purchasing VRI in response to highly variable soils offers little opportunity to increase profitability in comparison with optimally managing the entire field uniformly for the problem soils.

6.4 Situations Where VRI Can Conserve Water and Improve Profitability

VRI will be more profitable if the costs of water, or the marginal opportunity cost of lost water, is high. The marginal opportunity cost of lost water is greatest when growing high-value crops and water is already very limited. Below are some situations where VRI can conserve water and possibly improve profitability, depending on the economics of the specific situation.

6.4.1 Non-cropped areas

VRI can save water and agrochemicals and reduce maintenance problems by completely shutting the water off in areas of the field that should not be irrigated (Sadler *et al.*, 2005). These might include rock piles, ponds, streams, waterways or roads that cross through the field, or areas under the irrigation system that are otherwise not farmable. Sometimes pivots overlap. Shutting the water off on one of these pivots in the overlapped areas will reduce overwatering those areas. These constant, unchanging prescriptions where the water is turned off completely will result in the largest water and power savings at the lowest long-term management costs. Consequently most VRI systems being sold are primarily being used in this application (Evans *et al.*, 2012). Avoiding off-target application of agrichemicals or liquid wastes is another large driver for the adoption of VRI in these situations. The water savings are directly related to the proportion of the area that is not irrigated.

6.4.2 Areas of the field getting water from other sources

VRI can conserve water by applying less water to areas of the field where the crops are getting water from other sources. These may be either a high water table, or an area where water is ponding in the field due to runoff from suboptimal operation of the pivot or from water running on to the field from outside sources. Watering these areas less can reduce over-irrigation, saturation of soils, losses of nitrates through leaching, and losses of yield due to waterlogging (Sadler *et al.*, 2005). It is often necessary to modify the VRI prescription (variable irrigation map or plan) throughout the season to expand or contract the affected area, or to irrigate these areas more or less because the alternative sources of water may not be constant or able to keep up with ET throughout the entire season.

6.4.3 Different crops in the same field

VRI will allow growing different crops in different areas of the field and managing the water for these areas separately. This may be especially useful to those who cannot or do not want to plant in pie-shaped sections. It may be especially useful for researchers or seed growers who have a wide variety of different plots, crops, or water treatments under the same pivot.

6.4.4 Overwatering the inside span

The sprinkler flow rates required on the first span of a center pivot (nearest the pivot point) are so low that these small nozzle sizes can get plugged with small debris in the water. Because of this, many pivot dealers put on larger nozzles than necessary and over-irrigate the inside of the pivot circle. The area underneath these inside spans is relatively minor compared with the area underneath the outer spans and so many growers and pivot designers do not worry about it as much. However, variable zone irrigation could be used to periodically shut these nozzles off to avoid over-watering these two inside spans. Allowing the canopy in these areas to dry more often may reduce plant diseases and therefore disease spreading to other areas of the field.

6.4.5 Variations in crop water use (ET)

If there is a large variation in crop water use across the field (ET), applying less water to the areas where the ET is lower might conserve this water. These lower ET rates may be due to disease or pest pressure, among other possibilities. Because these areas are using less water, applying the same amount of water in these areas as to the rest of the field may result in water losses to deep percolation in these areas. This might be counterintuitive, because most people want to water areas that are not doing well more, not less.

6.4.6 Use of pivot as a variable rate sprayer

VRI may come in very handy when it is used for chemigation or fertigation and there is a need to apply these agrichemicals at a variable rate. This can be especially beneficial for applying liquid wastes.

6.4.7 Control for uniform dry down

In some instances it may be desirable for the crops in all areas of a field with highly spatially variable soils to experience water stress at the same time. In this case it may be desirable to restrict irrigation water to areas that have greater water-holding capacity (deep silts or clays) sooner, so that the soil profile will be depleted at about the same time as the areas with lower water-holding capacities (shallow or sandy soils).

6.4.8 Leaving room in the soil to capture rainfall

In humid areas where there is significant in-season rainfall, periodically shutting the water off to the areas of the field with larger water-holding capacities will leave space in the soil to capture and hold anticipated rainfall. The sandy areas will still have to be irrigated on a regular basis to avoid stress, because of their small water-holding capacity; however, the soil water in the silty or clay areas can be partially depleted. Then, during significant rainfall events, there will be capacity to hold this rainfall in the silt or clay areas of the field. At these events there will be unavoidable rainwater losses to deep percolation in the sandy areas. Doing this accurately requires additional data collection of the soil water content in the different areas of the field, good irrigation scheduling techniques, and in-season modifications to the VRI prescription in response to timing and depth of the precipitation events. This is demonstrated in Figs 6.9, 6.10 and 6.11 and summarized in Table 6.3.

Figure 6.9 shows the soil water content over time in the sandy area of the field where 9 mm (0.35 in) of water was applied every time that there

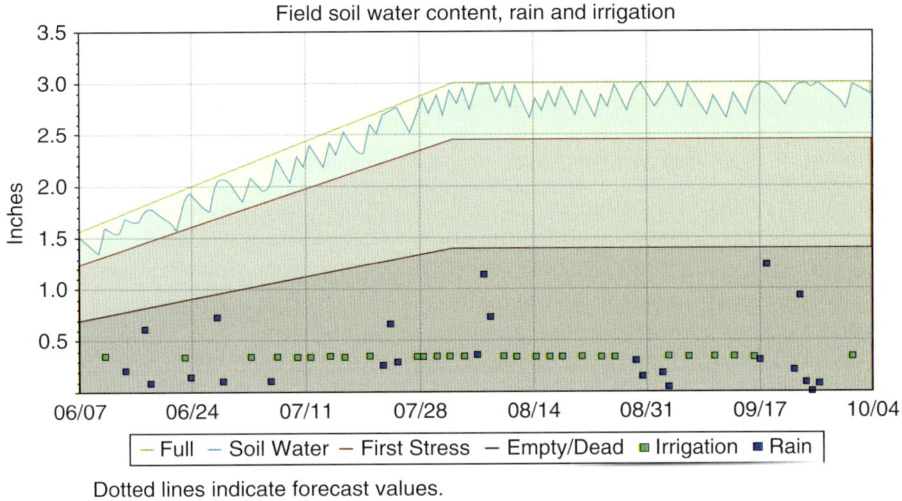

Fig. 6.9. Sandy soil managed to minimize stress in a humid region with significant in-season rainfall. The large rainfall events resulted in excessive deep percolation.

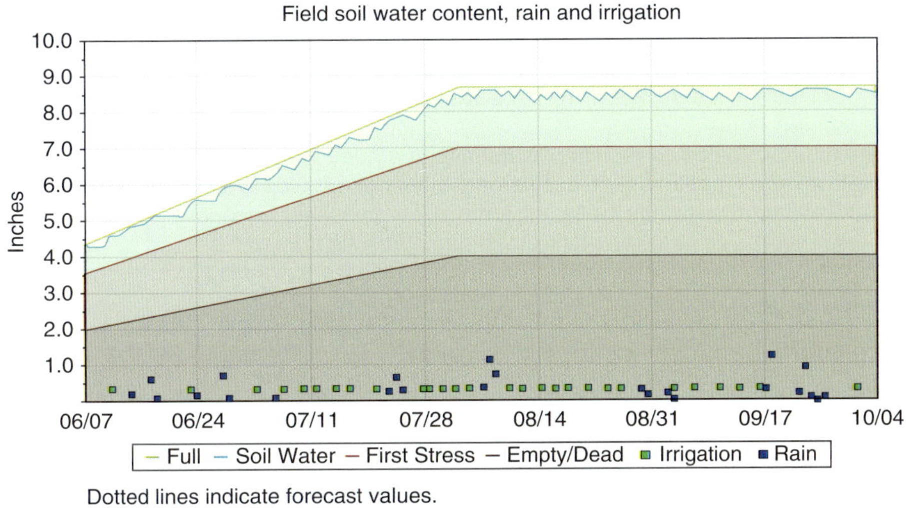

Fig. 6.10. Silt loam soil in the field managed to minimize stress in the sandy soil in a humid region with significant in-season rainfall. The large rainfall events also resulted in excessive deep percolation.

was capacity in the soil to hold it to ensure that the soil water content remained high. Figure 6.10 shows how the silt loam soil would fare if the field were managed uniformly under this scenario. It can be seen from the season totals in Table 6.3 that both of these scenarios resulted in the same amount of total crop ET and total irrigation amounts, and that no crop water stress was experienced by the crop throughout the season. However, there was a lot of water loss in all areas of the field to deep percolation,

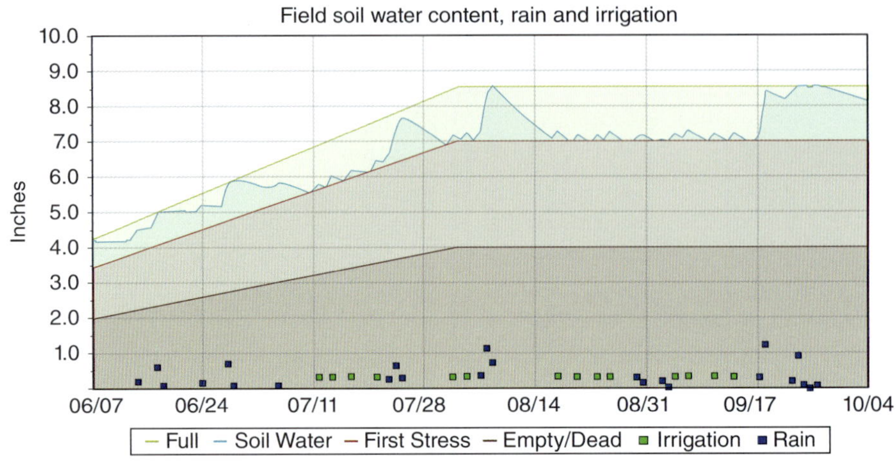

Fig. 6.11. Silt loam soil in the field managed to leave space in the soil to absorb significant in-season rain events in a humid region with significant in-season rainfall. VRI was used to withhold irrigations for these areas until the soil water content was near the first stress line. Because there was excess room in the soil it was possible to absorb the large rainfall events to avoid losses to deep percolation, and irrigation water was conserved while waiting for the soil to dry down again to near the first water stress point.

Table 6.3. Simulated season total water use in a humid region with significant in-season rainfall. A sandy area of the field is compared with the silt loam soil of the field if the whole field is managed uniformly for the silt loam soil with deeper irrigations applied less frequently. All water depths are in millimeters. Deep percolation is primarily caused by rainfall.

Scenario	Figure ref.	Season total evapo-transpiration (ET)	Total irrigation	Total rainfall	Deep percolation	Yield loss
Sandy soil managed to limit water stress	Fig. 6.9	321	240	223	145	0
Silt soil managed uniformly for the sand	Fig. 6.10	321	240	223	145	0
Silt soil managed with VRI to maintain space for in-season rainfall	Fig. 6.11	321	120	223	32	0

because the soil water content was kept high and there was little available space to capture this water. So, although in this simulation the rainfall infiltrated into the soil, that excess water was lost to deep percolation.

If, instead, the irrigation system is shut off over the areas of the field with larger water-holding capacities (deep silt or clay soils) until the crop is just about to experience water stress (Fig. 6.11), then there is much more capacity in the soil to absorb the in-season rainfall.

Table 6.3 shows that, although the crops did equally well in both areas, the silty areas of the field in the VRI scenario were able to make use of 120 mm of rainfall and consequently had much less water lost to deep percolation. Hedley *et al.* (2009) also found large water-savings potential under similar circumstances and that the water savings from VRI were related to rainfall events throughout the season (Hedley *et al.*, 2010).

However, managing for these kinds of results takes very sophisticated management practices, because:

- maintaining the soil water content right next to the water stress point leaves little margin for error;
- soil water-holding capacities are continuously variable across the field, requiring variable water restart periods for the spatially variable areas of the field;
- accurate decisions on how long to leave the water off and when to restart the irrigations to the different areas of the field would be complicated and vital to avoid crop yield losses to water stress;
- these decisions have to be re-evaluated, and new prescriptions uploaded to the VRI machine on a frequent basis throughout the season; and
- this would all be further complicated in the event that rainfall did not completely refill the soil.

The complexity of implementing this scenario may be a deterrent until more sophisticated data collection and decision support systems are available to help analyze the data and upload prescriptions that vary in both time and space.

6.5 Creating and Modifying VRI Prescriptions

This is not a trivial matter. The off-the-shelf VRI systems sold by pivot manufacturers and third-party dealers have been shown to be effective at applying the targeted amounts of water to the desired locations in the field (Dukes and Perry, 2006; O'Shaugnessy *et al.*, 2013; Higgens *et al.*, 2015a). In other words, the control systems and hardware work well and the equipment's ability to apply variable rates across the field is not a barrier to the adoption of VRI. The primary barrier is developing and modifying VRI prescriptions in a way that improves the overall profitability.

Prescriptions are the maps, or plans, for how the irrigation amounts will be varied in the different areas of the field. These are often developed based on experience, GPS or geographic information system (GIS) mapping, and/or GPS-referenced soil sampling. Electrical conductivity (EC) mapping, which is often used to indicate the differences in soil texture or water-holding capacity throughout the field, is also widely used. This data collection is often time consuming, expensive, and plagued by high degrees of uncertainty and sources of variability (Higgens *et al.*, 2015b). In addition it must be done by fairly educated and skilled (i.e. expensive to employ) personnel who are often hired consultants. Once the data that characterizes the variations in the field has been collected, it is not always clear how to vary irrigation amounts and timing in response to this data. Additional research is ongoing on these topics.

Further, irrigation decisions must be re-evaluated many times over a season. Crop performance relative to other areas of the field, the soil surface conditions that affect infiltration rates, and the various alternative sources of water (size of the pond in the field) rarely remain constant throughout a growing season. Using variable rate irrigation to leave space in soils with larger water-holding capacities to take advantage of water from anticipated rainfall events requires in-season modifications to avoid stressing the lower water-holding capacity areas and to adjust for the fact that the anticipated rainfall may not materialize. Therefore, it may be necessary to modify the prescriptions many times throughout the season. Such modifications can be especially challenging with continuously variable soils. This greatly increases the amount of data collection, analysis, decision-making, and modifications made to the VRI prescriptions throughout the season. This can be time consuming, complex, and therefore expensive. However, in the future it may be possible to completely automate this kind of decision making.

If the specific on-farm conditions allow the use of a consistent VRI prescription over time, then significant savings in management time and costs can be achieved and will likely result in considerable water savings. For instance, when there are non-cropped areas that can be left non-irrigated, or if the crops are getting water from a consistently high water table, then the VRI prescription need not be changed over time, and therefore these scenarios have the greatest potential for long-term implementation and measurable water savings.

6.6 What Other Researchers Have Found

Because the conditions under which VRI can be profitable do not apply to all fields, VRI does not always save water or conserve power (Stone et al., 2010). Israeli researchers using simulation models found that adopting practices to increase infiltration and using irrigation systems with high uniformity increased total yields per unit of applied water, but that the impacts of VRI were ambiguous (Feinerman and Voet, 2000). They also found that increasing the number of management units in a field did not necessarily result in more optimal water use, and that VRI did not guarantee savings and in many cases could yield the opposite result.

Several researchers used computer simulations to show that using VRI on center pivot fields with large differences in water-holding capacities in humid regions with frequent heavy rainfall during the growing seasons had the potential to save significant amounts of water and reduced deep percolation (Hedley *et al.*, 2009, 2010). These simulated benefits depend on the baseline, which might be suboptimal (see discussion of Figs 6.5, 6.6 and 6.7). Hedley *et al.* (2010) also found that larger water savings were related to years with rainfall events during the irrigation period. These studies showed that large differences in the water-holding capacities in the field and frequent large rainfall events strengthen the potential savings of VRI from rainfall capture. While computer simulations show potential benefits of VRI, published in-field testing results demonstrating similar benefits are still lacking.

Adoption of VRI has been generally limited and its use by early adopters has not always been sustained (Evans *et al.*, 2012). The complexity of installing, maintaining, and effectively managing VRI systems has been a significant barrier to adoption. In many instances the economic returns from adopting these technologies have not been easy to demonstrate consistently (Feinerman and Voet, 2000; Berne *et al.*, 2015). However, increased costs of water and energy, and severe water limitations will likely increase the financial incentives to adopt VRI (Evans *et al.*, 2012).

6.7 Summary

Variable rate irrigation gives a grower the ability to vary the amount of water that is applied to different areas of the field. On center pivots this can be done fairly simply and relatively inexpensively using variable speed irrigation. However, the spatial variations are limited to pie-shaped wedges. There are several other applications of variable speed irrigation besides VRI that can provide benefits in certain fields. Variable zone irrigation includes the ability of the system to vary both the speed and the amount of water applied along the lateral. It is more sophisticated, and flexible, but also much more expensive.

In-field variations in soil water-holding capacities and infiltration rates can be largely mitigated by proper water management for the entire field as a whole for the problem soils. If the whole field is irrigated to avoid deep percolation and water stress in the soils with the lowest water-holding capacity, the rest of the field will be fine. Likewise, managing the field as a whole to limit runoff in certain problem areas has little negative effects on the rest of the field.

Variable rate irrigation may provide water and power savings or crop yield benefits in the following circumstances: withholding irrigation in non-cropped areas; not irrigating areas of the field that are getting water from other sources; keeping the soil water content at a level so that rainfall can be captured (rainfall harvesting); varying irrigation for the different water needs of different crops in the same field; responding to spatial variations in crop water use (ET) due to crop health variations; using pivot as a variable rate sprayer or waste disposal system; or to avoid overwatering the inside span of the pivot.

The VRI systems currently being sold can fairly accurately implement uploaded VRI prescriptions. However, the data collection, analysis, and creation of optimal VRI prescriptions for a specific field's needs can be complex, time consuming and expensive, especially since many field situations require these prescriptions to vary both in time and in space. This is currently a significant barrier to the profitable use of VRI.

Variable speed irrigation currently has greater potential to be a good investment. Variable zone irrigation systems that are used to consistently avoid irrigating non-cropped areas are likely to be the most manageable and beneficial, especially when injecting agrichemicals or waste products and it is unlawful to apply these to non-cropped areas.

References

Berne, D., Valent, P. and Whitty, K. (2015) *Agricultural Irrigation Initiative: Grower Experience*. Research report #E15-005. Northwest Energy Efficiency Alliance (NEEA), Portland, Oregon. Available at: http://neea.org/docs/default-source/reports/grower-experience.pdf?sfvrsn=4 (accessed 3 June 2017).

Chavez, J.L., Pierce, F.J. and Evans, R.G. (2010) Compensating inherent linear move water application errors using a variable rate irrigation system. *Irrigation Science* 28, 203–210.

Dukes, M.D. and Perry, C. (2006) Uniformity testing of variable-rate center pivot irrigation control systems. *Precision Agriculture* 7, 205–218.

Evans, R.G., LaRue, J., Stone, K.C. and King, B.A. (2012) Adoption of site-specific variable rate sprinkler irrigation systems. *Irrigation Science* 31, 871–887.

Feinerman, E. and Voet, H. (2000) Site-specific management of agricultural inputs: an illustration for variable-rate irrigation. *European Review of Agricultural Economics* 27, 17–37.

Hedley, C.B., Yule, I.J., Tuohy, M.P. and Bogeler, I. (2009) Key performance indicators for simulated variable-rate irrigation of variable soils in humid regions. *Transactions of the ASABE* 52, 1575–1584.

Hedley, C.B., Bradbury, S., Ekanayake, J., Yule, I.J. and Carrick, S. (2010) Spatial irrigation scheduling for variable rate irrigation. *Proceedings of the New Zealand Grassland Association* 72, 97–102.

Higgens, C., Bar, C., Hillyer, C., Kelley, J. and Whitty, K. (2015a) *Agricultural Irrigation Initiative: Precision Water Application Test*. Research report #E15-009. Northwest Energy Efficiency Alliance (NEEA), Portland, Oregon. Available at: http://neea.org/docs/default-source/reports/precision-water-application-test.pdf?sfvrsn=4 (accessed 3 June 2017).

Higgens, C., Kelley, J., Liu, Z. and Hillyer, C. (2015b) *Agricultural Irrigation Initiative: Using Soil Electrical Conductivity Mapping for Precision Irrigation in the Columbia Basin*. Northwest Energy Efficiency Alliance (NEEA), Portland, Oregon. Research report #E15-010. Available at: https://neea.org/docs/default-source/reports/using-soil-electrical-conductivity-mapping-for-precision-irrigation-in-the-columbia-basin.pdf?sfvrsn=4 (accessed 3 June 2017).

Milton, A.W., Perry, C.D. and Khalilian, A. (2006) Status of Variable-Rate Irrigation in the Southeast. *Proceedings of the ASABE International Meeting in Portland*, Oregon. 9–12 July, 2006. Paper number 061075. American Society of Agricultural and Biological Engineers, St Joseph, Michigan. Available at: http://elibrary.asabe.org/azdez.asp?JID=5&AID=20599&CID=por2006&T=2 (accessed 3 June 2017).

O'Shaughnessy, S.A., Urrego, Y.F., Evett, S.R., Colaizzi, P.D. and Howell, T.A. (2013) Assessing application uniformity of a variable rate irrigation system in a windy location. *Applied Engineering in Agriculture* 29, 497–510.

Peters, R.T. (2014) *Irrigation Scheduler Mobile User's Manual and Documentation.* Available at: http://weather.wsu.edu/ism/ISMManual.pdf (accessed 3 June 2017).

Sadler, E.J., Evans, R.G., Stone, K.C. and Camp, C.R. (2005) Opportunities for conservation with precision irrigation. *Journal of Soil and Water Conservation* 60, 371–379.

Stone, K., Bauer, P., Busscher, W., Millen, J., Evans, D. and Strickland, E. (2010) Variable-rate irrigation management for peanut in the Eastern Coastal Plain. Paper IRR10-8977. In: Dukes, M. (ed.) *Proceedings of the 5th National Decennial Irrigation Conference*, Phoenix, Arizona, 5–8 December, 2010, Vol. 2. American Society of Agricultural and Biological Engineers (ASABE), St Joseph, Michigan, pp. 582–588.

7 Precision Technologies for Pest and Disease Management

Lav Khot[1]*, Gwen-Alyn Hoheisel[1], Yasin Osroosh[1] and Reza Ehsani[2]

[1]*Washington State University, Prosser, Washington, USA;* [2]*University of California, Merced, California, USA*

7.1 Introduction

Precision pest and disease management has become increasingly important in tree fruit production with the rising incidents of invasive insects, pests and pathogenic infestation. As consumers are becoming technology savvy, demand for quality produce that can readily be traced back to the source has been growing steadily. Regulatory agencies are also pushing for best management practices and traceability to effectively address produce safety and environmental concerns. Integrated pest and disease management (IPDM) is vitally important in such efforts.

The success of IPDM is heavily dependent on effective infestation monitoring. Monitoring is carried out with the purpose of spotting pest and disease damage and identifying pests and their key natural enemies, also known as beneficial organisms. As a basis for decision making, monitoring provides valuable information about the population densities and developmental stage of insects and their natural enemies. Over the years, technology has enabled us to transition from traditional labor-intensive pest and disease monitoring to digital 'smart' insect traps, pest modeling with orchard weather data, and remote sensing for disease infection mapping.

Analogously, variants of application technologies are being used for chemical and nutrient applications for increased plant protection and fruit quality. For example, in 'Honeycrisp' apple production, growers apply foliar calcium applications to improve fruit quality and reduce the incidence of bitter pit disorder. Similarly, application of foliar nutrients

* Corresponding author, email: lav.khot@wsu.edu

has been one of the strategies followed in Florida citrus production where such applications aim to maintain the health and viable yield of trees infected with huanglongbing (HLB) until a more permanent HLB disease management solution is found. In general, changes in canopy architecture and increased within-canopy variability due to varying pest and disease pressure have increased the demand to develop next-generation application technologies for site-specific and need-based chemical applications.

This chapter discusses aspects of integrated pest and disease management with a focus on: ways to monitor insects and pests using traditional as well as new technologies; advances in spatial and temporal microclimatic measurement techniques and the role of such data in pest monitoring; and overview of contact and non-contact sensing technologies for disease monitoring. This chapter also provides a state-of-art overview of the application technologies used in tree fruit production management, with a focus on the transition from traditional to variable-rate chemical application technologies, emerging future application technologies, and pertinent standardization issues.

7.2 Pest Monitoring Technologies

Monitoring insect and pest populations is a key activity in successful IPDM as it governs the effectiveness of the control strategy. Monitoring provides valuable information about the population density and developmental stages of insects and their natural enemies. The following sections discuss some of the conventional and emerging technologies used to monitor insects and pests.

7.2.1 Conventional pest monitoring techniques

In conventional monitoring, various objective sampling methods can be used to track pest and beneficial organism populations, depending on the pest type. Sampling methods include visual inspection of plants using an optical lens, manual collection of insects from plants, and the use of insect traps (e.g. sticky, pheromone-based, spore) (Fig. 7.1). Leafhoppers, caterpillars, leaf miners, aphids, psyllids, mites and most of their enemies are some of the pests that can be assessed with visual counts or manual collection through the use of a sweep net or beating sheet.

Regardless of the design (e.g. cylinder-shaped, wing, or tent types), pheromone-based traps slowly release chemical attractants to collect insects. Pheromones – classified as sex, aggregation, or feeding – are volatiles used to communicate among the same species of insect. Pheromone-based traps allow collection of valuable information on the life history of the insects which is required to run phenology models and to evaluate the efficacy of control programs (Beers, 1993). While these traps are a common and effective method for monitoring pest populations, they are often labor intensive and expensive to monitor and bait (Ding and Taylor, 2016).

Fig. 7.1. Traditional pest monitoring: (a) pheromone-baited and (b) sticky paper-based insect traps (photo courtesy of Lav Khot, Washington State University).

7.2.2 Emerging pest monitoring technologies

Recent advances have allowed the incorporation of digital imaging sensors, wireless transceiver-based communication and associated electronics into the traditional pheromone-based traps, i.e. to develop an 'electronic trap'. It monitors insect activity using an imaging sensor and allows automated insect count and classification. Such a system can be customized to transmit field images from the grower sites to multiple computing or mobile devices through cloud computing. Robust image processing algorithms can discriminate several types of pests from such field data. Experts can also use such imagery to identify the type of pest infestation. Quality of images in varied light conditions and reliable data transfer technology are key to successful adaptation of electronic traps. Ideally, integration of such traps into existing IPDM decision support systems (DSS) may lead to a significant economic saving through a reduction in the number of field scout hours and need-based pesticide applications. A broad range of camera-based insect traps have been developed in recent years. Tirelli *et al.* (2011) used a distributed imaging device integrated into a wireless sensor network (WSN) to automatically monitor remote insect activity and generate an insect population alarm. Selby *et al.* (2014) retrofitted a camera system used for monitoring mammals into a camera-based insect trap for observing the plum curculio population.

Several commercial low-cost pheromone-based camera traps are also available in the market. For example, Spensa Technologies Inc. (West Lafayette, Indiana) has commercialized the Z-Trap, which is an automatic wireless insect counting system. The Z-Trap uses a pheromone attractant to lure insects into impedance sensors calibrated to identify the codling moth, omnivorous leaf roller, or oriental fruit moth, common in orchard environments. Similarly, in Canada, SemiosBio Technologies Inc. (Vancouver, BC) has a pheromone-based camera trap system that relies on mesh network topology to collect information from several electronic traps. The system comes with a software program which allows for correlation of degree-days with trap catches and can be configured to activate pheromone puffers deployed in the field.

7.2.3 Advances in microclimatic measurements

Broadly, the microclimate is defined as the climate of a small area surrounding individual plants. Microclimate is the immediate environment upon which pest survival depends (Hatfield and Thomason, 1982). Air temperature, relative humidity, and surface wetness are some of the microclimatic variables commonly used for disease management (Sutton *et al.*, 1984). Knowing the relationship between weather and the epidemiology of pathogens, scientists have explored the possibility of reducing pest infestations through modifying the microclimate around crops (McDonald *et al.*, 2013; Pangga *et al.*, 2013; Gogo *et al.*, 2014).

7.2.3.1 Open field microclimate measurement

Site-specific management of crop pests and vector-borne diseases requires precise information about temporal and spatial variability of a microclimate (Matese *et al.*, 2014; Kotchi *et al.*, 2016). Such microclimate measurements on needed temporal resolution can be acquired by automated agricultural weather stations (AWSs). Typically, an AWS measures air and soil temperature, air relative humidity, solar radiation, wind speed and direction, and leaf wetness. AWSs can act as stand-alone units capable of recording measurements for a long period of time or can be part of a regional agricultural weather network (AWN). One example of a regional AWN is the AgWeatherNet managed by Washington State University (WSU-AgWeatherNet, Prosser, Washington). It has over 170 operational AWSs in eastern and central Washington.

Typically, a spatial range greater than 1 km is desired for most on-farm decision-making purposes. Considering the fact that such resolution is highly unlikely from most AWNs, researchers have suggested alternative approaches that do not rely on on-site weather stations (Magarey *et al.*, 2001). For example, gridded weather data from public or private organizations can be a feasible alternative to the data from local weather stations. Weather information with 1 km or higher resolution can provide the necessary input for a climate-based disease-warning system (Hong *et al.*, 2015; Rowlandson *et al.*, 2015).

Airborne remote sensing technologies such as Doppler radar or satellite images can provide accurate maps with spatial and temporal scales meeting the specific needs of IPDM (Wood *et al.*, 2003; Workneh *et al.*, 2005). Surface temperature plays an important role in pest risk assessment because of its relationship with microclimate variables such as air temperature (Da Silva *et al.*, 2015; Kotchi *et al.*, 2016). With the advent of airborne thermal infrared sensors, the estimation of surface temperature at a very fine scale has become possible (Deng and Wu, 2013). Surface temperature can be estimated using different satellites such as Landsat-8/TIRS, NOAA/AVHRR, and Aqua/MODIS.

7.2.3.2 In-field microclimate measurement

Due to small gradients in plant canopies, microclimatic measurements within the canopy require sensors that do not disrupt the natural pest habitat (Hatfield and Thomason, 1982). In recent years, a wide range of low-cost microclimate monitoring units have been developed for the purpose of phenological modeling. In-field microclimatic measurements can be carried out using mobile or stationary set-ups.

Leaf wetness can result from dew, rainfall, or irrigation events. Leaf wetness duration (LWD) is an important variable in plant disease epidemiology and can be used to quantify the exposure of pathogens to free water (Rowlandson *et al.*, 2015). Many disease warning systems use LWD data (Brown and Sutton, 1995; Llorente *et al.*, 2000; Peres and Timmer, 2006; Duttweiler *et al.*, 2008) to determine critical times to spray crops against diseases. Surface wetness duration is often measured indirectly using in-field sensors. Leaf wetness sensors (LWSs) measure either the resistance or dielectric constant of a printed grid of interlacing copper wires and relate it to the presence of water on the surface. Amongst several commercial LWSs available, the two commonly used are the 237-L and LWS-1 manufactured by Campbell Scientific, Inc. (Logan, Utah) and Decagon Devices, Inc. (Pullman, Washington), respectively. Both companies also produce microclimate monitoring stations which offer leaf wetness measurements. Although different studies have confirmed the impact of painting on 237-L sensitivity (Lau *et al.*, 2000; Sentelhas *et al.*, 2004), it is often deployed unpainted. WatchDog (Spectrum Technologies, Inc., Aurora, Illinois), and LW (RainWise, Inc., Trenton, Maine), both resistance-type, and LW10 (Environdata, Warwick, Queensland, Australia) are also among the commercially available sensors.

Besides the flat plate LWSs, cylindrical types have also been developed for non-commercial use. Sentelhas *et al.* (2007) compared flat and cylindrical LWS types in four different environments. Results showed that the cylindrical sensor overestimated LWD in a highly humid climate and detected wetness earlier than flat-type sensors. Reliability of LWSs has thus been argued by many researchers. There is no standardized alternative method for the calculation of LWD and several studies have compared physical and empirical models with visual observations (Rowlandson *et al.*, 2015). For example, Sentelhas *et al.* (2008) developed a simple empirical model for the estimation of LWD based solely on air relative humidity. Magarey *et al.* (2005, 2006) assessed surface wetness models based on a combination of water and energy balance approaches, and atmospheric variables and plant physical properties.

7.2.3.3 Climate data-driven decision systems

To make management decisions that are environmentally and economically sound, the plant disease risk and associated costs need to be assessed under different scenarios (Gent *et al.*, 2011). A climate-based DSS is a component of IPDM that enables growers to make informed decisions about pest and disease control practices based on weather variables rather than solely on phenological stages. To assist growers in plant disease management, a large number of weather-based DSSs have been developed. WSU-DAS (Washington State University Decision Aid System) and NAPPFAST (North Carolina State University) are some of the examples of well-established systems in the USA (Magarey *et al.*, 2002, 2007; Chambers *et al.*, 2011).

Similar to the above systems, Hong *et al.* (2015) developed and validated a Generic Pest Forecast System (GPFS), with observations of the oriental fruit fly, *Bactrocera dorsalis* (Hendel). The GPFS model was used for spatial and temporal simulation of the relative populations of non-indigenous arthropod pests. The model used hourly weather data as one of the inputs. The GPFS model simulations were compared with field survey data in three locations: Bangalore, India; Hawaii, USA; and Wuhan, China. The GPFS successfully captured major pest population peaks and the initial outbreaks of the pest. The GPFS model also proved to be informative in terms of relative abundance prediction. Kang *et al.* (2010) developed a web-based high spatial resolution information system for forecasting disease (scab and rust) outbreaks for pear growers in Gyeonggi-do, Korea. The system used spatially interpolated temperature, relative humidity, rainfall, and leaf wetness data. The system generates hourly or daily warnings with a spatial resolution of disease forecast that is high enough to estimate infection risks of individual farms.

Despite a large number of DSSs developed for disease forecasting, few are actually used by growers. This is simply because of the fact that agricultural product management requires a holistic view of the pests and diseases for a given crop and within a chosen field site, while most DSSs are intended for specific disease problems (Magarey *et al.*, 2002). In many cases, growers have a different perception of risk and risk management. Such perception often is a hindrance to the adoption of DSSs and IPDM (Gent *et al.*, 2011). In addition, lack or scarcity of site-specific weather data has been a major challenge in the application of climate-based disease management DSSs (Magarey *et al.*, 2001).

7.3 Disease Monitoring Technologies

7.3.1 Contact-type sensors

Over the years, molecular techniques (López *et al.*, 2003) have evolved into the most robust tool for detecting the presence of plant diseases. Enzyme-linked immunosorbent assay (ELISA) and polymerase chain reaction (PCR) are the molecular techniques most commonly used for disease detection (Saponari *et al.*, 2008). These techniques require tedious sample preparation protocols and often are labor intensive and time consuming. Moreover, the molecular techniques are moderately expensive, require specific instrumentation and are limited to well known diseases only (Sankaran *et al.*, 2010b).

As the fastest growing technology, biosensors are expected to replace ELISA in the near future (Lazcka *et al.*, 2007). Biosensors are chemical sensors that incorporate a biological recognition element for selective sensing, making it highly desirable as a basis for analysis of complex mixtures in real time (Velasco-Garcia and Mottram, 2003). A number of different aspects, including cost, durability of the biological element, sensitivity, and reproducibility, need to be improved upon for practical commercial applications of biosensors (Ruiz-Altisent *et al.*, 2010).

7.3.2 Non-contact-type sensors

Non-contact sensing is a rapid, non-destructive, cost-effective measurement method that allows for taking an unlimited number of samples. It is a promising technology for the detection of disease and can help in taking measures to prevent physiological stresses and physical damage caused by pathogens (Ushaa and Singh, 2013). To date, many non-invasive techniques have been developed for plant disease detection, including imaging and spectroscopic techniques, and volatile organic compounds (VOC) profiling-based techniques (Lee *et al.*, 2010; Sankaran *et al.*, 2010b). The spectroscopic and imaging techniques encompass a broad range of methods such as fluorescence spectroscopy, fluorescence imaging, hyperspectral imaging, and visible–infrared (vis–IR) spectroscopy (Belasque *et al.*, 2008; Guo *et al.*, 2009; Sundaram *et al.*, 2009). In fluorescence spectroscopy (Lins *et al.*, 2009), the object (vegetation) is excited with a beam of shortwave ultraviolet light and the emitted fluorescence is measured as a response. In fluorescence imaging, fluorescence images are taken using a camera (Chaerle *et al.*, 2007). In hyperspectral imaging, the spectral reflectance is acquired for a range that may include the visible and infrared regions of the electromagnetic spectra. Visible–near-infrared (vis–NIR) spectroscopy, mid-infrared spectroscopy, and hyperspectral imaging have been used in a number of studies for the detection of HLB (greening) in citrus orchards (Li *et al.*, 2006; Kumar *et al.*, 2010; Sankaran *et al.*, 2010a, 2011; Sankaran and Ehsani, 2012).

VOC profile-based disease detection uses an electronic nose or gas chromatography–mass spectrometry (GC–MS) analysis of volatile metabolites released by plants (healthy and diseased) to identify diseases. An electronic nose is made up of a series of gas sensors allowing for discrimination of a range of organic compounds that may be present in the air (Sankaran *et al.*, 2010b). Electronic nose systems have been successfully used for identifying plant diseases (Spinelli *et al.*, 2006; Moalemiyan *et al.*, 2006, 2007; Markom *et al.*, 2009).

Variants of spectroscopic, imaging, and VOC profiling-based techniques have shown a high degree of accuracy in non-contact detection of plant diseases. Depending on application and payload, a range of ground- and aerial-based platforms can be used for deploying proximal and remote non-contact sensors to detect crop diseases (Lee *et al.*, 2010). However, remote sensing techniques have mainly been used to evaluate the extent of disease damage and early crop disease detection is often difficult or even impossible using such methods (Lee *et al.*, 2010). The VOC profile naturally varies within plant species and can mask the changes due to the existence of diseases or stresses. To cope with this problem, distinct volatile biomarkers specific to a particular disease have to be identified. The spectral reflectance from the vegetation is also affected by environmental conditions (Griffin and Burke, 2003). This can be possibly resolved by identifying vegetative indices or wavelength ranges that are not sensitive to environmental changes (Sankaran *et al.*, 2010b).

7.3.3 Sensing technology adoption challenges

Multispectral and hyperspectral imaging systems have been integrated with ground and aerial platforms to study crop health. Airborne multispectral and hyperspectral imagery and high-resolution satellite imagery from commercial satellites (e.g. IKONOS, QuickBird, and SPOT 5) can provide image data at spatial resolutions fine enough for within-field plant variability mapping (Lee *et al.*, 2010). Airborne and satellite imagery have been successfully used for detecting crop diseases (Du *et al.*, 2004; Lee *et al.*, 2008; Shafri and Hamdan, 2009). Despite the advances in remote sensing technology, real-time automated monitoring of plant diseases under field conditions remains a challenge (Sankaran *et al.*, 2010b).

A key issue in direct application of remote sensing techniques is that multiple biotic and abiotic stressors may coincide and result in similar spectral reflectance. In such cases, additional information about the diseases and other sources of stress are required to determine which one is responsible for which morphological and/or physiological changes in the crop. In practical applications, the stress detection algorithm must monitor the soil, crop, and diseases simultaneously, otherwise the technique may fail to discriminate plant conditions (Lee *et al.*, 2010). Selection of remote sensing techniques, quality of the raw data, and robustness of the data-processing algorithm govern the usefulness of a particular plant and disease detection system under field conditions (Lee *et al.*, 2010; Sankaran *et al.*, 2010b).

7.4 Agricultural Application Technologies

7.4.1 Traditional orchard sprayers

Application technologies are critical for effective pest and disease management in tree fruit production. Orchard sprayers have evolved over a century of use from horse-drawn wagons with hand booms to tractor-pulled sprayers with sensors. Change has occurred because of increased integrated pest management, increased demand by consumers for quality produce, dramatic changes in horticulture, and the development of new technologies. Following passage of the 'Food Quality Protection Act', the US Environmental Protection Agency began phasing out widely used broad-spectrum and toxic insecticides (e.g. azinphosmethyl) that had been used to manage insect populations in spite of poor coverage. Most modern chemicals are far more specific in their mode of action and are applied at much lower doses than traditional broad-spectrum chemicals. Consequently, growers must use newer, safer materials with precise timing and excellent coverage to get effective control and to avoid crop damage and economic losses. Only a few newer, systemic pesticides that translocate through the plant can achieve adequate control with minimal regard to spray coverage.

Since 1949, the most widely used spray design has been that of an axial fan sprayer, more commonly known as an airblast sprayer. When the airblast sprayer was introduced, trees were grown in the iconic apple tree structure, i.e. 6 m tall, with many wide spanning branches that form almost a shade tree. They were planted in rows that were 4–6 m wide with the same spacing between trees. An airblast sprayer was an appropriate machine for such canopy architecture, because the large volume of air could push chemical-laden droplets 9 m in the air (Fox *et al.*, 2008).

The problem with large round or vase-shaped canopies is that they do not produce the highest quality of fruit and can be difficult to harvest and manage. Mature leaves should serve as a source of energy for the plant by absorbing sunlight and converting it to energy through photosynthesis. Leaves not exposed to the sun serve as energy sinks, meaning they consume energy to grow and live, but are not producing any more through photosynthesis. Studies have shown that dense canopies (i.e. more leaf layers) absorb less light (Wagenmakers, 1991; Wagenmakers and Callesen, 1995) and affect fruit quality. Essentially the lower layers of leaves which do not receive direct sunlight are energy sinks rather than sources, detracting from the plant being able to produce large fruit with high sugar quantity. Because of this and challenges in managing tall trees, advances in horticulture have focused on the implementation of dwarf rootstocks. These have allowed for modern orchard trees to be 4–5 m tall with 3–4 m between rows, and up to 1 m between trees. Dwarf rootstocks and these high-density plantings have also created the need for trellising fruit trees in which the limbs are shorter and thinner. This change in horticulture shows a change in tree architecture from trees that were 8 m tall and 3 m wide to 4 m tall and 1 m wide arranged in tight and more uniform canopies. This transition has driven the change in sprayer design and adoption by farmers.

The defining configuration or feature of an airblast sprayer is a single fan at the rear of the machine that pulls air in and redistributes it upward into the canopy. The volume and direction of air are critical factors related to spray coverage and deposition. The volume of air should be matched to the volume of air in the canopy. Therefore, larger, taller canopies are more appropriate for larger volumes of air while the reverse is true for smaller canopies. Traditionally, airblast sprayers create large volumes of air, but the volume of air is directly related to the size of the fan, and the speed or revolutions per minute (rpm) is adjusted by the power take-off (PTO) and throttle of the tractor. Newer designs have attempted to reduce the air output through manual louvers (e.g. Fede Sprayer) or automated louvers (Khot *et al.*, 2012, 2014), yet so far neither has been widely adopted by farmers. Nearly half of the pesticides applied with an airblast sprayer can be lost to the atmosphere above the canopy and to the ground via sedimentation, especially when used in smaller canopies (Herrington *et al.*, 1981; Raisigl *et al.*, 1991; Vercruysse *et al.*, 1999; Brown *et al.*, 2008).

The direction in which the air pushes spray droplets affects coverage. Axial fan sprayers are pulled on the ground with the nozzles no more than 1.7 m from the ground. Spray is directed upwards in a fan-shaped pattern. Alternatives to this sprayer design have focused on directing air horizontally into the canopy. Tower sprayers can range from 2.0 m to 3.7 m tall with nozzles arranged up the tower. This design typically provides a more uniform spray deposition with less off-target, above-canopy drift than conventional airblast sprayers, but orchard access can be limited in minimally managed canopies such as almond trees, where canopies merge across rows. Shorter canopies, such as grapes, have allowed more designs where spray heads and nozzles can be arranged on both sides of the canopy (i.e. 'over-the-row'), allowing for spray to be directed horizontally into the canopy from both directions. A design, such as this, that has opposing air direction can minimize drift.

Nozzles are another component that can impact canopy coverage. The majority of all canopy sprayers use a two-piece nozzle called a disc–core, which is made from either ceramic, stainless steel, or brass. There is a 'disc' that is the outer part of the nozzle, with an orifice for liquid, and an inner 'core' that spins to create a cone pattern. The output (e.g. liters per minute) of each nozzle is determined by the size of the orifices in the disc and core as well as operating pressure of the machine. These nozzles can be easily changed so that many configurations of output can be achieved on a single sprayer and more closely match the shape of the canopy. For example, nozzles with all the same output can be used on a tower sprayer in uniform-shaped canopies. In contrast, on a large round canopy, higher-output nozzles can be used on the top of the sprayer where the canopy is denser. However, the problem with disc–core nozzles is that they can wear and clog easily, requiring routine maintenance. As a solution, air-shear nozzles were developed. These 'nozzles' actually have no true nozzle tip, but rather rely on a 'Venturi' method to use air to push through a stream of liquid and create droplets. The benefit is that there is no nozzle to change, but very fine droplets (< 50 μm) can be created. While this will create a high-resolution pattern of coverage, droplets this small will drift more.

One nozzle solution to drift has been to evolve the disc–core nozzle into a one-piece air-induction (AI) nozzle. AI nozzles have a small opening in the top or bottom of the nozzle to pull air in and create bubbles within a droplet of pesticide. The droplet that emerges is large and does not drift as far. When it hits the leaf the bubbles pop, creating a splatter effect on the leaf. The advances in nozzle technology have not been adopted widely in the USA. Still the most common nozzle is disc–core. AI nozzles are more commonly used in herbicide applications and less in canopies. Air-shear nozzles are not interchangeable with other sprayers, meaning that a single sprayer can use either air-shear or standard nozzles, but not both.

Other common features of an orchard sprayer are pumps, tanks, and agitators. Each of these parts has multiple options regarding design and materials used. For example, a tank capacity can vary from 50 l to 1500 l and be made from either stainless steel, fiberglass, or plastic, each of which has limitations and advantages. It is the individualization of these mechanical parts that customizes the performance of a machine.

Traditionally, the spray applications in tree fruit crop production are based on the application rate per unit area covered by the ground vehicle. Row and tree spacing were used to estimate the agrochemical application rate per unit area. Spray application rates accounted for neither canopy height/width nor variability in canopy growth stages (e.g. leafing, half and fully developed canopies). In due course, with economic and environmental implications becoming a concern to growers and the community, several efforts have been made to optimize the spray coverage to complement canopy attributes, as summarized in the following section.

7.4.2 Variable rate technology: concept and implementation

Variable rate technology (VRT) utilizes different sensors and control instrumentation techniques mounted on agricultural vehicles for crop canopy-specific spray, fertilizer, and other crop inputs applications. Precision spraying pertinent to tree fruit production management can ideally be defined as the integrated technology that adjusts the chemical application rate based on the variability in the tree canopy, pest and disease pressure, and/or local microclimate.

Although knowing the spatial field variability is the key to the application of VRT (Sawyer, 1994), such variability needs to be perceived differently in tree fruit production. It could simply be the localized mapping of gaps in the canopy or canopy volume and density to the specific optical sensor-based georeferenced pest and disease infestation maps. Orchard topography and temporally varying microclimate factors such as wind speed and direction, temperature, humidity, and resulting atmospheric inversions can also be integrated with spatial canopy variability maps to decide the variable chemical application rates.

Recent advances in field-portable rugged sensors for non-contact canopy sensing, georeferenced remote variability mapping, spray droplet generation technologies, and embedded systems have helped the industry to realize and practically implement the precision sprayer concept (Gonzalez-de-Soto *et al.*, 2016). A typical precision sprayer (Fig. 7.2) can

Fig. 7.2. Orchard airblast sprayer retrofitted with variable rate spray and air-assist technology; C1 and C2 represent the air-assist louver in normally closed and open position, respectively (for additional details see Khot *et al.*, 2012).

encompass modules to (i) map spatial and climatic variability, (ii) computerize data acquisition and control (DAQ), and (iii) adjust spray rates and air-assist the droplets movement. Simplistic implementation of the above concept is through a two-dimensional (2D) or three-dimensional (3D) light detection and ranging (lidar) sensor and a global positioning system (GPS) receiver mounted on the front of an agricultural vehicle (i.e. tractor) for georeferenced canopy mapping. The collected information can be used for real-time actuation of individual spray nozzles through pulse-width modulation (PWM) for targeted spray applications. Precision sprayers may also use additional sensors in the feedback loop to perform real-time diagnosis of various sprayer components and optimize the sprayer performance accordingly. For example, a precision sprayer can have flow rate sensors to monitor liquid flowing through sections of spray manifolds as well as spray tank pressure and level monitoring sensors.

A spatial and climatic variability mapping module can use a range of optical sensors. Spatial variability in a canopy can be mapped by either an ultrasonic sensor or lidar sensor, both of which work on the time-of-flight measurement principle. Real-time climatic variability can be mapped using anemometers mounted on a tractor or sprayer unit that provides the components of wind and air temperature data. The attributes of a variability map are governed by the type of sensor. For example, an ultrasonic sensor can provide only a single data point (i.e. distance to canopy) and an array of sensors would be needed to derive approximated canopy height or volume. Lidar, on the other hand, can generate a high-resolution point cloud in the sensor field-of-view. Such point cloud data can then be used for extracting gaps in the canopy or canopy height, width, volume, and density. Aside from the cost, the range and resolution of optical sensors are some of the attributes that need to be considered when choosing the appropriate sensor for canopy mapping.

Small unmanned aerial system (sUAS) integrated optical sensors or other remote sensing techniques may also be useful in mapping the canopy variability. Through sUAS flights over the orchard blocks, growers can potentially generate either base canopy attributes (via a simple RGB imaging technique) or have canopy vigor and health maps using specific multispectral imaging sensors. Such georeferenced maps can be an input to the precision sprayer which then can vary the application rate based on sprayer location in the orchard.

The computerized DAQ (or embedded system) plays a central role in synthesizing spatial and climatic variability maps and actuation of the individual nozzle or array of nozzles to vary the application rates. Available commercial solutions use PWM-based solenoid valve assemblies that, when integrated with appropriate nozzle body and tips, can adjust the spray (liquid) output rates. Real-time variable air-assist for precise delivery of spray droplets can be achieved either via actuator-enabled louvers that divert excessive air from the air blast (Khot *et al.*, 2012) or through localized air-assist for a section of nozzles.

Overall, precision sprayer or VRT has not transitioned at the rate with which orchards have been transitioned from traditional to modern tree fruit architectures. Nonetheless, a few precision sprayer prototypes for use in nurseries, citrus, apple, and vineyard production management have been developed and tested by researchers (Chen *et al.*, 2013; Escolà *et al.*, 2013; Khot *et al.*, 2014). Researchers have been working on developing precision sprayers that use either map-based (Pérez-Ruiz *et al.*, 2011) or sensor-based (Chen *et al.*, 2011; Jeon and Zhu, 2012; Khot *et al.*, 2012) variability mapping and then adjust the application rates.

7.4.3 Future application technologies

All of the advances in rate controllers and sensors still do not solve several major problems with airblast sprayer-based applications. Because sprayers are only recommended to be driven at 3–6 km/h, applications can take a long time on large acreage. In row-crop systems, this has been solved by creating larger booms, but this is not possible in orchards. Vineyards, which are shorter and allow for over-the-row systems, have the largest multiple row systems, capable of spraying five rows simultaneously. Since sprays can take weeks to cover larger farms, timing of chemical applications can sometimes miss the optimal time for pest or disease control. Weather can also affect spray quality. High heat and low humidity can cause droplets to evaporate and reduce in size.

Small droplets will tend to drift off-target more and are affected by wind speed. Studies of airblast applications show differences in canopy deposition and off-target drift under changing wind speed and direction as it relates to canopy row orientation (Tsai, 2007; Armstrong *et al.*, 2013a, b). Finding ideal weather conditions for the entire duration of a spray can be quite challenging. Lastly, standard orchard or farm operations can interfere with spray applications. Large farm crews often need to be in the field working on horticultural issues such as pruning, thinning fruit, or other maintenance. Pesticide exposure from drift is a true risk to these workers if applications are being conducted anywhere nearby. Furthermore, for the reasons discussed earlier, old tree architectures are being replaced with modern vertical and Y-trellised tree architectures. Such modernization offers a unique opportunity for the industry to research and adopt novel chemical application techniques.

A well managed modern tree fruit architecture has a somewhat uniform canopy, and growers can benefit greatly if they adopt a precision spraying approach in such canopies. In one such approach, chemicals can be applied precisely into the canopies by a solid set canopy delivery system (SSCDS). Variants of SSCDS, also termed a fixed spraying system, have been studied in the USA (Carpenter *et al.*, 1985; Agnello and Landers, 2006; Sharda *et al.*, 2015) and Europe (Verpont *et al.*, 2015). An SSCDS constitutes an 'array of microsprayers positioned within the canopy for precise delivery of inputs in high-density orchards' (Grieshop *et al.*, 2013). Besides solid sets within the orchard block, the system also encompasses a chemical tank, pumping unit and air compressor that needs to be connected to a fixed set of microsprayers for chemical application.

Current systems can deliver chemical applications in a timely manner with sub-acre research blocks taking less than 1 min to spray. This would allow for applications during ideal weather conditions and minimize accidental human exposure risk. Existing studies have shown variable deposition in the canopy primarily due to the arrangement of the emitters in the canopy (Sharda *et al.*, 2013, 2015).

If designed and optimized properly, an SSCDS can offer several advantages (Agnello and Landers, 2006; Grieshop *et al.*, 2013; Verpont *et al.*, 2015), such as: (i) large acreage applications within a short time; (ii) possibility of microclimate-driven targeted applications; (iii) reduced chemical exposure to spray applicators; and (iv) reduced spray drift. In terms of adoption, as an SSCDS needs a fixed system installation within the orchard blocks, initial investment may be higher. Also, though SSCDS has been found suitable for short-length rows and for somewhat flat orchard topologies, more research is needed to optimize such systems in highly undulating orchard blocks that may have rows up to several hundred meters long. The future challenges also include developing systems for large acreage that do not interfere with other orchard and vineyard operations such as mechanical pruning and harvesting, as well as creating a fault detection system that shows when certain emitters are not working.

Aerial precision (or surgical) spraying could be another possible approach which can complement effective crop protection. In modern canopy architectures, emerging ground and remote sensing tools can be used to understand the pest and disease pressure (i.e. spatial variability).

Growers can then use ground or UAS integrated spray application techniques for surgical spraying instead of uniform chemical applications of the entire block. For example, Giles and Billing (2015) tested such a mid-sized UAS (model: RMax; Yamaha Motor Corp., Japan), with a platform weight greater than 25 kg (55 lb), for surgical spray applications in California vineyards. With optimized altitude and flight parameters for a given canopy architecture, mid-sized UASs (Fig. 7.3) offer advantages such as: (i) reduced chemical use with surgical applications; (ii) large acreage applications within a short time; (iii) relatively safe applications as such systems are unmanned; (iv) possibility of microclimate-driven targeted applications that are independent of row orientation; (v) possibility of site topography-driven applications instead of orchard tree row orientations; and (vi) no soil compaction related issues.

As mid-sized UASs is an emerging technology in agriculture, such systems need further refinements toward on-board intelligence, autonomous take-off and landing, improved platform stability and georeferencing, and improved payload lift capabilities (Ollero and Merino, 2004; Kotchi *et al.*, 2016). Such systems also need to be designed to integrate

Fig. 7.3. (a) Mid-sized unmanned helicopter and (b) aerial system-based surgical aerial spray applications in apple production (photo courtesy of Lav Khot, Washington State University).

real-time wind speed and direction to optimize the flight paths (Faiçal *et al.*, 2014) for targeted spray applications. The cost of ownership, timely maintenance, availability of a skilled workforce for operation and maintenance, and country-specific regulations may be some issues that will govern adoption of such technology in tree fruit production management.

7.4.4 Standardization issues

In the USA, configuration and operation of a sprayer can be challenging for all the reasons already mentioned, but also because of the lack of standardization for manufacturing processes. There are minimum standards set by the UN Food and Agriculture Organization (FAO) and similar entities for sprayer manufacturing, but they do not address the issues of repeatability amongst units. Sprayers are often manufactured manually, which means there are small differences in the assembly process. While machines can look similar, small changes in fan orientation or deflector alignment can lead to differences in air direction and spray patterns. This essentially means that while sprayers can perform similarly, they all perform as individuals.

Furthermore, other choices, such as nozzles, can be complex for farmers. Traditionally, droplet size categories (i.e. fine, coarse) from nozzles could not be compared across different manufacturers. With the regulation of ASABE S-572, nozzle manufacturers needed to classify the volume median diameter (VMD) of droplets into one of the eight defined categories with color codes. Now it is possible for a farmer to know that a nozzle producing 'fine' droplets meant the VMD was in the range of 106–235 microns, no matter the manufacturer. In Europe, manufacturers also follow standards set by the International Standards Organization (ISO). The nozzles are color coded according to ISO standards so that a single color will deliver the same output regardless of nozzle type. For example, a red hollow-cone nozzle will put out the same number of liters per minute as a red AI nozzle. This allows farms to easily switch among nozzles to optimize the application. In the USA, nozzle colors are not standardized unless the manufacturer chooses to use ISO colors. These inconsistencies in manufacturing of the machines and parts leads to uncertainties in the operation and optimization of the sprayers.

7.5 Summary

Pest management in the absence of a proper monitoring method usually leads to an excessive use of pesticides. Scouting is currently the main approach for monitoring tree and produce health. Conventional methods for monitoring pests and disease are costly, labor-intensive, and time-consuming. Better management of pests and diseases could alleviate these challenges by using techniques that enable the continuous monitoring of insect activity and health issues in tree crops. Emerging technologies such as pheromone-baited insect traps (smart traps) integrated into WSNs and advanced microclimatic monitoring systems with high spatial and temporal resolution have opened up new doors to an era of automatic pest monitoring. Advanced sensing tools for detecting crop diseases such as space-based satellite imagery, airborne remote sensing, ground-based sensor systems, and a broad range of non-invasive radiometric-based sensors are available today that need to be tested and customized for specific applications. There is a need for large-scale real-time plant disease detection tools that, under field conditions, are rapid, cost-effective, specific to a particular disease, and sensitive for detection at the early onset of the symptoms. More research is needed to advance data-driven algorithms for the discrimination of concurrent diseases and other plant stresses. Besides algorithms and sensing techniques, current research on insects and disease monitoring also includes electronic applications for smartphones, web-based decision support systems integrated with WSN systems, and sensor integration with autonomous vehicles and robots that monitor and control other aspects of crop and soil management. Pertinent to application technology, extensive research and extension efforts are needed to drive the precision spraying concepts from theory to field-level adoption.

References

Agnello, A. and Landers, A. (2006) Current progress in development of a fixed-spray pesticide application system for high-density apple planting. *New York Fruit Quarterly* 14(4), 22–26.

Armstrong J., Fenske, R.A., Yost, M.G., Galvin, K., Tchong-French, M. and Yu, J. (2013a) Presence of organophosphorus pesticide oxygen analogs in air samples. *Atmospheric Environment* 66, 145–150.

Armstrong J., Fenske, R.A., Yost, M.G., Tchong-French, M. and Yu, J. (2013b) Comparison of polyurethane foam and XAD-2 sampling matrices to measure airborne organophosphorus pesticides and their oxygen analogs in an agricultural community. *Chemosphere* 92(4), 451–457.

Beers, E.H. (1993) *Orchard Pest Management: A Resource Book for the Pacific Northwest.* Good Fruit Grower, Yakima, Washington.

Belasque, L., Gasparoto, M.C.G. and Marcassa, L.G. (2008) Detection of mechanical and disease stresses in citrus plants by fluorescence spectroscopy. *Applied Optics* 47(11), 1922–1926.

Brown, E.M. and Sutton, T.B. (1995) An empirical model for predicting the first symptoms of sooty blotch and flyspeck of apples. *Plant Disease* 79, 1165–1168.

Brown, D.L., Giles, D.K., Oliver, M.N. and Klassen, P. (2008) Targeted spray technology to reduce pesticide in runoff from dormant orchards. *Crop Protection* 27, 545–552.

Carpenter, T.G., Richard, D.L. and Wilson, S.M. (1985) Design and feasibility of a permanent pesticide application system for orchards. *Transactions of the ASAE* 28(2), 350–355.

Chaerle, L., Lenk, S., Hagenbeek, D., Buschmann, C. and Van Der Straeten, D. (2007) Multicolor fluorescence imaging for early detection of the hypersensitive reaction to tobacco mosaic virus. *Journal of Plant Physiology* 164(3), 253–262.

Chambers, U., Petit, B. and Jones, V.P. (2011) WSU-DAS – the online pest management support systems for tree fruits in Washington State. *Acta Horticulturae* 1068, 27–33.

Chen, Y., Zhu, H., Ozkan, H.E., Derksen, R.C. and Krause, C.R. (2011) *An Experimental Variable-rate Sprayer for Nursery and Orchard Applications*. ASABE Paper No. 1110497. American Society of Agricultural and Biological Engineers, St Joseph, Michigan.

Chen, Y., Zhu, H. and Ozkan, H.E. (2013) Development of a variable-rate sprayer with laser scanning sensor to synchronize spray outputs to tree structures. *Transactions of the ASABE* 55(3), 773–781.

Da Silva, J.R.M., Damásio, C.V., Sousa, A.M.O., Bugalho, L., Pessanha, L. and Quaresma, P. (2015) Agriculture pest and disease risk maps considering MSG satellite data and land surface temperature. *International Journal of Applied Earth Observation and Geoinformation* 38, 40–50.

Deng, C. and Wu, C. (2013) Estimating very high resolution urban surface temperature using a spectral unmixing and thermal mixing approach. *International Journal of Applied Earth Observation and Geoinformation* 23, 155–164.

Ding, W. and Taylor, G. (2016) Automatic moth detection from trap images for pest management. *Computers and Electronics in Agriculture* 123, 17–28.

Du, Q., French, J.V., Skaria, M., Yang, C. and Everitt, J.H. (2004) Citrus pest stress monitoring using airborne hyperspectral imagery. In: *Conference Proceedings of the International Geoscience and Remote Sensing Symposia*, Vol. VI, 3981–3984. Institute of Electrical and Electronics Engineers (IEEE), Piscataway, New Jersey.

Duttweiler, K.B., Gleason, M., Dixon, P.M., Sutton, T.B., McManus, P.S. and Monteiro, J.E.B. (2008) Adaptation of an apple sooty blotch and flyspeck warning system for the Upper Midwest United States. *Plant Disease* 92, 1215–1222.

Escolà, A.J., Rosell-Polo, R., Planas, S., Gil, E., Pomar, J. *et al.* (2013) Variable rate sprayer. Part 1 – Orchard prototype: design, implementation and validation. *Computers and Electronics in Agriculture* 95, 122–135.

Faiçal, B.S., Costa, F.G., Pessin, G., Ueyama, J., Freitas, H. *et al.* (2014) The use of unmanned aerial vehicles and wireless sensor networks for spraying pesticides. *Journal of Systems Architecture* 60(4), 393–404.

Fox, R.D., Derksen, R.C., Zhu, H., Brazee, R.D. and Svensson, S.A. (2008) A history of airblast sprayer development and future prospects. *Transactions of the ASABE* 51(2), 405–410.

Gent, D.H., De Wolf, E.D. and Pethybridge, S.J. (2011) Perceptions of risk, risk aversion, and barriers to adoption of decision support systems and integrated pest management: an introduction. *Phytopathology* 101, 640–643.

Giles, D.K. and Billing, R.C. (2015) Deployment and performance of a UAV for crop spraying. *Chemical Engineering Transactions* 44, 307–312.

Gogo, E.O., Saidi, M. and Ochieng, J.M. (2014) Microclimate modification and insect pest exclusion using Agronet improve pod yield and quality of French bean. *HortScience* 49(10), 1298–1304.

Gonzalez-de-Soto, M., Emmi, L., Perez-Ruiz, M., Aguera, J. and Gonzalez-de-Santos, P. (2016) Autonomous systems for precise spraying – evaluation of a robotized patch sprayer. *Biosystems Engineering* 146, 165–182.

Grieshop, M., Brunner, J. and Agnello, A. (2013) Progress report on 'development and optimization of solid-set canopy delivery system for resource-efficient, ecologically sustainable apple and cherry production'. Available at: http://www.canopydelivery.msu.edu/tag/progress/ (accessed 5 October 2016).

Griffin, M.K. and Burke, H.K. (2003) Compensation of hyperspectral data for atmospheric effects. *Lincoln Laboratory Journal* 14(1), 29–54.

Guo, T.T., Guo, L., Wang, X.H. and Li, M. (2009) Application of NIR spectroscopy in classification of plant species. In: *International Workshop on Education Technology and Computer Science*, Wuhan, Hubei, China, 3, pp. 879–883.

Hatfield, J. and Thomason, I.J. (eds) (1982) *Biometeorology in Integrated Pest Management*, 1st edn. Academic Press Inc., New York.

Herrington, P.J., Mapother, H.R. and Stringer, A. (1981) Spray retention and distribution on apple trees. *Pesticide Science* 12(5), 515–520.

Hong, S.C., Magarey, R.D., Borchert, D.M., Vargas, R.I. and Souder, S.K. (2015) Site-specific temporal and spatial validation of a generic plant pest forecast system with observations of *Bactrocera dorsalis* (oriental fruit fly). *NeoBiota* 27, 37–67.

Jeon, H. and Zhu, H. (2012) Development of variable-rate sprayer for nursery liner applications. *Transactions of the ASABE* 55(1), 303–312.

Kang, W.S., Hong, S.S., Han, Y.K., Kim, K.R., Kim, S.G. and Park, E.W. (2010) A web-based information system for plant disease forecast based on weather data at high spatial resolution. *Plant Pathology Journal* 26(1), 37–48.

Khot, L.R., Ehsani, R., Albrigo, G., Larbi, P.A., Landers, A., Campoy, J. and Wellington, C. (2012) Air-assisted sprayer adapted for precision horticulture: spray patterns and deposition assessments in small-sized citrus canopies. *Biosystems Engineering* 113(1), 76–85.

Khot, L.R., Ehsani, R., Maja, J.M., Campoy, J.M., Wellington, C. and Al-Jumaili, A. (2014) Evaluation of deposition and coverage of an air-assisted sprayer, and two airblast sprayers in citrus orchards. *Transactions of the ASABE* 57(4), 1007–1013.

Kotchi S.O., Barrette, N., Viau, A.A., Jang, J.D., Gond, V. and Mostafavi, M.A. (2016) Estimation and uncertainty assessment of surface microclimate indicators at local scale using airborne infrared thermography and multispectral imagery. In: Imperatore, P. and Pepe, A. (eds) *Geospatial Technology – Environmental and Social Applications*. Intech Publishers, Rijeka, Croatia, pp. 99–141.

Kumar, A., Lee, W.S., Ehsani, R., Albrigo, L.G., Yang, C. and Mangan, R.L. (2010) Citrus greening disease detection using airborne multispectral and hyper spectral imaging. In: *Proceedings of the 10th International Conference on Precision Agriculture*, 18–21, Denver, Colorado. Paper No. 427.15. Precision Agriculture Laboratory, Department of Agricultural and Biological Engineering, University of Florida, Gainesville, Florida.

Lau, Y., Gleason, M.L., Zriba, N., Taylor, S.E. and Hinz, P.N. (2000) Effects of coating, deployment angle, and compass orientation on performance of electronic wetness sensors during dew periods. *Plant Disease* 84, 192–197.

Lazcka, O., Campo, F.J.D. and Muñoz, F.X. (2007) Pathogen detection: a perspective of traditional methods and biosensors. *Biosensors and Bioelectronics* 22(7), 1205–1217.

Lee, W.S., Ehsani, R. and Albrigo, L.G. (2008) Citrus greening disease (huanglongbing) detection using aerial hyperspectral imaging. In: *Proceedings of the 9th International Conference on Precision Agriculture*, July 20–23, Denver, Colorado. Precision Agriculture Laboratory, Department of Agricultural and Biological Engineering, University of Florida, Gainesville, Florida.

Lee, W.S., Alchanatis, V., Yang, C., Hirafuji, M., Moshou, D. and Li, C. (2010) Sensing technologies for precision specialty crop production. *Computers and Electronics in Agriculture* 74, 2–33.

Li, W., Hartung, J.S. and Levy, L. (2006) Quantitative real-time PCR for detection and identification of *Candidatus liberibacter* species associated with citrus huanglongbing. *Journal of Microbiological Methods* 66(1), 104–115.

Lins, E.C., Belasque, J. Jr, and Marcassa, L.G. (2009) Detection of citrus canker in citrus plants using laser induced fluorescence spectroscopy. *Precision Agriculture* 10, 319–330.

Llorente, I., Vilardell, P., Bugiani, R., Gerardi, I. and Montesinos, E. (2000) Evaluation of BSPcast disease-warning system in reduced fungicide use programs for management of brown spot of pear. *Plant Disease* 84, 631–637.

López, M.M., Bertolini, E., Olmos, A., Caruso, P., Gorris, M.T. *et al.* (2003) Innovative tools for detection of plant pathogenic viruses and bacteria. *International Microbiology* 6, 233–243.

Magarey, R.D., Seem, R.C., Russo, J.M., Zack, J.W., Waight, K.T., Travis, J.W. and Oudemans, P.V. (2001) Site-specific weather information without on-site sensors. *Plant Disease* 85(12), 1216–1226.

Magarey, R.D., Travis, J.W., Russo, J.M., Seem, R.C. and Magarey, P.A. (2002) Decision support systems: quenching the thirst. *Plant Disease* 86, 4–14.

Magarey, R.D., Russo, J.M., Seem, R.C. and Gadoury, D.M. (2005) Surface wetness duration under controlled environmental conditions. *Agricultural and Forest Meteorology* 128, 111–122.

Magarey, R.D., Russo, J.M. and Seem, R.C. (2006) Simulation of surface wetness with a water budget and energy balance approach. *Agricultural and Forest Meteorology* 139, 373–381.

Magarey, R.D., Fowler, G.A., Borchert, D.M., Sutton, T.B., Colunga-Garcia, M. and Simpson, J.A. (2007) NAPPFAST: an internet system for the weather-based mapping of plant pathogens. *Plant Disease* 91(4), 336–345.

Markom, M.A., Md Shakaff, A.Y., Adom, A.H., Ahmad, M.N., Hidayat, W., Abdullah, A.H. and Ahmad Fikri, N. (2009) Intelligent electronic nose system for basal stem rot disease detection. *Computers and Electronics in Agriculture* 66(2), 140–146.

Matese, A., Crisci, A., Di Gennaro, S.F., Primicerio, J., Tomasi, D., Marcuzzo, P. and Guidoni, S. (2014) Spatial variability of meteorological conditions at different scales in viticulture. *Agricultural and Forest Meteorology* 189, 159–167.

McDonald, M.R., Gossen, B.D., Kora, C., Parker, M. and Boland, G. (2013) Using crop canopy modification to manage plant diseases. *European Journal of Plant Pathology* 135, 581–593.

Moalemiyan, M., Vikram, A., Kushalappa, A.C. and Yaylayan, V. (2006) Volatile metabolite profiling to detect and discriminate stem-end rot and anthracnose diseases of mango fruits. *Plant Pathology* 55, 792–802.

Moalemiyan, M., Vikram, A. and Kushalappa, A.C. (2007) Detection and discrimination of two fungal diseases of mango (cv. Keitt) fruits based on volatile metabolite profiles using GC/MS. *Postharvest Biology and Technology* 45, 117–125.

Ollero, A. and Merino, L. (2004) Control and perception techniques for aerial robotics. *Annual Reviews in Control* 28(2), 167–178.

Pangga, I.B., Hanan, J. and Chakraborty, S. (2013) Climate change impacts on plant canopy architecture: implications for pest and pathogen management. *European Journal of Plant Pathology* 135, 595–610.

Peres, N.A. and Timmer, L.W. (2006) Evaluation of the alter-rater model for spray timing for control of *Alternaria* brown spot of *Murcott tangor* in Brazil. *Crop Protection* 25, 454–460.

Pérez-Ruiz, M., Agüera, J., Gil, J.A. and Slaughter, D.C. (2011) Optimization of agrochemical application in olive groves based on positioning sensor. *Precision Agriculture* 12, 564–575.

Raisigl, U., Felber, H., Siegfried, W. and Krebs, C. (1991) *Comparison of different mist blowers and volume rates for orchard spraying.* BCPC Monograph 46, 185–196. British Crop Protection Council, Farnham, UK.

Rowlandson, T., Gleason, M., Sentelhas, P., Gillespie, T., Thomas, C. and Hornbuckle, B. (2015) Reconsidering leaf wetness duration determination for plant disease management. *Plant Disease* 99(3), 310–319.

Ruiz-Altisent, M., Ruiz-Garcia, L., Moreda, G.P., Lu, R., Hernandez-Sanchez, N. *et al.* (2010) Sensors for product characterization and quality of specialty crops – a review. *Computers and Electronics in Agriculture* 74, 176–194.

Sankaran, S. and Ehsani, R. (2012) Detection of Huanglongbing disease in citrus using fluorescence spectroscopy. *Transactions of the ASABE* 55(1), 313–320.

Sankaran, S., Ehsani, R. and Etxeberria, E. (2010a) Mid-infrared spectroscopy for detection of Huanglongbing (greening) in citrus leaves. *Talanta* 83, 574–581.
Sankaran, S., Mishra, A., Ehsani, R. and Davis, C. (2010b) A review of advanced techniques for detecting plant diseases. *Computers and Electronics in Agriculture* 72, 1–13.
Sankaran, S., Mishra, A., Maja, J.M. and Ehsani, R. (2011) Visible–near infrared spectroscopy for detection of Huanglongbing in citrus orchards. *Computer and Electronics in Agriculture* 77(2), 127–134.
Saponari, M., Manjunath, K. and Yokomi, R.K. (2008) Quantitative detection of *Citrus tristeza* virus in citrus and aphids by real-time reverse transcription-PCR (TaqMan). *Journal of Virological Methods* 147(1), 43–53.
Sawyer, J.E. (1994) Concepts of variable rate technology with considerations for fertilizer application. *Journal of Production Agriculture* 7(2), 195–201.
Selby, R.D., Gage, S.H. and Whalon, M.E. (2014) Precise and low-cost monitoring of plum curculio (Coleoptera: Curculionidae) pest activity in pyramid traps with cameras. *Environmental Entomology* 43(2), 421–431.
Sentelhas, P.C., Monteiro, J.E.B.A. and Gillespie, T.J. (2004) Electronic leaf wetness duration sensor: why it should be painted. *International Journal of Biometeorology* 48, 202–205.
Sentelhas, P.C., Gillespie, T.J. and Santos, E.A. (2007) Leaf wetness duration measurement: comparison of cylindrical and flat plate sensors under different field conditions. *International Journal of Biometeorology* 51, 265–273.
Sentelhas, P.C., Gillespie, T.J. and Santos, E.A. (2008) Spatial variability of leaf wetness duration in cotton, coffee and banana crop canopies. *Scientia Agricola* 65, 18–25.
Shafri, H.Z.M. and Hamdan, N. (2009) Hyperspectral imagery for mapping disease infection in oil palm plantation using vegetation indices and red edge techniques. *American Journal of Applied Sciences* 6(6), 1031–1035.
Sharda, A., Karkee, M. and Zhang, Q. (2013) *Fluid Dynamics of a Solid Set Canopy Spray Delivery System for Orchard Applications.* ASABE Paper No. 131620688. American Society of Agricultural and Biological Engineers (ASABE), St Joseph, Michigan.
Sharda, A., Karkee, M., Zhang, Q., Ewlanow, I., Adameit, U. and Brunner, J. (2015) Effect of nozzle type and mounting configuration around tree canopy on spray coverage using solid set canopy delivery system. *Computer and Electronics in Agriculture* 112(2015), 184–192.
Spinelli, F., Noferini, M. and Costa, G. (2006) Near infrared spectroscopy (NIRs): perspective of fire blight detection in asymptomatic plant material. Proceedings of the 10th International Workshop on Fire Blight. *Acta Horticulturae* 704, 87–90.
Sundaram, J., Kandala, C.V. and Butts, C.L. (2009) Application of near infrared (NIR) spectroscopy to peanut grading and quality analysis: overview. *Sensing and Instrumentation for Food Quality and Safety* 3(3), 156–164.
Sutton, J.C., Gillespie, T.J. and Hildebrand, P.D. (1984) Monitoring weather factors in relation to plant disease. *Plant Disease* 68, 78–84.
Tirelli, P., Borghese, N.A., Pedersini, F., Galassi, G. and Oberti, R. (2011) Automatic monitoring of pest insects traps by Zigbee-based wireless networking of image sensors. *Instrumentation and Measurement Technology Conference (I2MTC)*, IEEE, 10–12 May 2011.
Tsai, M.Y. (2007) The Washington Orchard Spray Drift Study: understanding the broader mechanisms of pesticide spray drift. Doctoral Dissertation, University of Washington School of Public Health, Seattle, Washington.
Ushaa, K. and Singh, B. (2013) Potential applications of remote sensing in horticulture – a review. *Scientia Horticulturae* 153, 71–83.
Velasco-Garcia, M.N. and Mottram, T. (2003) Biosensor technology addressing agricultural problems. *Biosystems Engineering* 84(1), 1–12.

Vercruysse, F., Steurbaut, W., Drieghe, S. and Dejonckheere, W. (1999) Off target ground deposits from spraying a semi-dwarf orchard. *Crop Protection* 18, 565–570.

Verpont, F., Favareille, J. and Zavagli, F. (2015) Fixed spraying system: a future potential way to apply pesticides in an apple orchard? In: Triloff, P. (ed.) *13th Workshop on Spray Application in Fruit Growing*, Lindau, Germany, 15–18 July 2015. Bundesforschungsinstitut für Kulturpflanzen (Julius Kühn-Institut), Quedlinburg, Germany, pp. 53–54.

Wagenmakers, P. (1991) Simulation of light distribution in dense orchard systems. *Agricultural and Forest Meteorology* 57(1–3), 13–25.

Wagenmakers, P. and Callesen, O. (1995) Light distribution in apple orchard systems in relation to production and fruit quality. *Journal of Horticultural Science* 70(6), 935–948.

Wood, G.A., Taylor, J.C. and Godwin, R.J. (2003) Calibration methodology for mapping within-field crop variability using remote sensing. *Biosystems Engineering* 84(4), 409–423.

Workneh, F., Narasimhan, B., Srinivasan, R. and Rush, C.M. (2005) Potential of radar-estimated rainfall for plant disease risk forecast. *Phytopathology* 95, 25–27.

8 Precision Nutrient Management

Gerry Neilsen* and Denise Neilsen

Agriculture and Agri-Food Canada, Summerland, British Columbia, Canada

8.1 Introduction

In horticultural production systems, increased precision in the application of both water and nutrients can lead to improvements in the efficiency of nutrient uptake by the crop, particularly when using simultaneous applications – a process that is commonly referred to as fertigation. Fertigation principles and practices have previously been reviewed (Haynes, 1985; Bar-Yosef, 1999). Fertigation has been applied to a wide range of perennial fruit crops, including apple (Neilsen *et al.*, 1999), sweet cherry (Neilsen *et al.*, 2004a), pecans (Wells, 2015), peach (Bussi *et al.*, 1991), grape (Treeby, 2008), blueberry (Vargas *et al.*, 2015), coffee (Bruna *et al.*, 2015) and citrus (Alva, 2008).

This chapter will focus on cumulative fertigation research undertaken primarily in high-density apple and cherry as a case study of the practical application of fertigation. Thus the information is most relevant to semi-arid fruit-growing regions, where irrigation is required to achieve production. However, the principles may also have relevance to more humid regions where supplemental irrigation is used as a buffer against erratic precipitation. Fertigation is of particular interest in orchards that are located on coarse-textured sandy loams, loamy sands or sandy soils. Such soils are susceptible to several management problems, including slow growth of newly planted orchards and the development of various nutrient imbalances attributed to low organic matter content and their generally poor nutrient and water retention capacities.

The precision control of nutrients and water applications via fertigation, primarily through low-pressure irrigation systems, will be considered in

* Corresponding author, email: Gerry.Neilsen@AGR.GC.CA

combination with automated atmometer-scheduled irrigation. Fertigation is particularly suited for low-pressure irrigation systems, such as drip, which tend to concentrate root development (and hence nutrient uptake) into a restricted wetted zone receiving nutrients (Bravdo and Proebsting, 1993).

8.2 Fertigation Methods

A typical automated fertigation injection system (Fig. 8.1) can be used to fertigate orchards. Fertigation relies on a method of injecting fertilizer into an irrigation system, suitable soluble fertilizers, effective irrigation scheduling, and periodic monitoring to verify functioning of the system.

8.2.1 Fertilizer injection

There is a range of injection systems commercially available to inject fertilizers into an orchard's irrigation system. They vary from low-cost Venturi systems that rely on the pressure drop created by the change in water velocity as it passes through a constriction (Fig. 8.1, component 6) to more expensive, electrically powered or water-driven injector pump systems. Practical considerations concerning choice of injectors and other details of fertigation adoption are often available from irrigation manuals designed for various regions and crop production systems. For example, much of the field research subsequently described was informed by the British Columbia Trickle Irrigation Manual (Van der Gulik, 1999).

1. Back flow preventer
2. Filter
3. Solenoid valve
4. Flow meter
5. Pressure regulator
6. Injection port
   i. Venturi – in line passive
   ii. Pumps – active injection
7. Pressure compensating emitters
8. Automated soil moisture or ET data

Fig. 8.1. Components of an automated fertigation system.

8.2.2 Irrigation scheduling

Precision water applications will greatly improve the efficiency of use of fertigated nutrients. There is a range of approaches (as discussed in Chapter 6) to improve the precision of irrigation that involve detecting water stress and computing crop water needs in order to schedule irrigation automatically (Osroosh *et al.*, 2016). The approach discussed in this chapter involved scheduling water applications to meet evaporative demand. Water was applied daily, based on the previous day's irrigation as measured by an electronic atmometer (Fig. 8.1, component 8; ET Gage Co., Loveland, Colorado) and modified according to a crop coefficient curve, based on seasonal canopy development (Parchomchuk *et al.*, 1996).

More recently, a crop coefficient curve was developed from the growth-stage crop coefficients for temperate fruit trees outlined in Allen *et al.* (1998). This has been modified to start at full bloom, which coincides with rapid shoot growth in both apple and cherry, and the crop coefficient was plotted against days after full bloom (DAFB) (Fig. 8.2a). Although measurements of total canopy development were not available, estimates were made from measurements of shoot growth at a number of sites for both apple and sweet cherry and compared with the generalized curve (Fig. 8.2a). The generalized curve approximates the timing of the peak of apple and cherry canopy development but is a better fit to the apple data, suggesting that individualized curves for each crop would be preferable. The differences between the two crops earlier in the growing season reflect the earlier bud break of sweet cherry. Another approach would be to use a temperature-based development curve (Fig. 8.2b). In this case, the growing degree-day base 5°C (GDD5) accumulation after full bloom was related to shoot growth data for 'Gala' apple ($R^2 = 0.89$) and 'Sweetheart' sweet cherry ($R^2 = 0.93$).

8.2.3 Fertilizer sources

Soluble fertilizers are required for fertigation and a range of fertilizers has been used as nutrient sources (Table 8.1). Caution is required when co-applying fertilizers in order to select compatible fertilizers that will not form insoluble precipitates within irrigation lines. For example, many phosphate (P) fertilizers will form precipitates when combined with soluble calcium (Ca) and magnesium (Mg) fertilizers. Similarly, irrigation waters may naturally contain high concentrations of dissolved calcium or magnesium which will precipitate with P-fertilizers. If the chemical composition of irrigation water is unknown, a simple test for potential precipitation can be performed by mixing the proposed fertigation solution and irrigation water in the same proportion as will occur in the irrigation line.

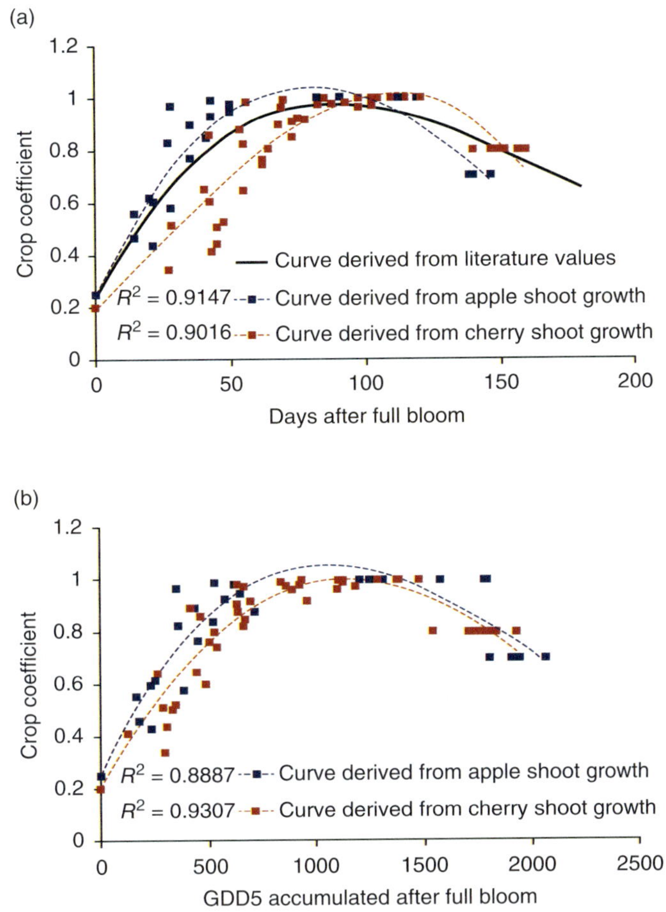

Fig. 8.2. Crop coefficient curves for use with automated fertigation systems derived from literature values (Allen *et al.*, 1998); apple and sweet cherry shoot growth data collected in 2009–2010 for orchard sites in the Okanagan Valley, BC, Canada and related to (a) days after full bloom and (b) GDD5 accumulated after full bloom.

8.2.4 Nutrient and moisture monitoring

It is advantageous to determine how nutrient availability in the soil is affected by fertigation. The distribution of nitrogen (N) and P forms has been measured post fertigation after analyses of soil samples collected at various depths and distances in both laboratory column studies (Bar-Yosef and Sheikholslami, 1976) and field studies in orchards (Klein and Spieler, 1987). In a study of 20 apple orchards, composite soil samples were collected for examination of soil chemical properties, including nutrient availability at 0–15 cm depth directly below drip emitters after NP-fertigation had occurred for 2–5 years (Neilsen *et al.*, 1995c). In this study, comparisons were made with samples collected from the same depth from

Table 8.1. Commonly fertigated nutrient sources for perennial fruit crops. (Various sources, including International Plant Nutrition Institute.)

Nutrient	Compound	Nutrient content	Solubility (20°C) g/l H_2O	Comment
Nitrogen	Ammonium nitrate	33–34% N	1900	Acidifying
	Ammonium sulphate	21% N, 24% S	750	V. acidifying
	Urea	45–46% N	1060	Acidifying
	Calcium nitrate	15.5% N	1310	
	Potassium nitrate	13–14% N, 44–46% K_2O	316	
	Urea solutions, various	20–23% N	Liquid	
	Urea-ammonia solutions, various	28–32% N	Liquid	
Phosphorus	Phosphoric acid	52–75% P_2O_5	5480	Acidifying
	Ammonium polyphosphate	8–11% N, 34% P_2O_5	Liquid	Acidifying
	Mono-ammonium phosphate	11% N, 50% P_2O_5	370	Acidifying
	Di-ammonium phosphate	18% N, 46% P_2O_5	588	Acidifying
Potassium	Potassium chloride	60–62% K_2O	344	
	Potassium sulphate	50% K_2O, 17% S	120	Ultra-fine grind
	Potassium-magnesium sulphate	22% K_2O, 11% MgO	240	
	Potassium thiosulphate	22% K_2O, 17% S	Liquid	
Boron	Sodium borate	20% B	110	

the unfertigated mid-row zone of the orchard. It was discovered that soil pH and extractable Mg, potassium (K) and boron (B) values were declining and preventive nutrient management strategies should be implemented. Useful information can therefore result from periodic destructive soil sampling but this is not feasible if soils are to be sampled frequently in order to determine, for example, changes in nutrient availability within season.

In contrast, permanently installed soil suction lysimeters have been used for repetitive monitoring of soil solution N, P, and K concentrations during the growing season in response to fertigation treatments (Klein and Spieler, 1987; Neilsen, *et al.*, 1998a). Once installed, suction lysimeters allow extraction and chemical analysis of soil solution with minimal disruption of the root zone. Suction lysimeters were extensively used to monitor soil nutrient changes within the growing season to understand system response to fertigation treatments in much of the subsequently reported research.

When using automated irrigation systems, it is always useful to have back-up confirmation of the performance of the irrigation system by monitoring soil moisture content (Fig. 8.1, component 8). There are several commercially available systems that allow automated recording of soil moisture to data loggers. In the bulk of the fertigated research cited from British Columbia, determination of volumetric soil moisture content was made using depth-integrated time domain reflectometry (TDR) (Topp and Reynolds, 1998; Tektronix, Beaverton, Oregon). The seasonal pattern of soil moisture variation over multiple growing seasons (Fig. 8.3) illustrates the degree of control of soil moisture content possible in an NPKB-fertigated apple orchard receiving daily irrigation at 100% evapotranspiration (ET) replacement (Neilsen *et al.*, 2016). The effects of high precipitation events (2007 and 2009) on soil moisture content can be readily discerned as well as the lack of information available when electronic recording equipment fails (2008).

8.3 Nutrient Requirements

It has been difficult to determine the annual nutrient requirements of perennial fruit crops in order to develop annual fertigation application rates. Successive destructive sampling and chemical analysis of representative trees are usually required in order to determine the nutrients needed for incremental growth of above-ground woody structures, buds, leaves and fruit and below-ground roots. In one such study involving 'Gala' apple on M.26 rootstock grown in sand culture with adequate nutrition, annual nutrient requirements were measured for essential major and minor nutrients to support a crop load of 52.5 t/ha (Cheng and Raba, 2009) (Table 8.2). It was observed that highest total nutrient requirements in a growing season were for N (62% contained in shoots and leaves) and K (77% contained in fruit) with minor requirements (< 1 kg/ha) for B, zinc (Zn), copper (Cu), manganese (Mn), and iron (Fe). Other studies have indicated for N that 40–50% of total tree N is required for fruit and senescent leaves which can be lost annually from the orchard (Neilsen and Neilsen, 2002). Estimates from these studies of the annual nitrogen required to replace that removed in fruit and senescent leaves are summarized for a series of experiments involving high-density apple trees planted on M.9 rootstock (Table 8.3).

In addition to quantity of nutrients required, other considerations when fertigating include: variation in demand for nutrients within the growing season and among fruit crops; effects on crop performance, including fruit quality; and efforts to improve nutrient use efficiency and minimize environmental impact. Experience with fertigation to address these issues will be subsequently discussed for individual nutrients, using apple and sweet cherry as model fruit crops. Results were derived from a series of case studies of fertigation trials established in the field over the past two decades.

8.3.1 Nitrogen

The effects of various fertilizer and water application strategies on root-zone N availability was monitored by the determination of soil solution

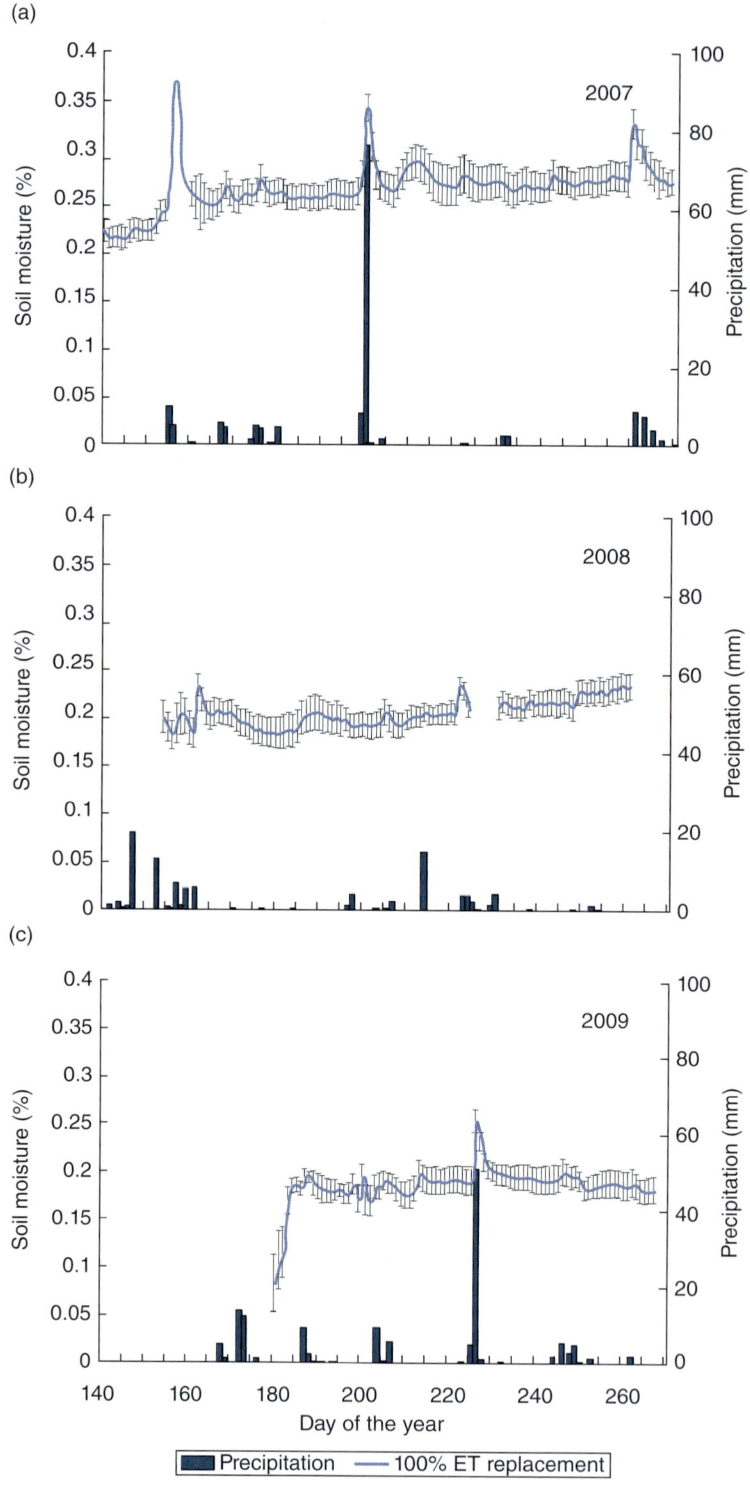

Fig. 8.3. Average daily volumetric soil moisture content in response to 100% ET replacement irrigation and daily precipitation: (a) 2007, (b) 2008 and (c) 2009. From Neilsen *et al*. (2016).

Table 8.2. Net accumulation of nutrients from bud break to fruit harvest for 6-year-old 'Gala'/M.26 rootstock[a] grown in sand culture. (Adapted from Cheng and Raba, 2009.)

	Accumulation			Accumulation	
Nutrient	(g/tree)	(kg/ha)	Nutrient	(mg/tree)	(kg/ha)
N	19.8	55.2	B	93.6	0.26
P	3.3	9.2	Zn	60.9	0.17
K	36.0	100.4	Cu	46.5	0.13
Ca	14.2	39.6	Mn	189.8	0.52
Mg	4.4	12.3	Fe	148.7	0.42
S	1.6	4.5			

[a]Tree spacing 1.07 × 3.35 m. Tree yield 52.5 t/ha.

Table 8.3. Estimates of annual nitrogen requirements of apple trees on 'Malling 9' (M.9) dwarfing rootstock; nitrogen removal in fruit and senescent leaves. (Adapted from Neilsen and Neilsen, 2002.)

		N content	
Expt.	Variety	g/tree	kg/ha[a]
1	'Golden Delicious'/M.9 end of year 1[b]	2.9	9.2
2	'Elstar'/M.9 end of year 4	10.2	33.1
3	'Gala'/M.9 end of year 3	10.5	34.4
4	'Gala'/M.9 end of year 6	12.2	39.8

[a]Assumes density of 3300 trees/ha
[b]Leaves only

nitrate-N (NO_3^--N) concentrations from permanently installed suction lysimeters at 30 cm depth directly beneath drip emitters within the root zone of high-density apple (Neilsen *et al.*,1998a) (Fig. 8.4). Nitrate-N concentration in the soil solution beneath the drip emitters remained higher over more of the growing season for weekly fertigation and daily drip irrigation, compared with single broadcast fertilizer application and sprinkler irrigation (Fig. 8.4a). With daily calcium nitrate fertigation and daily drip irrigation, N concentrations increased and decreased rapidly with both the onset and end of fertigation, remained relatively constant during the intervening period and varied depending up on the time of N application (N time 1 compared with N time 2, Fig. 8.4b).

Systematic soil solution monitoring also indicated that the volume of applied water affected soil solution NO_3^--N concentrations, with doubling the daily volume of water applied halving soil solution nitrate concentrations (Neilsen *et al.*, 1998a). Since quantities of water applied to the soil via irrigation or natural precipitation affect the concentration of NO_3^--N in the root zone, scheduling irrigation in response to evapotranspiration using an atmometer can be used to ensure that excess water is not applied to the soil when the weather is cool and wet. In a 3-year study in a young high-density apple orchard, the effects of scheduling irrigation to meet evaporative demand was compared with irrigation applied

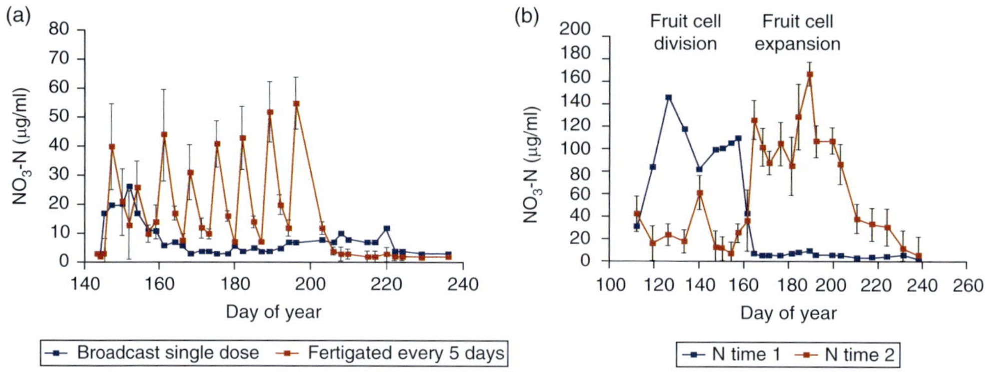

Fig. 8.4. Soil solution nitrate-N concentration in response to (a) either single broadcast fertilizer application or fertigation at 5-day intervals (from Neilsen *et al.*, 1998a) or (b) daily fertigation at different phenological stages (from Neilsen *et al.*, 2001).

at a fixed rate (unscheduled) in order to assess the effects on leaching loss of NO_3^--N (Neilsen and Neilsen, 2002). Water losses beneath the root zone, as measured by capillary wick samplers, were greater for fixed-rate compared with scheduled irrigation during the coolest months (May, June and September) of irrigation application. Thus any N fertigated during a period of unnecessarily high water applications would be susceptible to leaching. Greater N-use efficiency was measured for trees when irrigation was scheduled to meet evaporation demands rather than applied at a fixed rate year round. Oversupply of water resulted in deep drainage of fertigated N (and P) from both drip and microjet irrigation systems in an apple orchard even when irrigation was automatically applied twice daily in response to ET estimated from an electronic atmometer (Neilsen *et al.*, 2008a). Splitting total daily irrigation requirements into quarter applications applied four times every 6 h improved establishment and initial growth of sweet cherry on a dwarfing rootstock (Gisela 6) in an orchard with a coarse-textured loamy sand (Neilsen *et al.*, 2010b). More frequent irrigations but of smaller volume may improve fruit tree growth while reducing nutrient leaching.

Timing of N application is an important consideration when fertigating N. Annual fruit tree growth is supported by N remobilized from storage in spring as well as that taken up by roots. Labeled N studies have indicated that root uptake of soil N is negligible during early spring prior to bloom for several perennial fruit crops, including apple (Neilsen *et al.*, 2001), sweet cherry (Grassi *et al.*, 2003) and nectarine (Tagliavini *et al.*, 1999). As illustrated for apple, the start of rapid uptake of N from the soil is associated with the end of within-tree remobilization and coincides with the start of annual shoot growth (Fig. 8.5). This means that pre-bloom N fertilizer applications have often proven ineffective and the preferred time of N-fertigation is in the 4–6-week period of rapid shoot growth following petal fall.

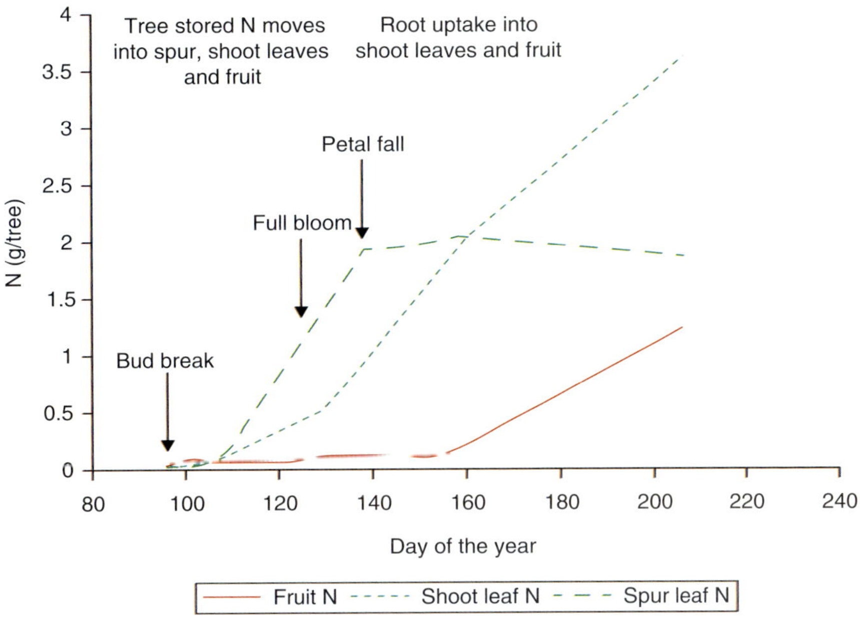

Fig. 8.5. Uptake of fertigated 15N-fertilizer into sequentially harvested 2-year-old 'Fuji'/M.9 relative to tree phenology. Adapted from Guak *et al.* (2003) and Neilsen *et al.* (2006).

The effects of N-fertigation on high-density apple is illustrated in a long-term experiment maintained for the first six fruiting seasons in an orchard containing five different cultivars ('Ambrosia', 'Cameo', 'Fuji', 'Gala', and 'Silken') on M.9 rootstock (Neilsen *et al.*, 2009). Fertigation treatments were a combination of two N rates and three times of N application. N was applied at low (28 mg N/l) or high (168 mg N/l) concentrations daily at 0–4, 4–8 or 8–12 weeks after full bloom (of the 'Ambrosia' cultivar). It was difficult to distinguish consistent effects of altering the timing of N-fertigation within the 0–12-week period following bloom, although there is a tendency to improve yield and fruit size by early application of N within 4 weeks post bloom while N applications closer to harvest (8–12 weeks following bloom) tend to delay apple maturity. Fruit firmness and color were unaffected by fertigation timing.

At high N rates, all cultivars had increased midsummer leaf and harvested fruit N concentrations, decreased fruit firmness, and, in heavy crop years, decreased percentage of red color (Table 8.4). Annual yield of all cultivars was significantly increased by N in a single year, but their cumulative yields were not different among treatments as a result of rate or timing (Table 8.4). It should be noted that the low N rate was equivalent to application of 25 kg N/ha/year and the high N rate equivalent to 125 kg N/ha/year. The high N rate was associated with fruit quality decline, including decreases in fruit firmness and, at high crop load, reductions in percentage of red color. Optimum fruit quality and adequate N availability would be achievable by maintaining fertigation concentrations at 42 mg N/l, which would have applied 40 kg N/ha by daily applications over a minimum of a 4-week period during the main growing season. This illustrates the potential of achieving good apple production at relatively low rates of N application per unit area of land.

Similar long-term fertigation trials for sweet cherry indicate that response to fertigation will differ among fruit crops (Neilsen *et al.*, 2004a, 2007). In sprinkler-fertigated experiments on 'Lapins' sweet cherry on Gisela 5 rootstock, yield was depressed when fertigated N rates exceeded 250 kg N/ha (Table 8.5). Also very large year-to-year decreases in yield were observed, as in 2005, when spring frosts occurred. Fruit size is an important quality attribute for sweet cherry and was frequently maximum at lowest rates of fertigated N and particularly high in low crop years.

Table 8.4. Average leaf and fruit concentration, yield, harvest fruit firmness and color for 'Ambrosia', 'Cameo', 'Fuji', and 'Gala' apple cultivars as affected by low and high N rates over a 6-year period (1999–2004). (Adapted from Neilsen *et al.*, 2009.)

Factor	Cumulative yield (kg/tree)	Leaf N (% dry weight)	Fruit N (mg/kg fresh weight)	Firmness (N)	Color (% red)
N rate[a]					
Low	52.6	2.25	364	84.6	85
High	55.6	2.45	483	82.3	80
Significant years	0	6	6	5	3

[a]Fertigated at either 28 mg N/l (low) or 168 mg N/l (high) as calcium nitrate (15.5N–0P–0K)

Table 8.5. Yield and average fruit size of 'Lapins' sweet cherry as affected by annual rate of fertigated N over the six fruiting years. (Adapted from Neilsen *et al.*, 2004a, 2007.)

N fertigation rate	2000	2001	2002	2003	2004	2005[a]
			Average tree yield (kg/tree)			
Low (63 kg N/ha)	1.8	13.5	8.6	37.8	46.5	12.5
Medium (125 kg N/ha)	2.0	13.2	8.7	34.0	37.0	10.7
High (254 kg N/ha)	3.0	13.3	12.2	21.5	23.0	10.1
	NS	NS	NS	****	***	NS
			Average fruit size (g/fruit)			
Low	12.6	11.0	10.0	11.2	9.7	14.9
Medium	12.0	10.0	9.0	9.9	10.1	14.4
High	12.3	9.6	9.0	8.5	9.4	13.8
	NS	*	*	**	NS	**

[a]Crop reduced by severe spring frost

8.3.2 Phosphorus

Despite a lack of response of apple trees to P-fertilization in early research work (Boyton and Oberly, 1966), several soil conditions have been identified when apples responded to P applications. These include times when apple root length is limited, as when trees are newly planted (Taylor and Goubran, 1975) and when replant disorders further inhibit growth (Neilsen and Yorston, 1991), or when low soil P availability limits P acquisition by roots (Cripps, 1987).

Fertigation increases P mobility in sandy soils. The improved mobility has been attributed to the movement of P by mass flow with irrigation waters after saturation of sorption sites near the point of application (Neilsen *et al.*, 1993a). Annual fertigation of 17.5 g P per apple tree as a single dose of ammonium polyphosphate increased extractable soil P concentrations at 10–20 cm depth directly beneath drip emitters to values approaching 100 mg/kg soon after application, with a decline later in the growing season (Neilsen *et al.*, 1997). This pattern during the growing season was also reflected in soil suction lysimeter values measured at the same depth in the same experiment (Fig. 8.6).

One benefit of first year P-fertigation is increased flowering of apple trees in the second year and improved vigor of trees planted in old orchard soil (Neilsen *et al.*, 1990). A single, annual application of 20 g P per tree at bloom as ammonium polyphosphate dissolved in irrigation water was beneficial for the performance, over the first five fruiting seasons, of a range of apple cultivars, including 'Fuji', 'Gala', 'Ambrosia', 'Silken', and 'Cameo' planted at high density on the dwarfing rootstock M.9 (Neilsen *et al.*, 2008b). These trees were otherwise receiving optimum fertigation recommendations for sandy soils, which included daily application of

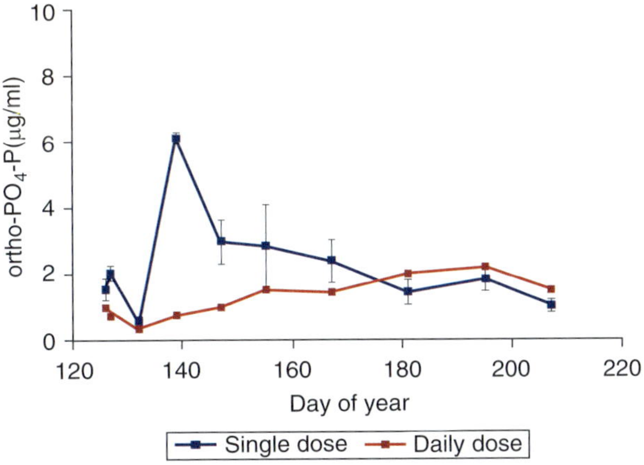

Fig. 8.6. Soil solution ortho-phosphate P concentrations in the early–mid-growing season in response to a single dose or daily doses of fertigated phosphoric acid. Adapted from Neilsen *et al.* (1999).

168 mg N/l as calcium nitrate, maintenance B applications (0.17 g B per tree as Solubor), both applied 0–4 weeks post bloom, and daily applications of K applied 8–12 weeks post bloom as potassium chloride (approximately 20 g K per tree). Cumulative yield of these trees was increased by about 20% over all tested apple cultivars during the first five growing seasons. These yield increases were associated with maintenance of standard fruit quality parameters, including fruit size, soluble solids content (SSC), titratable acidity (TA) and proportion of red coloration (for red-skinned cultivars) at harvest. Reduction in incidence of water core at harvest, resistance to browning of cut slices, reduced membrane leakage and elevated antioxidant content of fruit after cold air storage indicated a role for P in maintenance of apple fruit membrane stability. Cumulative results from this research also suggested that optimum mid-summer leaf P concentrations should exceed 2.2 mg/g dry weight (dw) for fruiting, fertigated, high-density apple orchards, much in excess of current recommendations of 1.5 mg/g dw. Associated recommended harvest fruit P concentrations would then be 100–120 mg/kg fresh weight (fw).

Fertigation of a single annual dose of P in early spring at 20 g P per tree, through sprinkler irrigation lines, did not increase leaf and fruit P concentration of 'Lapins' sweet cherry on Gisela 6 rootstock nor meaningfully affect tree performance (Neilsen *et al.*, 2004a). However, when a single application of P at the same rate was applied through drip emitters, yield was increased in the first three fruiting years for 'Skeena' sweet cherry on Gisela 6 rootstock (Neilsen *et al.*, 2010b). Subsequently, the yield benefit of fertigated P disappeared and P-fertigation was associated with delayed fruit color development and hence a potential for later harvest (Neilsen *et al.*, 2014).

8.3.3 Potassium

Potassium can be effectively applied as potassium chloride via fertigation, as indicated by increases in soil solution concentrations at 30 cm depth directly below drip emitters over a 4-year period when fertigated from mid-June to mid-August (Fig. 8.7) (Neilsen and Neilsen, 2008). Potassium fertigation was able to maintain leaf K concentrations above deficiency levels and increase fruit K concentrations, unlike the treatment not receiving fertigated K. Furthermore, K-fertigation increased fruit yield, size, titratable acidity and red color at harvest for the apple cultivars 'Gala', 'Fuji, 'Fiesta' and 'Spartan' which were grown at this site. Drip fertigation tends to concentrate root development in the wetted zone with nearly half of roots located within 30 cm depth and lateral distance of emitters after 8 years of fertigation (Neilsen *et al.*, 2000). In coarse-textured soils, as at this site, K-deficiency developed after 5 years without K-fertigation, resulting in the necessity of applying K to correct the deficiency (Neilsen *et al.*, 1998b). Annual fertigation of K is recommended for coarse-textured soils to maintain adequate K-nutrition, vigor and yield even when supplemental K may be available from use of mulches and cover crops which recycle additional K to orchards (Hogue *et al.*, 2010).

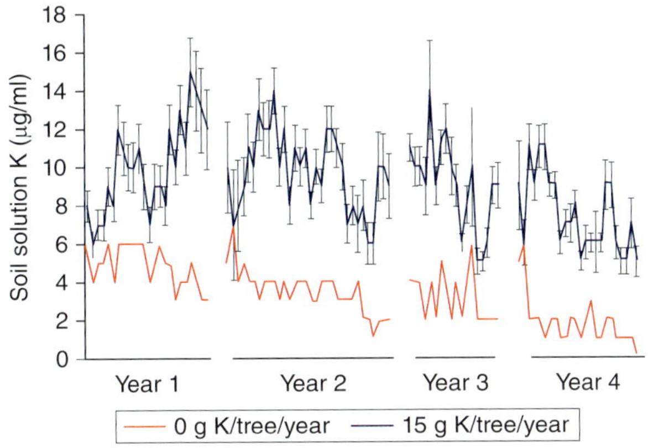

Fig. 8.7. Soil solution potassium (K) concentration at 30 cm depth directly beneath the drip emitter as affected by daily fertigation on a sandy loam orchard soil. Adapted from Neilsen and Neilsen (2008).

Leaf and fruit concentrations of 'Lapins' sweet cherry on Gisela 5 rootstock were consistently lower when irrigation and NP-fertigation were applied via drip as opposed to micro sprinkler irrigation (Fig. 8.8). Leaf K concentrations were deficient by the 5th year of production for the drip-irrigated trees, which were also smaller with reduced yield relative to the sprinkler-irrigated trees receiving the same NP-fertigation regime (Neilsen *et al.*, 2007). Thus, similar to apple, NPK-fertigation has been applied to sweet cherry grown on coarse-textured soil to avoid the development of K-deficiency (Neilsen *et al.*, 2010b). Potassium chloride was fertigated to apply 20 g/tree for 4–5 weeks prior to harvest when K requirements were large for rapidly growing cherry fruit.

Subsequent research involving 'Jonagold' apple on M.9 indicated there were few differences among K-fertilizers including KMag (potassium, magnesium sulphate), sulphate, chloride and thiosulphate forms indicating that a range of soluble materials would be suitable fertigants for improving inadequate K-nutrition (Neilsen and Neilsen, 2006). Use of potassium magnesium sulphate was effective at increasing extractable soil Mg content but relative effects on leaf and fruit Mg concentrations were slight, indicating the difficulties of ameliorating inadequate leaf Mg via fertigation when co-applying K. Short-term differential effects of treatments on leaf and fruit calcium (Ca) concentrations were minimal and there was no increase in bitter pit disorders (Table 8.6). The sensitivity of the orchard production system to long-term disruption in Ca nutrition was indicated by a general decrease in soil Ca availability after fertigation of K, especially in association with Mg additions and low pH (Table 8.7), and by increased whole fruit Ca/K ratios at harvest. Thus application of significant amounts of K to coarse-textured soils, via irrigation water, when the K nutritional status of apple trees is unknown, should be accompanied by vigilance to ensure Ca concentration of harvested fruit is optimum.

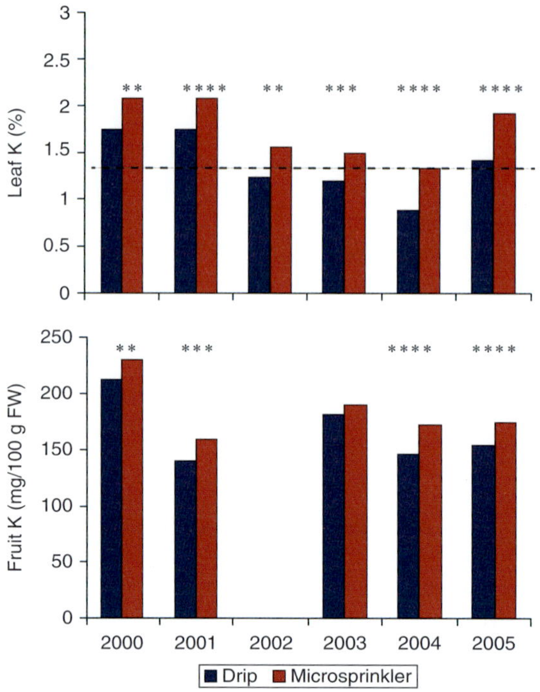

Fig. 8.8. Leaf and fruit potassium concentration for 'Lapins' sweet cherry on Gisela 5 rootstock in response to irrigation method. Columns within years with symbols above significantly different at P = 0.01 (**), P = 0.001 (***), P = 0.0001 (****) or not significantly different (blank). Adapted from Neilsen *et al*. (2007).

Table 8.6. Leaf and fruit Ca concentrations and incidence of bitter pit as affected by K-fertigation treatment for 'Jonagold' on M.9 rootstock, 2000–2002. (Adapted from Neilsen and Neilsen, 2006.)

	Leaf Ca (% dry weight)			Fruit Ca (g/kg fresh weight)			Bitter pit (%)		
Fertigation treatment	2000	2001	2002	2000	2001	2002	2000	2001	2002
Control (no K)	1.23	1.35	1.15	32.6	24.7	26.0	8	15	17
KCl (15 g K/tree)	1.21	1.29	1.25	34.8	23.8	26.3	2	13	18
KCl (30 g K/tree)	1.18	1.18	1.19	36.3	23.7	24.2	5	5	18
KMag (18 g K/tree)	1.21	1.22	1.15	35.6	23.2	26.2	8	10	27
KMag (30 g K/tree)	1.24	1.18	1.10	34.8	22.0	22.5	7	8	20
K_2SO_4 (30 g K/tree)	1.24	1.22	1.15	32.8	23.5	24.6	2	5	10
KTS (30 g K/tree)	1.24	1.20	1.18	33.2	23.6	24.6	8	4	13
Contrasts									
Control vs all	NS	**	NS	NS	NS	NS	NS	NS	NS
K-form	NS	NS	*	NS	NS	NS	NS	NS	NS
K-rate	NS	NS	NS	NS	NS	**	NS	NS	NS

KCl: Potassium chloride; KMag: Potassium, magnesium sulphate; K_2SO_4: Potassium sulphate; KTS: Potassium thiosulphate; Statistical comparison between contrasts significantly different at P=0.05 (*), P=0.01 (**) or not significantly different (NS).

Table 8.7. Effect of K-fertigation treatment on exchangeable Ca concentration at 0–10 cm, 10–20 cm and 20–30 cm directly beneath the drip emitters after three growing seasons of treatment. (Adapted from Neilsen and Neilsen, 2006.)

Factor	Exchangeable Ca (cmol (+)/kg)		
Depth (cm)	0–10 cm	10–20 cm	20–30 cm
Treatment			
Control (no K)	3.44a	2.64a	2.16a
KCl	2.90b	2.46ab	1.92b
KMag	2.91b	2.10bc	1.28c
K_2SO_4	3.20ab	2.57a	2.07ab
KTS	2.85b	1.43c	1.38c
Significance	*	**	****

All K treatments applied at 30 g K/tree annually for 3 years;
KCl: Potassium chloride;
KMag: Potassium, magnesium sulphate;
K_2SO_4: Potassium sulphate;
KTS: Potassium thiosulphate;
Statistical comparison between contrasts significantly different at P=0.05(*), P=0.01 (**) or P=0.001(***).

8.3.4 Other Nutrients

Several important nutrients are traditionally more effectively applied via foliar sprays rather than by applications to the soil. These include Ca, which is critical for ensuring optimum apple fruit harvest and storage quality (Vang-Petersen, 1980). British Columbia production recommendations for orchard Ca management are similar to those for many fruit-growing regions of the world and usually involve, for blocks susceptible to Ca disorders, three to six sprays of soluble Ca salts (food-grade calcium chloride preferred) applied during mid- to late-growing season when the fruit is large. Cultivars such as 'Braeburn' and 'Honeycrisp' have been found to respond to foliar-applied Ca earlier in the growing season prior to the cessation of shoot growth (Peryea *et al.*, 2007). There has been a limited response of apple trees to Ca application via fertigation. One exception is a comparative study in a low pH (high leaf Mn) orchard where fruit Ca in 'Jonagold' apple was higher when fertigated with calcium nitrate rather than urea or ammonium nitrate (Table 8.8) (Neilsen *et al.*, 1993b).

Also many micronutrients are preferentially applied by foliar sprays. For example, leaf concentrations of the micronutrient Zn was effectively increased via foliar application in a long-term trial on apple (Neilsen *et al.*, 2004b). Fertigation of Zn as soluble zinc sulphate was ineffective at increasing leaf Zn concentration (Fig. 8.9). Foliar B applications also augmented leaf B concentrations and ameliorated deficiency symptoms associated with 'blossom blast' in the spring and fruit corking and cracking at harvest in 1994 after 2 years without B application in a B-deficient soil (Fig. 8.10). In contrast to Zn, fertigated B was mobile within the soil (Fig. 8.11) and it was relatively easy to increase leaf and fruit B concentrations via fertigation of modest rates of 0.34 g B per tree. The ready response of tissue

Table 8.8. Effect of form of N on 'Jonagold' fruit Ca concentration during the second and third growing seasons. (Adapted from Neilsen *et al.*, 1993b.)

Fertilizer treatment	Fruit Ca (mg/kg fresh weight) Year 2	Year 3
Form of N[a]		
Urea (46-0-0)	62b[b]	42
Ammonium nitrate (34-0-0)	53b	41
Calcium nitrate (15.5-0-0)	75a	44
Significance	**	NS

[a]All fertigated at rate of 57 g N/tree in eight weekly applications during May and June each year.
[b]Means within columns followed by different letters significantly different at $P = 0.01$ (**).

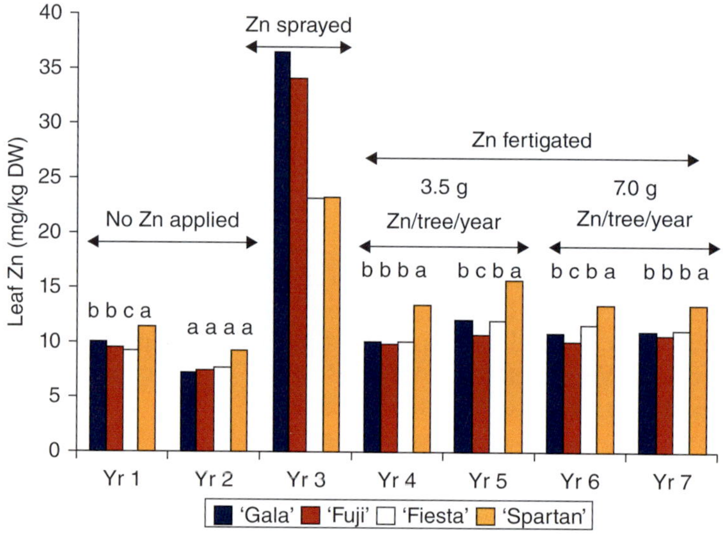

Fig. 8.9. Leaf zinc (Zn) concentration of four apple cultivars in response to Zn application method over the first seven growing seasons. Columns within years with different letters above significantly different at P = 0.01. Adapted from Neilsen *et al.* (2004b).

B concentrations to changes in rate of fertigated B (Fig. 8.10) suggest that caution is required to select moderate B application rates in order to avoid toxicity when fertigating B (Neilsen *et al.*, 2004b).

8.4 Fertigation Challenges

8.4.1 Crop load

Variation in crop load has long been recognized as having a major influence on apple tree and fruit concentration (Hansen, 1980). Heavy cropping is known to increase overall tree K-uptake but also to depress leaf K concentrations due to strong demand for K by fruit (Jadczuk and Lenz, 1998). It is therefore important when fertigating nutrients to be aware of

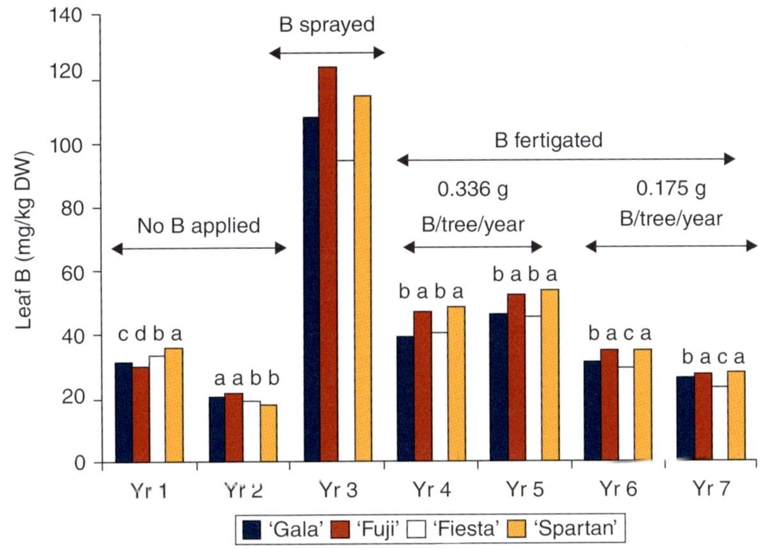

Fig. 8.10. Leaf boron (B) concentration of four apple cultivars in response to B application method over the first seven growing seasons. Columns within years with different letters above significantly different at P = 0.01. Adapted from Neilsen *et al*. (2004b).

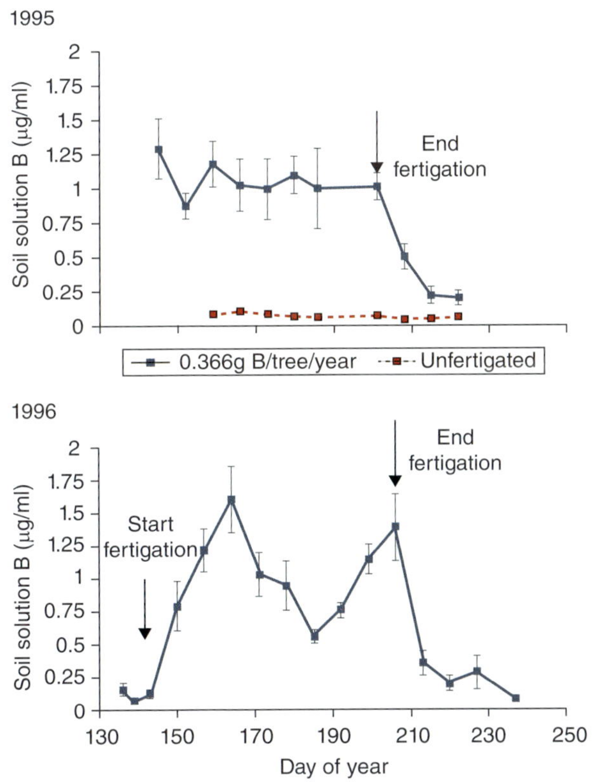

Fig. 8.11. Soil solution B concentration at 30 cm depth directly beneath the drip emitter as affected by daily fertigation on a sandy loam orchard soil. Adapted from Neilsen *et al*. (2004b).

an increase in annual nutrient demands with a large crop, especially for K, which occurs at the highest concentration in fruit. Some of the consequences of crop load variation on leaf and fruit nutrient concentrations are illustrated from the first year of a multi-year fertigation trial (Neilsen *et al.*, 2010a, 2015). Low, standard and high crop loads (2.5, 5 and 10 fruits/cm^2 trunk cross-sectional area (TCSA), respectively) were maintained on 'Ambrosia' apple on M.9 rootstock over a range of irrigation treatments.

To ensure that trees were not nutrient limited, N was fertigated daily as calcium nitrate ($Ca(NO_3)_2$) for 6 weeks after bloom to provide 75 g N/tree, P was fertigated 1 day after full bloom before the start of N applications as ammonium polyphosphate at 20 g P/tree, and K was fertigated daily 3–6 weeks after full bloom as potassium chloride (KCl) at 20 g K/tree. Nevertheless, as crop load increased, fruit concentration and total tree partitioning to fruit of N, P, K, Mg and B and leaf concentrations of K and P decreased, suggesting some limitations in availability to the fruit when demand was high from a large crop (Table 8.9). High crop load did not result in negative consequences to fruit Ca concentration, which affects fruit storage and harvest fruit quality, possibly due to smaller apples occurring when crop load was high. However, the results suggest that variation in crop load should be considered when fertigating, with the possibility (after examination of leaf concentrations) of increasing P and K fertigation rates when a large crop is anticipated.

8.4.2 Soil acidification

A problem that has arisen from continued fertigation of ammonium-containing N fertilizers into restricted soil volumes is soil acidification resulting from nitrification of the applied ammonium (NH_4-N) (Edwards *et al.*, 1982). Declines of two pH units have been measured in fertigated orchards within a single year. Soil acidification has been sufficient to release toxic quantities of Mn and aluminum (Al) (Ross *et al.*, 1985). Soil pH below 5.5 has been associated with tree stunting and the development of internal bark necrosis on susceptible apple cultivars such as 'Delicious'

Table 8.9. Effects of crop load on leaf and fruit nutrition of fruiting 'Ambrosia' apple on M.9 rootstock. (Adapted from Neilsen *et al.*, 2010a.)

	Leaf (% dry weight)		Fruit (mg/kg fresh weight)					
Cropload[a]	K	P	N	P	K	Ca	Mg	B
Low	1.67a[b]	0.20a	635a	132a	1426a	32c	62a	3.3a
Standard	1.59b	0.20a	570b	109b	1252b	37b	58b	3.1ab
High	1.42c	0.19b	518c	89c	1060c	47a	54c	2.7b
Significance	****	**	****	****	****	****	****	*

[a]Crop load adjusted 6 weeks after full bloom to low, standard and high crop loads (2.5, 5 and 10 fruit/cm^2 trunk cross-sectional area).
[b]Means within columns followed by different letters significantly different at indicated probability.

(Hoyt and Neilsen, 1985). In addition to reduced tree growth from Mn and Al toxicity, other nutrient deficiencies can be induced, including for K (Neilsen *et al.*, 1994). The sensitivity of orchard soils to fertigation-induced acidification varies with the buffering capacity of the soil and can be estimated by an acidification resistance index to assess soil susceptibility to acidification (Neilsen *et al.*, 1995b). Soils with little resistance to acidification should receive N fertigated in nitrate form rather than ammonium form. The problem can be particularly serious in regions where soil pH is already low.

8.4.3 Nutrient balance

Fertigation frequently involves application of a single nutrient at a time, due to the necessity of avoiding precipitation reactions between multiple nutrients within irrigation lines and the variation in seasonal demand for individual nutrients. Nutritional problems can concern several nutrients in the field. The relationship between Ca, Mg and K is of particularly importance for apple tree nutrition because of the critical importance of Ca to fruit quality (Faust, 1989), and the association of deficient fruit Ca concentrations with several fruit disorders, including bitter pit (Ferguson and Watkins, 1989). Excessive Mg and particularly K supply have been identified as potential problems for achieving sufficiently high fruit Ca concentrations to optimize fruit quality (Vang-Petersen, 1980). However, there is limited evidence that annual application of both K and Mg as K, Mg sulphate at 200 kg K/ha for 3 years decreased fruit Ca concentrations or increased incidence of Ca disorders of apples despite relative enrichment of the surface 20 cm depth of the orchard with K and Mg relative to Ca (Neilsen and Neilsen, 2011); regular monitoring of fruit Ca concentration was, however, recommended.

For leaves, a strong antagonism exists between leaf Mg and K as illustrated by high leaf Mg concentration when K concentrations are at deficiency levels (Neilsen and Neilsen, 2008). Similarly it has proven difficult to increase leaf Mg concentrations when fertigating K (Neilsen and Neilsen, 2006). Acidic soil conditions which can result from fertigation of ammonium-N fertilizers can create nutrient imbalances when Mn uptake is favored at the expense of other micronutrients such as Cu and Zn (Neilsen *et al.*, 1994). Vigilance is thus required for long-term successful fertigation programs to avoid the development of apparently anomalous nutrient disorders. Annual leaf sampling and analysis will help to maintain balanced nutrition.

8.5 Future Fertigation Management Developments

Future developments may improve the precision of automation and facilitate the adoption of fertigation. Currently, fertigation is particularly well suited for semi-arid regions that have traditionally been irrigated. The challenges of implementing precision irrigation (and hence fertigation) in humid climate regions has recently been discussed (Daccache *et al.*, 2015).

Advances in precision irrigation will improve retention of soluble fertigated nutrients in the root-zone of irrigated orchards. Traditionally fertigation has ignored soil and crop variability by making uniform applications across an orchard. Improved understanding of soil variability and use of geostatistical techniques may allow creation of several irrigation management zones within fields, allowing variable rate irrigation and improving the overall match between water required and applied. Furthermore, decision support systems may be developed that can monitor the spatial variability of tree water status. For example, thermal sensors installed on unmanned airborne vehicles also provide field measurements to refine variable rate irrigation methods (Gonzalez-Dugo *et al.*, 2015). The potentials and limitations of ground-based thermal remote sensing to estimate evapotranspiration and drought stress have been reviewed (Maes and Steppe, 2012). In an ideal world it would be desirable to develop a system that would allow individual nutrient and water applications to be made to single trees according to their requirements. Engineering limitations, including imprecision in emitter discharge rates, fluid travel times, sloping orchard soil surfaces and the high cost of wireless networked sensors and valves, prevent the use of a high number of simultaneously operating hydrozones (Coates *et al.*, 2012).

Another limitation to improving fertigation is incomplete knowledge of the annual and seasonal nutrient demands of orchard trees. Where vegetables are grown in hydroponic systems, relationships are more readily established between biomass increase and nutrient requirements by successive harvest and chemical analyses during the growth cycle, but this is impossible under field conditions. Annual fruit nutrient removal rates can be more readily obtained as the product of fresh-weight yield and fruit nutrient concentration (determined on a fresh-weight basis). Calibration curves could then be developed to calculate nutrient demands as yield increases. A rational approach could therefore be used for increasing fertigation nutrient application rates at high yield, especially for nutrients, such as K, that occur at high concentrations in fruit.

To determine leaf N non-destructively, the use of radiation sensors to determine leaf chlorophyll *in situ* has been proposed (Hunt and Daughtry, 2014). SPAD-502 meter readings have been used to determine the N status of apple trees (Neilsen *et al.*, 1995a). The technique showed promise as a means of identifying the need for fertigated N but not the quantity required. It is likely that any sensing technology developed for rapid field determination of nutrients in order to allow in-season adjustment of fertigation rates would have to undergo standardization of methods prior to evaluation with respect to potential to refine fertigation strategies.

A decision support system has recently been proposed to optimize fertilizer costs for fertigation by taking into account composition of irrigation waters, fertilizer availability and application rate and water flow (Pagan *et al.*, 2015). The details of fertigation will evolve in order to adjust methods for new crops and cultivars, to incorporate novel soil amendments, including soluble organics, and to account for different soil management techniques, including cover crops and mulches (Bryla and Strik, 2015).

8.6 Summary

Fertigation can be an important aspect of improving precision nutrient management, particularly for production of perennial fruits in semi-arid regions requiring irrigation. The automated injection of essential nutrients can be timed to match crop nutrient demand more closely. Coupled with automated irrigation systems that schedule water applications according to plant water demands, excessive leaching of nutrients from the root-zone can also be minimized. The research cited here has focused on strategies developed in a series of field experiments conducted in apple and sweet cherry orchards. Daily irrigation was automatically controlled based on the previous day's evapotranspiration, with quantity applied modified by a crop coefficient which varied within the growing season as the crop canopy expanded. To avoid underperformance or catastrophic failure, it is desirable to be continuously aware of the actual performance of any automated system. Continuous measurement of the soil moisture regime, as by TDR, is useful to avoid water stress. Soil solution monitoring is particularly useful for determining the dynamics of soil N availability.

The efficiency of N-fertigation is improved by avoiding irrigation in excess of tree water requirements, thereby reducing N-leaching and by minimizing N-fertigation when tree N-requirement is largely met by remobilization of stored N. Suitable production can be achieved at relatively low rates relative to broadcast applications when N is fertigated. For apple 40 kg N/ha/year and for sweet cherry 60 kg N/ha/year were sufficient when applied over a 4–6-week period post bloom when rapid shoot growth and fruit cell division occur. Detrimental effects of excess N differ by fruit species, with high N associated with decreased fruit firmness and red coloration for apple and excessive shoot vigor and decreased fruit size for sweet cherry. Fertigation of a single application of 15–20 g P/tree around bloom increased P mobility within the soil profile, increased leaf and fruit P concentration and initial bloom and yield of both apple and sweet cherry. K-fertigation can be effective at preventing the development of K-deficiency in sandy orchard soils which are prone to restrict tree K uptake when drip-irrigated and may have inadequate tree K in heavy crop years. Soluble K forms are effectively applied for 4–6 weeks late in the growing season as fruit reaches its maximum size. Some nutrients are better supplied via foliar sprays rather than by fertigation, including Ca to the fruit and most micronutrients (Mn, Fe, Cu and Zn) to foliage or bark. An exception is B, which is highly mobile within the soil and plant; maintenance applications can be achieved at modest rates of several kilograms per hectare which would otherwise be difficult to broadcast uniformly on the soil surface.

Long-term maintenance of fertigation programs requires avoidance of excessive soil pH decline associated with overreliance on use of ammonium-rich N fertilizers, which acidify the soil as ammonium is converted to nitrate-N. The problem is particularly serious for soils with limited capacity to buffer against soil pH change and can result in solubilization of potentially toxic levels of Mn and Al. Furthermore, the long-term requirements for balanced plant nutrition requires attention to possible detrimental consequences of continuous K-fertigation on leaf Mg and fruit Ca concentrations.

Future research developments hold the promise of improving fertigation application. Improved automatic sensing of water stress allowing adjustment of irrigation will improve the efficiency of water and hence nutrient application. Tailoring nutrient applications on an individual tree basis or through multiple 'uniform hydro-zones' within an otherwise uniformly treated orchard block may be stimulated by continued evolution of cheap sensors and valves that can be linked via wireless networks. Improved knowledge of orchard annual and seasonal nutrient demands has the potential to refine fertigation application rates by, for example, adjusting annual fertigation according to anticipated crop load. There will always be scope to adjust fertigation according to improved knowledge of new crops, different cultivars and soil management techniques and the availability of novel soluble fertilizers and organic adjuvants.

References

Allen, R.G., Pereira, L.S., Raes, D. and Smith, M. (1998) *Crop Evapotranspiration Guidelines for Computing Crop Requirements.* FAO Irrigation and drainage Paper 56. Food and Agriculture Organization of the United Nations, Rome.

Alva, A.K. (2008) Role of fertigation in horticultural crops: Citrus. In: Imus, P. and Price, M.R. (eds) *Fertigation: Optimizing the Utilization of Water and Nutrition.* International Potash Institute, Horgen, Switzerland, pp. 76–88.

Bar-Yosef, B. (1999) Advances in fertigation. *Advances in Agronomy* 65, 1–76.

Bar-Yosef, B. and Sheikholslami, M.R. (1976) Distribution of water and ions in soils irrigated and fertilized from a trickle source. *Soil Science Society of America Journal* 40, 575–582.

Boyton, D. and Oberly, G.H. (1966) Apple nutrition. In: Childers, N.F. (ed.) *Nutrition of Fruit Crops.* Horticultural Publication, New Brunswick, New Jersey, pp. 1–50.

Bravdo, B. and Proebsting, E.L. (1993) Use of drip irrigation in orchards. *HortTechnology* 3, 44–49.

Bruna, I.P., Reichardt, K., Bortolotto, R.P., Pinto, V.M., Bachii, O.O.S., Dourado-Neto, D. and Unkovich, M.J. (2015) Nitrogen balance and fertigation use efficiency in a field coffee crop. *Journal of Plant Nutrition* 38, 2055–2076.

Bryla, D.R. and Strik, B.C. (2015) Nutrient requirements, leaf tissue standards, and new options for fertigation of northern highbush blueberry. *HortTechnology* 25, 464–470.

Bussi, C., Huguet, J.-G. and Defrance, H. (1991) Fertilization scheduling in peach orchards under trickle irrigation. *Journal of Horticultural Science and Biotechnology* 66, 487–483.

Cheng, L. and Raba, R. (2009) Accumulation of macro- and micronutrients and nitrogen demand-supply relationship of 'Gala'/'Malling 26' apple trees grown in sand culture. *Journal of the American Society for Horticultural Science* 134, 3–13.

Coates, R.W., Sahoo, P.K., Schwankl, L.J. and Delwiche, M.J. (2012) Fertigation techniques for use with multiple hydrozones in simultaneous operation. *Precision Agriculture* 13, 219–235.

Cripps, J.E.L. (1987) Response of apple trees to soil applications of phosphorus, nitrogen and potassium. *Australian Journal of Experimental Agriculture* 27, 909–914.

Daccache, A., Knox, J.W., Weatherhead, E.K., Daneshkhah, A. and Hess, T.M. (2015) Implementing precision irrigation in a humid climate. Recent experiences and on-going challenges. *Agricultural Water Management* 147, 135–143.

Edwards, J.H., Bruce, R.R., Horton, B.D., Chesness, J.L. and Wehunt, E.J. (1982) Soil cation and water distribution as affected by NH_4NO_3 applied through a drip irrigation system. *Journal of the American Society for Horticultural Science* 107, 1142–1148.

Faust, M. (1989) *Physiology of Temperate Zone Fruit Trees.* John Wiley and Sons, New York.

Ferguson, I.B. and Watkins, C.B. (1989) Bitter pit in apple fruit. *Horticultural Reviews* 11, 289–355.

Gonzalez-Dugo, V., Goldhammer, D., Zarco-Tejada, P.J. and Fereres, E. (2015) Improving the precision of irrigation in a pistachio farm using an unmanned airborne thermal system. *Irrigation Science* 33, 43–52.

Grassi, G., Millard, P., Gioaccchini, P. and Tagliavini, M. (2003) Recycling of nitrogen in the xylem of *Prunus avium* trees starts when spring remobilization of internal reserves declines. *Tree Physiology* 23, 1061–1068.

Guak, S., Neilsen, D., Millard, P., Wendler, R. and Neilsen, G.H. (2003) Determining the role of N remobilization for growth of apple (*Malus domestica* Borkh.) trees by xylem-sap N flux. *Journal of Experimental Botany* 54, 2121–2131.

Hansen, P. (1980) Crop load and nutrient translocation. In: Atkinson, D., Jackson, J.E., Sharples, R.O. and Waller, W.M. (eds) *Mineral Nutrition of Fruit Trees.* Butterworth, London, pp. 201–212.

Haynes, R.J. (1985) Principles of fertilizer use for trickle irrigated crops. *Fertilizer Research* 6, 235–255.

Hogue, E.J., Cline, J.A., Neilsen, G. and Neilsen, D. (2010) Growth and yield responses to mulches and cover crops under low potassium conditions in drip-irrigated apple orchards on coarse soils. *HortScience* 45, 1866–1871.

Hoyt, P.B. and Neilsen, G.H. (1985) Effects of soil pH and associated cations on growth of apple trees planted in old orchard soil. *Plant Soil* 86, 395–401.

Hunt, E.R. Jr and Daughtry, C.S.T. (2014) Chlorophyll meter calibrations for chlorophyll content using measured and simulated leaf transmittances. *Agronomy Journal* 106, 931–939.

Jadczuk, E. and Lenz, F. (1998) Effect of water supply and fruit load on uptake and distribution of potassium in 'Golden Delicious' trees. *Gartenbauwissenshaft* 63, 193–196.

Klein, I. and Spieler, G. (1987) Irrigation of apples with nitrate or ammonium nitrogen under drip irrigation. II: Nutrient distribution in the soil. *Communication in Soil Science and Plant Analysis* 18, 323–339.

Maes, W.H. and Steppe, K. (2012) Estimating evapotranspiration and drought stress with ground-based thermal remote sensing in agriculture; a review. *Journal of Experimental Botany* 63, 4671–4712.

Neilsen, D. and Neilsen, G.H. (2002) Efficient use of nitrogen and water in high-density apple orchards. *HortTechnology* 12, 19–25.

Neilsen, G.H. and Neilsen, D. (2006) Response of high density apple orchards on coarse-textured soil to form of potassium applied by fertigation. *Canadian Journal of Soil Science* 86, 749–755.

Neilsen, D. and Neilsen, G. (2008) Fertigation of deciduous fruit trees: apple and sweet cherry. In: Imus, P. and Price, M.R. (eds) *Fertigation: Optimizing the Utilization of Water and Nutrition*. International Potash Institute, Horgen, Switzerland, pp. 76–88.

Neilsen, G.H. and Neilsen, D. (2011) Consequences of potassium, magnesium sulphate fertilization of high density Fuji apple orchards. *Canadian Journal of Soil Science* 91, 1013–1027.

Neilsen, G.H. and Yorston, J. (1991) Soil disinfection and monoammonium phosphate fertilization increase precocity of apples on replant problem soils. *Journal of the American Society for Horticultural Science* 116, 651–654.

Neilsen, G.H., Hogue, E.J. and Parchomchuk, P. (1990) Flowering of apple trees in the second year is increased by first-year P fertigation. *HortScience* 25, 1247–1250.

Neilsen, D., Parchomchuk, P. and Hogue, E.J. (1993a) Soil and peach seedling responses to soluble phosphorus applied in single or multiple doses. *Communication Soil Science and Plant Analysis* 24, 881–898.

Neilsen, G.H., Parchomchuk, P., Wolk, W.D. and Lau, O.L. (1993b) Growth and mineral composition of newly planted apple trees N and P. *Journal of the American Society for Horticultural Science* 118, 50–53.

Neilsen, G.H., Parchomchuk, P., Hogue, E.J., Wolk, W.D. and Lau, O.L. (1994) Response of apple trees to fertigation induced soil acidification. *Canadian Journal of Plant Science* 74, 347–351.

Neilsen, D., Hogue, E.J., Neilsen, G.H. and Parchomchuk, P. (1995a) Using SPAD-502 values to assess the nitrogen status of apple trees. *HortScience* 30, 508–512.

Neilsen, D., Hoyt, P.B., Parchomchuk, P., Neilsen, G.H. and Hogue, E.J. (1995b) Measurement of the sensitivity of orchard soils to acidification. *Canadian Journal of Soil Science* 75, 391–395.

Neilsen, G.H., Hoyt, P.B. and Neilsen, D. (1995c) Soil chemical changes associated with NP-fertigated and drip irrigated high-density apple orchards. *Canadian Journal of Soil Science* 75, 307–310.

Neilsen, G.H., Parchomchuk, P. and Neilsen, D. (1997) Distribution of soil P and K as affected by NP-fertigation in high density apple orchards. *Acta Horticulturae* 448, 439–447.

Neilsen, D., Parchomchuk, P., Neilsen, G.H. and Hogue, E.J. (1998a) Use of soil solution monitoring to determine the effects of irrigation management and fertigation on nitrogen availability in high-density apples. *Journal of the American Society for Horticultural Science* 123, 706–713.

Neilsen, G., Parchomchuk, P., Meheriuk, M. and Neilsen, D. (1998b) Development and correction of K-deficiency in drip-irrigated apple. *HortScience* 33, 258–261.

Neilsen, G.H., Neilsen, D. and Peryea, F. (1999) Response of soil and irrigated fruit trees to fertigation or broadcast application of nitrogen, phosphorus, and potassium. *HortTechnnology* 9, 393–401.

Neilsen, G.H., Parchomchuk, P., Neilsen, D. and Zebarth, B.J. (2000) Drip-fertigation of apple trees affects root distribution and development of K deficiency. *Canadian Journal of Soil Science* 80, 353–361.

Neilsen, D., Millard, P., Herbert, L.C., Neilsen, G.H., Hogue, E.J., Parchomchuk, P. and Zebarth, B.J. (2001) Remobilization and uptake of N by newly planted apple (*Malus domestica*) trees in response to irrigation method and timing of N application. *Tree Physiology* 21, 513–521.

Neilsen, G., Kappel, F. and Neilsen, D. (2004a) Fertigation method affects performance of 'Lapins' sweet cherry in Gisela 5 rootstock. *HortScience* 39, 1716–1721.

Neilsen, G.H., Neilsen, D., Hogue, E.J. and Herbert, L.C. (2004b) Zinc and boron nutrition management in fertigated high density apple orchards. *Canadian Journal of Plant Science* 84, 823–828.

Neilsen, D., Neilsen, G.H., Herbert, L., Millard, P. and Guak, S. (2006) Allocation of dry matter and N to fruit and shoots in dwarf apple in response to sink size and N availability. *Acta Horticulturae* 721, 33–40.

Neilsen, G., Kappel, F. and Neilsen, D. (2007) Fertigation and crop load affect yield, nutrition, and fruit quality of 'Lapins' sweet cherry on Gisela 5 rootstock. *HortScience* 42, 1456–1462.

Neilsen, D., Neilsen, G.H., Gregory, D., Forge, T. and Zebarth, B. (2008a) Drainage losses of water, N and P from micro-irrigation systems in a young, high density apple planting. *Acta Horticulturae* 792, 483–490.

Neilsen, G.H., Neilsen, D., Toivonen, P. and Herbert, L. (2008b) Annual bloom-time phosphorus fertigation affects soil phosphorus, apple tree phosphorus nutrition, yield and fruit quality. *HortScience* 43, 885–890.

Neilsen, G.H., Neilsen, D. and Herbert, L. (2009) Nitrogen fertigation concentration and timing of application affect nitrogen nutrition, yield, firmness and color of apples grown at high density. *HortScience* 44, 1425–1431.

Neilsen, D., Neilsen, G.H., Herbert, L. and Guak, S. (2010a) Effect of irrigation and crop load management on fruit nutrition and quality for Ambrosia/M.9 apple. *Acta Horticulturae* 868, 63–71.

Neilsen, G.H., Neilsen, D., Kappel, F., Toivonen, P. and Herbert, L. (2010b) Factors affecting establishment of sweet cherry on Gisela 6 rootstock. *HortScience* 45, 939–945.

Neilsen, G.H., Neilsen, D., Kappel, F. and Forge, T. (2014) Interaction of irrigation and soil management on sweet cherry productivity and fruit quality at different crop loads that simulate those occurring by environmental extremes. *HortScience* 49, 215–220.

Neilsen, G.H., Neilsen, D., Guak, S. and Forge, T. (2015) The effect of deficit irrigation and crop load on leaf and fruit nutrition of fertigated 'Ambrosia'/'M.9' apple. *HortScience* 50, 1387–1393.

Neilsen, D., Neilsen, G.H., Guak, S. and Forge, T. (2016) Consequences of deficit irrigation and crop load reduction on plant water relations, yield, and quality of 'Ambrosia' apple. *HortScience* 51, 98–106.

Osroosh, Y., Peters, R.T., Campbell, C.S. and Zhang, Q. (2016) Comparison of irrigation automation algorithms for drip-irrigated apple trees. *Computers in Electronics in Agriculture* 128, 87–99.

Pagan, F.J., Ferrández-Villena, M., Fernández-Pacheco, D.G. and Rosillo, J.J. (2015) Optifer: an application to optimize fertiliser costs in fertigation. *Agricultural Water Management* 151, 19–29.

Parchomchuk, P., Berard, R.C. and Van der Gulik, T.W. (1996) Automatic irrigation scheduling using an electronic atmometer. In: Camp, C.R., Sadler, E.J. and Yoder, R.E. (eds) *Evapotranspiration and Irrigation Scheduling. Proceedings of International Conferences, San Antonio, Texas.* American Society of Agricultural Engineers, St Joseph, Michigan, pp. 1099–1104.

Peryea, F.J., Neilsen, G.H. and Faubion, D. (2007) Start-timing for calcium chloride spray programs influences fruit calcium and bitter pit in 'Braeburn' and 'Honeycrisp' apples. *Journal of Plant Nutrition* 30, 1213–1227.

Ross, G.J., Hoyt, P.B. and Neilsen G.H. (1985) Soil chemical and mineralogical changes due to acidification in Okanagan apple orchards. *Canadian Journal of Soil Science* 65, 347–355.

Tagliavini, M., Millard, P., Quattieri, M. and Marangoni, B. (1999) Timing of nitrogen uptake affects winter storage and spring remobilization of nitrogen in nectarine (*Prunus persica* var. *nectarina)* trees. *Plant Soil* 211, 149–153.

Taylor, B.K. and Goubran, F.H. (1975) The phosphorus nutrition of the apple tree. Influence of rate of application of superphosphate on the performance of young trees. *Australian Journal of Agricultural Research* 26, 842–853.

Topp, G.C. and Reynolds, W.D. (1998) Time domain reflectrometry: a seminal technique for measuring mass and energy in soil. *Soil Tillage Research* 47, 125–132.

Treeby, M. (2008) Manipulating grapevine annual shoot growth, yield and composition of grapes using fertigation. In: Imus, P. and Price, M.R. (eds) *Fertigation: Optimizing the Utilization of Water and Nutrition*. International Potash Institute, Horgen, Switzerland, pp. 89–102.

Van der Gulik, T.W. (1999) *BC Trickle Irrigation Manual*. Resource Management Branch, British Columbia Ministry of Agriculture and Food, Abbotsford, British Columbia, Canada.

Vang-Petersen, O. (1980) Calcium nutrition of apple trees: a review. *Scientia Horticulturae* 12, 1–9.

Vargas, O.L., Bryla, D.R., Weiland, J.R., Strik, B.C. and Sun, L. (2015) Irrigation and fertigation with drip and alternative micro irrigation systems in northern highbush blueberry. *HortScience* 50, 897–903.

Wells, L. (2015) Growth and nitrogen status of young pecan trees using fertigation. *HortScience* 50, 904–908.

9 Precise Crop Load Management

Caixi Zhang[1]* and Du Chen[2]

[1]Shanghai Jiaotong University, Shanghai, China; [2]China Agricultural University, Beijing, China

9.1 Introduction

The development of fruit size and quality depends on many factors, such as the leaf–fruit ratio, genetic and climatic factors, position in the canopy, tree age, water and nutrient supply, source–sink relationship and crop load (Dennis, 2003). The management of crop load is one of the most important areas of orchard management that growers face each year, because most fruit species often set more fruit than necessary if growing conditions are optimal (Westwood, 1993). An excess of fruits with respect to vegetative growth may lead to low fruit size and to irregular or alternate bearing in many perennial crops, particularly in apple, pear, plum, olive, and citrus (Monselise and Goldschmidt, 1982). For most tree fruit species, the alternation of large and small crops is caused by competition between the current season's crop and the coming season's flower buds. To ensure good size and fruit quality, it is of great importance to manipulate the source–sink relationship and the balance between vegetative and reproductive growth of a fruit tree. In sweet cherries, manipulation of the number of fruits on the tree, and leaf area, can be used to encourage larger and sweeter fruit through balanced carbohydrate supply and demand (Whiting and Lang, 2004).

When a plant develops a heavy fruit load, the fruits seem to have a priority for the photosynthate from most leaves: both the direction and pathway of assimilate transport change in favor of fruit growth (Ho, 1996). Therefore, a high fruit load decreases the distribution of assimilates to the roots and other permanent plant organs; the lack of assimilates have negative effects on fruit production in the following years (Lenz, 2009)

* Corresponding author, email: acaizh@sjtu.edu.cn

and lead to alternate bearing. Over-thinning also carries economic perils and fruit size will be excessively large, with reduced fruit quality due to reduced flesh firmness, reduced color and a much reduced post-harvest life in apple (Robinson *et al.*, 2013). Thus, management of crop load is a balancing act between reducing crop load sufficiently to achieve optimum fruit size and adequate return bloom without reducing yield excessively. On the other hand, low fruit set usually results from the unsuitable environmental conditions, including declining honeybee numbers; therefore, crop pollination has become an important issue for crop load management in many fruit species.

Orchardists use a variety of methods for tree fruit crop load management, including fruit thinning, pruning and artificial pollination (WSU, 2016). However, the conventional approaches of crop load management are labor intensive. Recently, innovation technologies in precise crop load management have been developed and show a great potential for the future. There are many benefits to precision crop load management, such as less risk for over-thinning and alternate bearing, reduced costs for hand thinning, pruning and harvest, and increased profits due to matching fruit size to the markets.

9.2 Manual Approaches for Crop Load Management

Crop load management is the single most important yet difficult management strategy that determines the annual profitability of orchards. The number of fruit that remain on a tree directly affects yield, fruit size, and fruit quality, which largely determine crop value. There are three primary management practices that have a large effect on crop load: pruning, chemical thinning, and hand thinning.

9.2.1 Pruning

In most orchards, there are more flower buds than needed every year. Pruning is the first approach to reduce flower bud load to maintain the balance of tree growth in the next year. In fruit crops, pruning has long been considered an art, requiring the skill of a surgeon and the insights of a poet (McFerson, 2012). In general, the pruning usually includes summer pruning and dormant pruning. Summer pruning refers to shoot tipping and removing water shoots or branches during the growing season, in order to expose fruit to adequate sunlight and modulate sink–source relationships for better photosynthate accumulation in the fruit. Dormant pruning refers to the annual removal of wood during the dormant season, and it is the most important and most expensive orchard management practice. In addition, artificial spur extinction (ASE) uses hand thinning of whole buds in late dormancy to reach targeted floral bud densities on every limb. Manipulation of floral bud density has beneficially altered carbon availability within floral spurs during early-season development. Increasing labor costs and unavailability are threatening the long-term economic viability of orchards. In the past, the lack of uniformity of semi-dwarf trees and the massive number of buds on a tree made accurately counting buds impractical. However, with the adoption of high-density cropping systems in apples and cherries, it becomes practical to count the number of flower buds on representative trees. Therefore, precision pruning is a possible practice to reduce the flower bud number per tree to a pre-defined number.

In the wine grape cultivar 'Pinot gris', mechanical pruning provided 79% cost savings in labor operations compared with hand pruning alone in a warm growing region (Geller and Kurtural, 2013). In China and Japan, the pruning of table grapes usually is designed to establish and maintain the grapevine in a specific form, to reduce labor cost and improve operation efficiency. There are two principal pruning methods: short cane pruning (severe spur-pruning); and long cane pruning. The choice of pruning method depends on the varieties and regions, especially whether the basal nodes on the cane of the grapevine are fruitful. The objectives of short cane pruning are to leave only one to three buds on one lateral shoot and maintain straight primary shoots (Fig. 9.1a). This kind of heavy pruning is easy for growers to carry out. On the other hand, in the long cane pruning method 1-year-old canes that elongated in the previous year are pruned leaving seven to nine buds. With the adoption of horizontal T-trellis systems and the technology of rooting-zone restriction, precision and mechanical pruning, bagging, thinning and harvest in grapevine will become practical in the near future (Fig. 9.1b). A flexible horizontal trellis that allows adjustment of the trellis angle to the stature of the grower has been developed to save labor hours and enhance management efficiency under protected culture.

9.2.2 Chemical thinning

Chemical thinning, which usually relies on compounds that are commonly referred to as hormones and chemicals, has been the primary method growers have used to achieve the proper crop load and consistent annual cropping during the past 50 years. However, its efficacy and performance remain unpredictable and are highly dependent on weather conditions, tree status, and application methods. Chemical blossom thinners cause damage to blossoms and reduce overall fruit set. Post-bloom thinners mimic plant hormones, causing physiological responses in trees, which cause fruit to drop and reduce the crop load. Many fruit growers consider chemical thinning to be the single most important and most challenging practice they perform each season.

Greene *et al.* (2013) developed an improved method of conducting chemical thinning that utilizes both the carbohydrate model and the fruit growth model in apple, namely 'precision chemical thinning'. With the

Fig. 9.1. A horizontal T-trellis grapevine with rooting-zone restriction under a protected culture in Shanghai, China: (a) after short cane pruning; (b) after cluster bagging.

variety-specific target of final fruit number per tree and the thinning task in mind, a precision thinning program is conducted by applying sequential thinning sprays followed by rapid assessment of the results in time to apply a subsequent thinning spray, and then an early reassessment, followed by another spray if needed until the final target fruit number for each variety is achieved. While pome fruits are relatively easy to thin, stone fruits and particularly apricot, cherry, and plum, fail to react to chemical thinning agents with caustic chemicals or hormones (Drkenda *et al.*, 1998). Thus, hand thinning is a necessary but costly management practice in peach production. Chemical thinning alone may not be sufficient to promote annual bearing for several commercially important apple cultivars, such as 'Golden Delicious' and 'Fuji', that possess a strong genetic tendency to alternate bearing (Robinson and Lakso, 2011; Robinson *et al.*, 2013). When compared with 'Gala', 'Jonagold' and 'Granny Smith', which have a low alternate bearing index, 'Rome' shows mild susceptibility to alternate bearing, and 'Delicious' and 'McIntosh' show moderate susceptibility to alternate bearing. There is evidence that alternate bearing has become a less significant problem for apples and pears than it once was (Monselise and Goldschmidt, 1982). This may be attributed in part to selection of annual bearing cultivars, but far more so to the commercial development of effective chemical thinners.

9.2.3 Hand thinning

Hand thinning can be done by selectively plucking individual fruits from a branch, a technique commonly used with larger-fruited species. Large-fruited species may be 'pre-thinned' using a combination of pruning and limb tapping, to reduce the cost of selective hand thinning to make the final crop load adjustment. Apples, pears, table grapes and Asian pears usually require thinning. Cherries may benefit from thinning as well.

9.2.3.1 Cluster trimming and berry thinning

Crop load management of the grapevine, including cluster trimming, gibberellic acid (GA) treatment, and berry thinning, are indispensable in order to produce high-quality table grape bunches in China, South Korea and Japan. Moreover, only one or two clusters are retained per cane, depending upon the density of the latter. Monta *et al.* (1995) developed a berry thinning end-effector, but this was not applied in table grape production thereafter. Recently, hand-held cluster trimmers for grapevine have been developed (Fig. 9.2) and adopted by the growers for their efficiency. However, most growers suffer the long, hard and intensive labor requirement for berry thinning, cluster trimming and thinning. Cluster thinning involves the removal of all clusters during the first non-fruiting year and removal of excess clusters during fruiting years to keep from over-cropping. An early cluster thinning can be done when the shoots are about 30 cm in length for table grapes and some wine grapes that exhibit poor fruit set. This early cluster removal should increase fruit set and berry size. Early cluster thinning is not recommended for cultivars that normally set tight clusters, due to the increased chance of tighter clusters causing bunch rots. A late cluster removal can also be done around the onset of ripening (veraison) for a final crop load adjustment and to promote ripening of wine grapes.

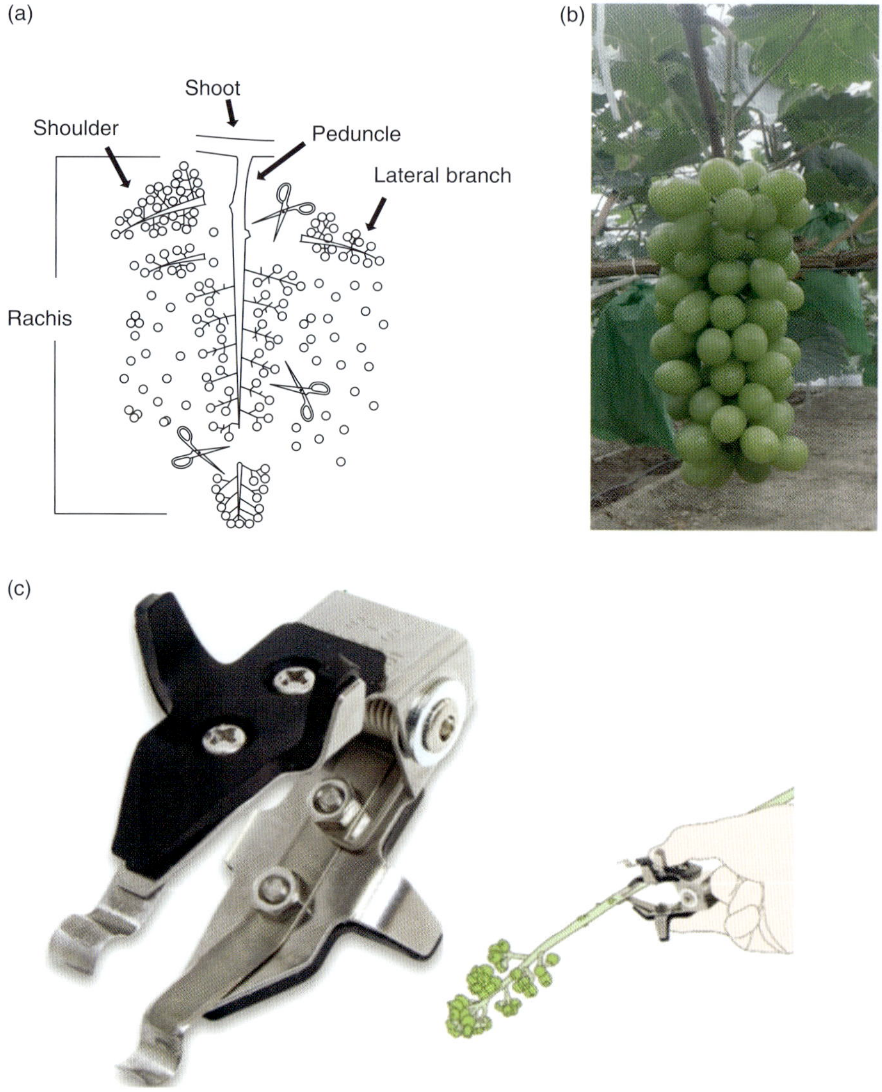

Fig. 9.2. Method for grape berry thinning (a, b) and cluster trimming (c) using hand-held cluster trimmers.

9.2.3.2 Blossom/fruit thinning

Blossom thinning is reported to mitigate alternate bearing in apple and increase final fruit size in both apple and stone fruit. It involves the removal of individual flowers from the five-flower apple cluster; otherwise, multiple flowers in a cluster lead to many small, hard, green and unripe fruits. In apples, the practice of hand thinning can be beneficial to increase fruit size and color. Hand thinning can take place anytime during the growing season between fruit set and harvest. Since king flowers usually develop into a high-quality fruit in apple, side flowers/fruitlets are removed as soon as possible. In Japan, 'Fuji' apple growers usually hand thin the side flowers in the cluster before bloom on 1-year-old shoots prior to fruitlet thinning within 4 weeks of petal fall. Hand blossom and/or green fruit thinning is the most labor intensive and costly operation in the production of apples and peaches.

Heavy crop loads result in smaller fruit than light cropping because of source limitation, and accordingly, fruit thinning is widely used to increase the final size and quality of the remaining fruit on the trees (Jackson, 2003). The earlier that fruit thinning is done, the better, in order to have a positive impact for carrying a larger crop, and for a good return bloom in the following year (Westwood, 1993). In apples, early hand thinning, within 6 weeks of bloom and before flower bud initiation, will help prevent biennial bearing and give the maximum improvement in fruit size. Hand thinning later in the growing season only helps to increase fruit size marginally and can be used to grade fruit by removing. Therefore, thinning must be done as early as possible and in a short period.

Hand thinning of blossoms and fruitlets is labor intensive and expensive but is used when other methods have not removed enough fruit, or more precise thinning is still required. This may particularly be the case where fruit size is important and the cultivar has a high market value, therefore justifying the added labor cost. In Japan, both hand thinning and hand pollination are routine approaches to balance the cropping in Japanese pear. Fruitlets are hand thinned twice until 6 weeks after pollination to limit crop load to 10–15 fruit/m^2. The third to fifth flowers from the bottom of the cluster are left and the other fruit are hand thinned. A second fruit thinning is conducted before bagging and only one fruit per spur is left to develop. In the whole tree, only one fruit per three to five clusters is left to bear. To produce larger fruit, the normal crop load levels (leaf–fruit ratio) of Japanese pear are usually adjusted to around 10 fruit/m^2 on the early- and medium-maturing cultivars and 5–6 fruit/m^2 on the late-maturing cultivars (Zhang *et al.*, 2005). At present, Japanese pear is hand thinned, which involves about 15–20% of the total labor time necessary for pear production each year. Japanese pear will flower and fruit prolifically and so there is a definite requirement for successful flower and fruit thinning to obtain good fruit size and to ensure an annual crop.

In recent years, fruit growers in America have relied primarily on chemical thinning to adjust crop load, with a lesser reliance on hand thinning, to reduce labor requirements. Robust apple bloom-thinning programs in Washington State, when preceded by appropriate pruning and followed by appropriate post-bloom spray programs, have dramatically reduced alternate bearing cycles and created a more reliable supply. The year-to-year fluctuations in apple crop production have been reduced from 20% to less than 5% (McFerson, 2012). In other countries, hand thinning is still the primary means of adjusting crop load. A few progressive growers have also begun to view pruning as a means to adjust crop load. Pruning removes bearing surface (fruit buds) and stimulates vegetative growth from remaining buds. This promotion of vegetative vigor prevents many of the remaining buds from becoming floral. Pruning is a non-selective mass-thinning technique, and therefore is reasonably labor efficient compared with hand thinning.

9.3 Case Study of Mechanical Crop Load Management

9.3.1 Continuously pruning technology for grapevine

Grapevines need to be pruned every year to improve yield as well as reduce diseases. This effective pruning work needs lots of skilled labor, the availability of which has become a very real problem. The labor cost has been increasing in recent years. To maintain crop profitability, it is necessary to seek alternatives to reduce the cultivation costs.

Mechanical pruning, as part of the whole mechanization system, can reduce labor cost in specialty crops by as much as 50% (Morris, 1998). Mechanization of canopy and crop load (Ravaz index (RI), i.e. ratio of yield to pruning weight) management in vineyards was shown to reduce labor costs by 44–80%, maintain yield and quality at the farm gate, and reduce the overheads associated with human resources (Morris, 2007; Kurtural *et al.*, 2012). This section will provide the basic knowledge and configuration of several continuously pruning technologies commonly used for grapevine pruning. A further technique with a robot system for selective pruning will also be covered to illustrate a promising solution for innovative growers.

Mechanical pruning includes non-selective and selective technologies. Non-selective mechanical pruning technology is initially adopted to achieve higher speed and make more savings compared with hand pruning (Sansavini, 1978). It has been introduced from forestry, where it is widely used because of high efficiency and less labor requirement. Particularly in wine grape production, both surface and inner pruners are used to remove surplus vines continuously.

The surface pruner is an implement that removes branches continuously by sickle bars or circular saws with a predetermined plan. The main objective of surface pruning is to control the size of vine cluster for better distribution of nutrients. Topping, hedging and skirting are three common types of surface pruning. The major difference among these pruning patterns is the cutting position according to diverse physiologic requirement. Topping is done to reduce the height of the grapevine and hedging is used to reduce the width. Compared with these pruning methods for canopy volume reduction, skirting is done to cut off lower non-productive branches.

This type of machinery can be attached to tractors, or self propelled with its own power unit. The sickle bar machine is particularly suitable for topping, usually removing shoots up to a diameter of 30–40 mm. The reciprocating moving bars can be driven easily and are relatively inexpensive. Another type of pruner with rotating blades has a flexible boom, more or less adjustable, which makes the operation more efficient in achieving the desired vine shape. The speed of the rotating blades ranges from 1500 to 3000 rpm. These pruners usually have a higher work capacity and require more energy input.

Although surface pruners can efficiently remove vines in short time, this would harm vines and should be used infrequently. The indiscriminate cutting made by the machine involves removing branches at right angles, leaving long stubs and producing excessive epicormic sprouts. These may result in yield reduction and plant decay.

Compared with surface pruners, inner pruners are designed to provide the ability to prune inside the vine in a pass. The purpose of inner pruning is to remove excessive branches in order to increase light penetration and air movement throughout the canopy. The consequential canopy density can be determined by adjusting the speed and size of the cutting blade. However, this non-selective cutting behavior can also remove healthy shoots and result in vine decay.

The inner pruner has similar mechanisms to the surface pruner, but these machines usually involve deep and slot pruning into the vine. Inner cutting can lower the canopy density, and remove dead and diseased branches within the plant. One type of inner pruner (Fig. 9.3) employs a pair of counter-rotating drums driven by hydraulic power on an over-row frame. The tractor-mounted device straddles the plant and removes both surface and internal branches by a series of mulching cutting disks. This

Fig. 9.3. Inner pruner for wine grapes.

over-the-row pruner is usually used for vineyard pruning and can reach a relatively higher efficiency compared with hand pruning.

In wine grape production, the major concern with non-selective mechanical pruners is the uncertainty of yield and fruit quality, which could affect the economic advantage. Also, instead of enhancing the form or character of plants, the uncontrollable pruning process made by the machine would destroy the plants. The limit of the non-selective mechanical pruner is that it is hard to make a plan in advance considering many factors, such as fruit variety, plant density, age and shape of grapevine. The principle of selective pruning has been proposed to remove branches with corresponding rules. In contrast to non-selective pruners, selective pruners are constructed to gain high-quality fruits rather than merely considering the work efficiency. Currently, several organizations (Landwise, 2015; Vision Robotics, 2016) are involved in developing robotic pruners, especially for premium wine production. The selective pruner is designed to accommodate the following objectives:

- ability to scan the entire plant and build a three-dimensional (3D) model of the plant quickly;
- ability to work out a pruning rule based on the 3D modeling results to guide the actuator; and
- ability to make the cut consistently and precisely.

Currently, several subsystems are included to complete this selective pruning operation: stereoscopic vision system for plant figure sensing; 3D reconstruction algorithm integrated with pruning rules to recognize diseased and non-productive branches that need pruning; and mechatronic actuator, such as hydraulic clippers, for branch removal. All these systems are under development for better efficiency and accuracy.

9.3.2 Practice of mechanical thinning in peach and sweet cherry

Blossom/fruit thinning is a costly but necessary practice to produce larger and higher-quality fruit in tree fruit production (Schupp *et al.*, 2008). Poor or inadequate thinning will reduce profitability in the current year and result in inadequate return bloom in the following year. Mechanical devices to aid in thinning have been developed, and research shows that mechanical thinning appears to be a promising technique for supplementing hand thinning in apple and peach trees. However, few have proven highly efficient and capable of completely replacing hand thinning. Narrow canopy training systems, such as upright fruiting offshoot (UFO) systems, and novel peach tree growth habits offer new opportunities to examine mechanical methods for thinning peach and cherry trees. This section illustrates applications of mechanical thinning techniques including a string thinning machine, a spiked-drum shaker, and a hand-held targeted thinning device used for blossom/fruit thinning in peach and sweet cherry production. They are adopted to sufficiently remove the extra blossom or fruits with high throughput ability compared with manual methods, which are very time-consuming and labor intensive tasks associated with high cost to producers.

9.3.2.1 Blossom thinning with string thinning machine and drum shaker in peach

To decrease production costs, a number of mechanical devices, including trunk shakers, low-frequency electrodynamic limb shakers, high-pressure water streams and rotating rope curtains, have been developed and evaluated (Berlage and Langmo, 1982; Baugher *et al.*, 1991; Glenn *et al.*, 1994). Among these techniques, string thinning machines and drum shakers have gained interest in the past few years and have shown promise in effective and practical blossom/fruit thinning during peach production (Miller *et al.*, 2011). Especially for peach production in North America, a rotating string thinner (Darwin 300, Fruit-Tec, Germany, as shown in Fig. 9.4) and a US Department of Agriculture (USDA) spiked-drum shaker have been tested and evaluated in Penn State University (Schupp *et al.*, 2008). The Darwin string thinning machine consists of a tractor-mounted frame with a 3.0 m tall vertical spindle mounted on a central free-rotating shaft. Attached to the spindle are 648 plastic cords, each 0.5 m long, which can knock out fruit buds, blossom buds or flowers. Thinning intensity can be adjusted by installing different numbers of strings on the spindle, and adjusting the rotation speed of the spindle or ground speed of the tractor. In addition, the Darwin thinning machine has a group of mounting arrangements, designed to operate between rows of fruit trees in orchards trained to 3D architectures

Fig. 9.4. Darwin string thinning machine mounted on a tractor.

or exposed canopies such as tall spindle, vertical axe and fruiting walls. The USDA machine is a vibrating direct-drive double spiked-drum shaker, which was originally designed for citrus harvesting and lightly modified for blossom thinning. The shaker consists of two rotating drums, each 2.4 m wide and 1.5 m high. Six whorls of 16 nylon rods, each 32 mm wide and 1.1 m long, are configured on each drum. The drums can freely rotate as they pass through trees to knock green fruits off the limbs.

Tests with the string thinner were conducted during bloom and those with the spiked-drum shaker were conducted before pit hardening when fruit ranged from 20 mm to 30 mm. Across all trials, the spiked-drum shakers removed an average of 37% green fruit. The Darwin string thinner at 60–80% full bloom reduced crop load by 21–50%. Current results obtained from experiments showed that both machines could effectively reduce crop load and follow-up hand thinning time, and harvested fruit size was often significantly larger in machine-thinned trees (Miller *et al.*, 2011). However, there were side effects of mechanical thinning and limitations of the current string thinning machine and spiked-drum shaker, such as:

- induced physical damage to tree limbs, fruiting spurs, and leaves;
- not applicable for multidimensional orchard systems with complex, random architecture; and
- not capable of selective thinning, resulting in over-thinning or uneven thinning.

9.3.2.2 Hand-held targeted thinning machine for sweet cherry

To fill the gap between manual and tractor-mounted mechanized thinning, the hand-held blossom thinner employing a targeted thinning approach is considered as a promising supplement to the machines (Wang, 2013). Compared with currently available tractor-mounted machines, the hand-held device has wider applicability as it can be used in any orchard system and it offers increased selectivity. Furthermore, operators can take a non-selective thinning spindle and place it selectively within the tree canopy so that only the right amount of the least desirable bloom is removed. Hand-held mechanical blossom thinning is also a practical transition to automated selective mechanical thinning methods and tools. Another advantage for a hand-held mechanical device is its low cost. For small-scale orchards, hand-held devices are more suitable than tractor-based machines because the cost of tractor-based machines is higher. Researchers from Washington State University have developed a hand-held mechanical thinning device for targeted thinning of blossoms on sweet cherry trees regardless of tree architecture. The following features of a hand-held device were listed (Wang *et al.*, 2013):

- capable of allowing users to effectively and selectively remove flowers with desired crop loads or strategies;
- capable of being used in any orchard training system;

- much lower product cost when compared with commercially available tractor-powered thinners, especially for small orchards; and
- convenience of operation with low weight and easiness of transportation.

To validate this suitability, a prototype hand-held targeted thinner was firstly developed based on a gasoline-powered weed trimmer (Model FS 09R, Stihl Inc., Waiblingen, Germany), weighing 5.8 kg and extending to 1.5 m long. Modifications included the replacement of the original trimming head by a few specially developed blossom-thinning spindles and the addition of a finer-tip speed controller for adjusting engine speed setting. This type of targeted thinning device could effectively remove blossom with three rows of 150 mm plastic strings for sweet cherry. The thinning spindle speed was found to be the most important variable affecting the effectiveness of blossom removal. Additionally, the string material was observed having an impact on the thinning effect. However, this prototype has been found to be heavy, noisy and difficult to control, considering its further commercial use. To solve these problems, a test bench was developed to investigate the key parameters and factors that affect the performance of the device for providing first-hand knowledge to both the manufacturer and the end-users. Experimental results revealed that spindle rotating speed, swiping speed, and thinning distance were crucial parameters influencing the working efficiency of the thinner. Stiffness of thinning strings also affected the controllability of the percentage of flower removal. Subsequently, a version with spindles driven by a direct-current (DC) motor and two batteries in a backpack worn by the worker was developed, as shown in Fig. 9.5. Consistency of rotating speed has been proven crucial for flower removal rate, considering the speed drop when suffering a power shortage. To develop a better hand-held mechanical blossom-thinning product for precision thinning, Wang (2013) suggested improving the ergonomic design of the current hand-held device, especially the physical comfort of the individual operator.

9.3.3 Electrostatic pollination in fruit tree crops

Bloom and fruit set are annual worries for orchardists, because fruit set determines crop potential for the year. Poor pollination may lead to lower yields and low fruit quality, which will influence the grower's profit. Most fruit tree crops are pollinated either by insects or by wind, which can be easily influenced by reduced bee populations and unsuitable climates. For bee pollination, timely beehive placement, strength of beehives, and activity during cool weather are all worries for growers. Providing bees for pollination does not directly influence alternate bearing, rather it reduces the risk of cropping irregularities, which could trigger alternate bearing. Hand pollination, with minimal pollen consumption and well distributed pollination, has been a commercial practice for pollination supplementation for at least 2500 years. In Japan, apple growers hand pollinate only the king flower in each cluster, since the king flower tends

Fig. 9.5. DC-powered targeted thinning machine used in sweet cherry orchard.

to grow larger and sweeter fruit. However, the drawback of manual pollination is that it requires large amounts of human labor. To provide more effective solutions, both ground- and aircraft-based applicators have been applied (Gan-Mor *et al.*, 2009). Mechanical pollination of fruit trees is of particular importance during years in which the environmental conditions adversely affect natural pollination. These devices could achieve fairly high working efficiency. However, several ineluctable problems, such as pollen drift, large amount of pollen usage and wind-induced inaccuracy deposition, need to be solved to obtain more ideal economic results. The electrostatic technique has been widely used in the industry, for the deposition and separation of small particles (Kawamoto, 2008). It provides a potential technique to achieve precise pollination for fruit trees, which could be an innovative solution for fruit growers. This section will introduce the basic principle of pollen deposition with electrostatic methods. Biological aspects and partial results of field trials will also be presented to show the adaptability of the commercial application of this technology.

The first prototype of a pollination system with electrostatic technology has been developed in Israel and tested in a local date orchard. To investigate the practicability of electrostatic pollination system configuration, mathematical models have been firstly developed based on a finite-element method. The following equation could be used to describe the electrostatic problem (Stremler *et al.*, 1990):

$$\frac{\partial}{\partial x}\left(\varepsilon\frac{\partial V}{\partial x}\right)+\frac{\partial}{\partial y}\left(\varepsilon\frac{\partial V}{\partial y}\right)=-\rho \tag{9.1}$$

where V = potential, ε = electric permittivity, and ρ = charge density.

The proof-of-concept study involved system modeling, simulation, laboratory tests and field trials. To validate the model and subsequent trajectory simulation results, a prototype pollination machine has been constructed based on the corona charging principle in the laboratory, consisting of an air-pressure control system, pollen-charging system and a 500 cc pollen-storage tank. The pressure control system could provide air pressure with a range of 20–500 kPa, which could be used to adjust the mixed pollen–air flow rate as charging. A high electrode located in the centre of an electrostatic nozzle could carry a potential up to 80 kV, which could provide several charging levels to the pollen cloud for experimental validation. Both simulation and laboratory results indicated that pollination with electrostatic charge could introduce much higher pollen density on the pistil surface in the flower envelope (Bechar *et al.*, 1999). In their further field investigations, the prototype provided a promising pollen deposition percentage on the stigma, also with an increasing fruit yield.

During the next few years, this prototype pollinator was upgraded to be utilized for a wide range of fruit tree pollination, such as almond, kiwi, date and pistachio (Gan-Mor *et al.*, 2003). Pollination machines were first used in New Zealand's kiwifruit production. At this stage, an electrostatic pollen applicator had been optimally designed and developed, which consisted of air–pollen mixer, feeder motor, feed-rate regulator and controller, air blower, high-voltage DC converter, and corona charging system. This applicator was referentially designed for almond pollination application, and needed partial modification for pistachio and kiwifruit pollination, and was considered for the treatment of large-scale orchards. For example, for kiwifruit pollination, the pollen applicator has been mounted on a John Deere 'Gator' with predetermined angle suitable for a trained structure. The advantage of this electrostatic applicator is the convenience of pollen feed-rate regulation with negligible fluctuations. Because of adjustable air stream, both the output air velocity and pollen–air ratio could be appropriately controlled in accordance with the fruit variety. Field trials were conducted to compare pollination efficiency under different operation configurations, including charged pollination, uncharged pollination, and open pollination treatment (Gan-Mor *et al.*, 2003). The yield and nut size were chosen as the evaluation standard from both horticultural and economic perspectives. The results showed different effects for diverse fruit varieties. For example, in almond production, electrostatic pollination increased the total yield, with slightly reduced average fruit size; in pistachio, it could both increase the yield and improve fruit quality. Although electrostatic pollination provides much greater efficiency on pollen deposition, accompanied with less pollen consumption and more economic benefits, it also has some defects during commercial application, such as uncertain fruit quality, high requirement of horticultural knowledge, and tree structure adaptability.

To overcome these challenges, another field trial was conducted to further evaluate the performance of an electrostatic applicator for date pollination. In this study, electrostatic pollination, considered as a precise load management method, could be utilized to control the amount of flowers and fruit yield (Gan-Mor *et al.*, 2009). Together with target thinning technology, it showed a promising mechanical solution for fruit tree load management, which has brought hardship to growers for a long time.

Similar approaches of mechanically spraying pollen are being tried around the world. Researchers are working to develop a 'Robobee' machine that could deposit pollen where it is needed by sensing female flowers. In Europe, scientists are studying artificial pollination by using orchard sprayers to improve fruit set of pears. Preliminary research at Washington State University indicates that mechanical pollination also shows promise for managing the crop load in cherry and apple orchards.

9.4 Challenges and Opportunities for Automation in Crop Load Management

Mechanical methods for fruit tree load management have been proven practical for specific plant varieties. However, the uncertain physiological effect on the fruit plants is still questionable for a diverse range of fruit varieties. Long-term studies should be conducted to understand the behavior of the plant under different load management approaches, especially the yield and fruit quality. Precision crop load management utilizes three management approaches to adjust crop load. It begins with precision pruning to leave a preset bud load on the tree, followed by precision chemical thinning to reduce initial flower number per tree to a preset fruit number, and ends with precision hand thinning to bring the load down to desired levels. Both accurate crop load estimation and established forecasting models are important for efficient crop load management.

With more accessible training systems and high-density planting in fruit crops, new technologies for mechanically pruning, blossom thinning and even pollination are being commercially used or tested today for more efficient and feasible crop load management. However, precision crop load management still remains largely conceptual for tree fruit. Pruning and thinning actually become more predictable, but are still done by hand and inconsistent. During the development of automation technology, consideration should be given to the varieties, rootstocks, tree age, phenology, and physiological and environmental factors when making crop load management decisions. More precise management of crop load will help growers achieve the optimum crop load and maximize crop value.

9.5 Summary

Fruit crops tend to set more fruit than necessary, thus adjusting crop load is important not only for annual bearing, but also for economic sustainability in the current season as well as the next. Three approaches of precision crop load management were discussed in this chapter. Several production practices can be used to reduce crop load, but the chief practices are fruit thinning and pruning. Obtaining the optimal crop load for fruit size, quality and return bloom with chemical thinning is challenging. The new technologies of mechanical thinning, pruning and pollination take us closer to managing fruit trees at the level required to maximize yields and profits.

References

Baugher, T.A., Elliott, K.C., Leach, D.W., Horton, B.D. and Miller, S.S. (1991) Improved methods of mechanically thinning peaches at full bloom. *Journal of American Society for Horticultural Science* 116, 766–769.

Bechar, A., Shmulevich, I., Eisikowitch, D., Vaknin, Y., Ronen, B. and Gan-Mor, S. (1999) Modeling and experiment analysis of electrostatic date pollination. *Transactions of the ASAE* 42(6), 1511–1516.

Berlage, A.G. and Langmo, R.D. (1982) Machine vs. hand thinning of peaches. *Transactions of the ASAE* 25, 538–543.

Dennis, F. Jr (2003) Flowering, pollination and fruit set and development. In: Ferree, D.C. and Warrington, I.J. (eds) *Apples: Botany, Production and Uses*. CAB International Publishing, Cambridge, Massachusetts, pp. 153–166.

Drkenda, P., Bertschinger, L. and Stadler, W. (1998) Fruchtbehang und Fruchtqualität tragwilliger Zwetschgensorten. *Schweiz Z Obstund Weinbau* 134, 156–158.

Gan-Mor, S., Bechar, A., Ronen, B., Eisikowitch, D. and Yaknin, Y. (2003) Electrostatic pollen applicator development and tests for almond, kiwifruit, date, and pistachio – an overview. *Applied Engineering in Agriculture* 19, 119–124.

Gan-Mor, S., Ronen, B., Vaaknin, Y., Glik, Y., Samocha, Y. and Eisikowitch, D. (2009) Further study on electrostatic date pollination – from the laboratory bench to field unit performance test. *Applied Engineering in Agriculture* 25, 643–646.

Geller, J.P. and Kurtural, S.K. (2013) Mechanical canopy and crop-load management of Pinot gris in a warm climate. *American Journal of Enology and Viticulture* 64, 65–73.

Glenn, D.M., Peterson, D.L., Giovannini, D. and Faust, M. (1994) Mechanical thinning of peaches is effective postbloom. *HortScience* 29, 850–853.

Greene, D.W., Lakso, A.N., Robinson, T.L. and Schwallier, P. (2013) Development of a fruitlet growth model to predict thinner response on apples. *HortScience* 48, 584–587.

Ho, L.C. (1996) The mechanism of assimilate partitioning and carbohydrate compartmentation in fruit in relation to the quality and yield of tomato. *Journal of Experimental Botany* 47, 1239–1243.

Jackson, J.E. (2003) Flowers and fruits. In: Jackson, J.E. (ed.) *Biology of Apples and Pears*. Cambridge University Press, New York, pp. 268–308.

Kawamoto, H. (2008) Some techniques on electrostatic separation of particle size utilizing electrostatic travelling-wave field. *Journal of Electrostatics* 66, 220–228.

Kurtural, S.K., Dervishian, G. and Wample, R.L. (2012) Mechanical canopy management reduces labor costs and maintains fruit composition in 'Cabemet Sauvginon' grape production. *HortTechnology* 22(4), 509–516.

Landwise (2015) Grape vine pruning robot. Available at: www.landwise.org.nz/precision-agriculture/grape-vine-pruning-robot/ (accessed 12 August 2016).

Lenz, F. (2009) Fruit effects on the dry matter and carbohydrate distribution in apple tree. *Acta Horticulturae* 835, 21–38.

McFerson, J.R. (2012) Applying precision agriculture to tree fruit. *American Fruit Grower*, 6 March 2012. Available from: www.growingproduce.com/fruits/applying-precision-agriculture-to-tree-fruit (accessed 5 June 2017).

Miller, S.S., Schupp, J.R., Baugher, T.A. and Wolford, S.D. (2011) Performance of mechanical thinners for bloom or green fruit thinning in peaches. *Journal of American Society for Horticultural Science* 46, 43–51.

Monselise, S.P. and Goldschmidt, E.E. (1982) Alternate bearing in fruit trees. In: Janick, J. (ed.) *Horticultural Review*, Volume 4. John Wiley & Sons, Inc., Hoboken, New Jersey, pp. 128–173.

Monta, M., Kondo, N. and Shibano, Y. (1995) Agricultural robot in grape production system. *Proceedings of the 1995 IEEE International Conference on Robotics and Automation* 3, pp. 2504–2509.

Morris, J.R. (1998) *Vineyard mechanization for juice and wine production for Missouri and Mid-America.* Paper presented at Midwest Regional Grape and Wine Conference, Lake-of-the-Ozarks, Missouri, January 1998. Available at Northwest Berry & Grape Information Network (posted 15 February 2007): www.berrygrape.org (accessed 5 June 2017).

Morris, J.R. (2007) Development and commercialization of a complete vineyard mechanization system. *HortTechnology* 17(4), 411–420.

Robinson, T.L. and Lakso, A.N. (2011) Advances in predicting chemical thinning response using a carbon balance model. *New York Fruit Quarterly* 19,15–20.

Robinson, T.L., Lakso, A.N., Greene, D. and Hoying, S. (2013) Precision crop load management. *New York Fruit Quarterly* 21, 3–9.

Sansavini, S. (1978) Mechanical pruning of fruit trees. *Acta Horticulturae* 65, 183–198.

Schupp, J.R., Baugher, T.A., Miller, S.S., Harsh, R.M. and Lesser, K.M. (2008) Mechanical thinning of peach and apple trees reduces labor input and increases fruit size. *Hortechnology* 18, 660–670.

Stremler, F.G., Klein, S.A., Luo, F.F. and Liao, Y. (1990) Numerical solutions and mapping of electrostatic fields using the Apple Macintosh computer. *IEEE Transactions on Education* 33, 104–110.

Vision Robotics (2016) Intelligent autonomous grapevine pruner. Available at: www.vision-robotics.com/vr-grapevine-pruner (accessed 12 August 2016).

Wang, M. (2013) A hand-held mechanical device for targeted blossom thinning in sweet cherry. PhD dissertation, Washington State University, Pullman, Washington.

Wang, M., Wang, H., Zhang, Q., Lewis, K.M. and Scharf, P.A. (2013) A hand-held mechanical blossom thinning device for fruit trees. *Applied Engineering in Agriculture* 29, 155–160.

Westwood, M.N. (1993) *Temperate Zone Pomology: Physiology and Culture.* Timber Press, Portland, Oregon.

Whiting, M.D. and Lang, G.A. (2004) Bing sweet cherry on the dwarfing rootstock Gisela 5: crop load affects fruit quality and vegetative growth but not net CO_2 exchange. *Journal of American Society for Horticultural Science* 129, 407–415.

WSU (2016) *Crop Load Management.* WSU Tree Fruit, Washington State University. Available at: http://treefruit.wsu.edu/orchard-management/crop-load-management/ (accessed 8 October 2016).

Zhang, C., Tanabe, K., Tamura, F., Itai, A. and Wang, S. (2005) Spur characteristics, fruit growth and carbon partitioning in two late-maturing Japanese pear (*Pyrus pyrifolia*) cultivars with contrasting fruit size. *Journal of American Society for Horticultural Science* 130, 252–260.

10 Mechanical Harvest and In-field Handling of Tree Fruit Crops

Manoj Karkee[1*], Abhisesh Silwal[1] and Joseph R. Davidson[2]

[1]*Washington State University, Prosser, Washington, USA;* [2]*Washington State University, Richland, Washington, USA*

10.1 Introduction

Through intensive automation and mechanization, agricultural productivity has substantially increased in the past century. Farming technologies commercially adopted over the course of the 20th century include equipment for field operations, such as tractors, planters, sprayers and combine harvesters, as well as irrigation systems, all of which have profoundly altered the structure of agriculture (Silwal *et al.*, 2016a). The production of row crops like corn and wheat has seen unparalleled reduction in labor use and improvement in crop yield and quality through the application of these technologies. However, commercial adoption of mechanization and automation technologies for fresh market tree fruit crops such as apples and pears is still limited.

A large workforce of seasonal laborers is currently used for fruit production operations such as training, pruning, thinning, and harvesting. Among all field operations in tree fruit production, harvesting is the most labor-intensive as well as time-sensitive task (Gallardo and Brady, 2015; Silwal *et al.*, 2016b). For example, Washington State alone employs an additional 36,425 seasonal agricultural workers during the peak tree fruit harvesting months (WSESD, 2013). Another study from Gallardo *et al.* (2010) reported that harvesting labor accounts for nearly one-third of the total annual variable costs of apple production, making it the most expensive orchard operation.

Typically, hand harvesting of tree fruit crops includes the use of ladders to access fruit higher in the canopy. This ladder use requires repetitive movements up and down the ladder and also to and from the collection

* Corresponding author, email: manoj.karkee@wsu.edu

bin or container with heavy loads of fruit. Such activities expose every fruit picker to the risk of falling from the ladder as well as other ergonomic injuries due to heavy lifting and repetitive hand actions (Fathallah, 2010; Elkins *et al.*, 2011). Orchard platforms have the potential to improve worker productivity and safety, but their adoption is highly limited. In Washington State, only 11% of producers are using mechanical platforms in their operations (Gallardo and Brady, 2015). Use of these platforms is also not widespread in growing regions in the eastern USA (Robinson *et al.*, 2013). Some of the major reasons cited for limited adoption of the technology are: (i) incompatibility of platforms with orchard architecture; (ii) non-uniform and variable tree structures; and (iii) slowing down of faster workers as a group of workers on the same platform have to work at the speed of the slowest worker (Elkins *et al.*, 2011; Robinson *et al.*, 2013; Gallardo and Brady, 2015). The worker compensation system based on individual piece rate, which is commonly used by growers, is also a limitation to platform adoption.

Past research and experience have shown that engineering design and horticultural modification need to go hand-in-hand to be successful in mechanical harvesting of fruit crops. If we try to develop a robotic system to harvest fruit from complex four-dimensional (4D) trees grown conventionally (Section 10.2.4), it is highly challenging to achieve a desirable level of harvesting efficiency. The machine is also going to be exceptionally complex (for example, a large number of degrees of freedom in a robotic manipulator), which leads to high cost for both acquisition and maintenance of the machine and thus limits its practical adoption. In the past few decades, tree canopy architectures have been improved increasingly towards more planar (two-dimensional) structures called 'fruiting wall' architectures (WSDA, 2011). These architectures have improved working environments and ergonomics for orchard workers and have provided opportunities to increase the adoption of assistive technology such as the mechanized platform. More importantly, such improvement has opened up opportunities for developing practically adoptable mechanized and automated tree fruit harvesting systems (Silwal *et al.*, 2017).

Although farmworkers play an important role in the agricultural industry in the USA and around the world, their wages and salaries account for less than 1% of total US wage and salary data (USDA ERS, 2012). A non-supervised farmworker earns about US$10–12/h. In addition, the demand for farm labor is highly seasonal in nature and migrant workers have to find alternative employment opportunities during the off-season. Mechanization and automation have the potential to reduce the dependency on migrant workers and 'end an era of importing poverty' (Gallardo and Brady, 2015). More mechanized and automated field operations will also provide higher paying and permanent jobs to a pool of farmworkers, thereby making a positive impact on the socio-economic status of rural agricultural communities.

The major goal of harvesting a crop is to gather the commodities from the field with consistent maturity at rapid speed while minimizing physical damage, crop loss, and harvesting cost (Kader, 2002). Fresh market tree fruit crops like cherries and peaches are delicate, have short harvesting windows, and may have a wide range of maturity. Because of these restrictions, the speed and throughput of harvesting systems are critically important for a tree fruit harvesting system. To reduce dependency on the seasonal labor force, accelerate the supply chain of tree fruit crops, and improve the long-term sustainability of the fruit industry, mechanization and automation of harvesting and handling systems are critical and have a clear need in today's economy. The major significances of a harvesting and handling system are that: (i) mechanical harvesting and handling systems have the potential to reduce risk from uncertain labor availability and increase harvest productivity; and ii) mechanical harvesting and handling systems enable sustainable production systems and accelerate the supply of agricultural commodities to the market at affordable cost.

Fully exploring the potential benefits of mechanization, research and development of tree fruit harvesting and handling systems has received a renewed focus in recent years. There are a number of research groups and private industries from around the world, including the USA, Europe, Asia, Australia, and New Zealand, engaged in research and development activities for fruit crop harvesting. Washington State University's Center for Precision and Automated Agriculture (WSU CPAAS) has also focused strongly in the area of tree fruit harvesting for the past 5 years. This chapter provides a general overview and description of various components of mechanized and automated fruit harvesting and handling systems. The chapter will also introduce two case studies: one in shake-and-catch cherry harvesting; and another in robotic apple harvesting. As the design of crop canopies is a critical consideration for the successful development and implementation of harvesting solutions, the chapter begins with a description of several relevant crop architectures and the suitability of each for mechanized or automated harvesting. The chapter ends with a brief discussion on the current challenges and limitations and potential directions for future research and development in tree fruit harvesting.

10.2 Crop Architecture

Godin *et al.* (1999) defined crop architecture as the organization of various canopy components such as the trunk (or leaders) and branches in a three-dimensional (3D) space, which evolves over time through growth and development as well as through training and pruning. Many different types of fruit crop are grown commercially around the world with numerous different tree canopy architectures. In this section, a brief summary of commonly used canopy architectures relevant to mechanization and automation is presented. Terms and descriptions used in representing these canopy architectures are not always consistent and there can be a lot of overlap between the use of these terms and definitions in different parts of the world. In this chapter, we have tried to employ some of the most widely used definitions, but there may exist some level of inconsistency in other literatures.

10.2.1 Central leader architecture

Central leader is a free-standing (non-trellised) tree architecture that consists of a strong leader or trunk in the center with a few sets of branches grown laterally from the central leader. A set of branches (also called a scaffold) at the lower level are grown longer compared with the branches in the higher level of the canopy, thus creating a pyramidal canopy structure (Fig. 10.1).

Generally, central leader trees are planted at 4.0–6.0 m row spacing and 1.8–3.0 m tree spacing within a row, leading to roughly 500–1500 trees/ha (Robinson *et al.*, 2013). Depending on the crop type and rootstock, tree height is controlled to 3.5–5.0 m. Many different types of crop such as apples, cherries (sweet and tart), pears, plums, and olives are grown in this architecture, sometimes with slight variations for specific crops, such as Vogel central leader for sweet cherries (Long *et al.*, 2015) and mini-central leader for olives (Anonymous, 2016). Central leader trees provide a good structure for mass harvesting of fruit with trunk shaking, as can be seen in commercially available tart cherry and olive harvesters. However, fruit catching can occur only below the lowest level of branches, creating a large fruit drop height as well as interception of fruit by branches as they fall down on to the catching surface. Because of these limitations, it is challenging to maintain desirable fruit quality for fresh market produce. If central leader canopies can be trained and pruned such that there is a clear gap between multiple scaffolds, a fruit-catching system could

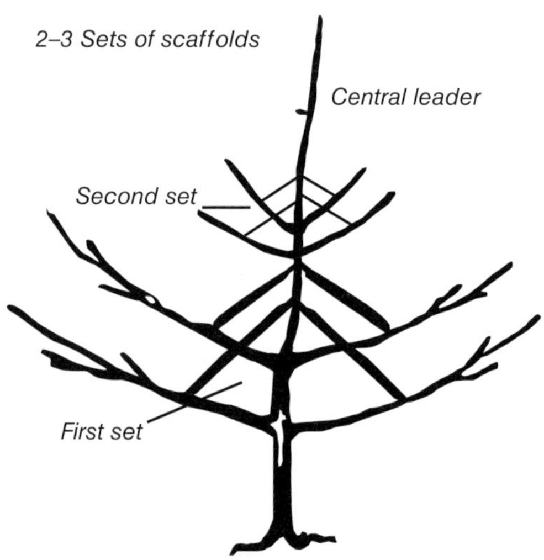

Fig. 10.1. A mature fruit tree structure trained to a central leader architecture (PSE, 2016).

be designed to be inserted into the canopy at multiple levels, potentially improving fruit quality with shake-and-catch harvesting. In recent years, central leader orchards have increasingly been replaced by some variations of SNAP (Simple, Narrow, Accessible, Productive) architecture (WSDA, 2011) such as super spindle systems (Section 10.2.2), which are friendlier for automated harvesting of fresh market fruit.

10.2.2 Spindle architecture

Spindle trees are also single-leader (trunk) trees with lateral branches. However, unlike central leader trees, the branches are maintained much shorter and the trees are planted closer together. There are a lot of variations and improvements of spindle tree architectures, including tall spindle apples (Fig. 10.2) and super slender cherries. The principle of training and pruning for these variations are similar. The major difference will be in how close together the trees are planted and how narrow the canopy is maintained. On one side of the spectrum from wide to narrow structures are canopies such as vertical axis and on the other side of it are canopies such as super spindle. For wider canopies such as vertical axis, lateral branches in the lower level of the canopy are generally tied down for better light distribution. Different types of trellising are used to support the canopies in spindle structures, particularly for those narrower architectures such as super spindle.

Apples, pears and sweet cherries are some of the crops grown in some variations of spindle architectures. A few major variations of spindle architecture are listed in Table 10.1 with a general range of parameters for each architecture. As can be seen in the table, the major difference between the spindle architectures is how closely the trees are planted. Slender spindle and tall spindle architectures, however, are separated primarily by the tree height. Slender spindle, one of the older spindle canopies developed and used in Europe, has a shorter canopy height (~2.5 m) compared with that in tall spindle architecture (Perry, 2010).

The narrower these spindle canopies are, the more visible the fruit, branches and other canopy components are to the sensors and actuators, which makes mechanization and automation easier. Both vertical and angled fruiting wall orchards (Section 10.2.4) can be developed with some of the narrower spindle structures such as super spindle trees.

10.2.3 Multi-leader canopies

Various types of multi-leader tree canopies (canopies with more than one trunk or stem) are grown across different types of crops, including peaches, pears, cherries and apples. A few common multi-leader architectures are discussed below.

Fig. 10.2. A tall spindle apple orchard with around 4500 trees per hectare.

10.2.3.1 Open vase architecture

Open vase trees, also called open center trees, are trained such that multiple, equally strong leaders/stems are grown in a certain orientation so as to create an opening in the center of the canopy. Peaches, apples and

Table 10.1. Geometric parameters of various spindle canopy architectures. Please note that these parameters vary substantially both between crops and within a crop, due to various biological (e.g. crop vigor) and management factors; and are provided here as only general guidelines. More discussion with specific parameters for many of these architectures can be found in Mika (1991), Sigler (2010), Robinson *et al.* (2013), Long *et al.* (2015).

Spindle canopy type	No. of trees/ha	Row spacing (m)	Tree spacing (m)	Canopy height (m)	Canopy depth (m)
Vertical Axis	1000–2000	~4.5	~2.0	3.5–4.5	~1.5
Slender Spindle	2000–3000	~3.0	~1.5	~2.5	~1.2
Tall Spindle	2300–4000	~3.0	0.8–1.6	3.0–4.0	~1.2
Super Spindle	4000–12,000	<3.0	<0.8	3.0–4.0	<0.9

plums are some of the crops grown in this freestanding architecture. In this architecture, about 400–600 trees are planted per hectare with a few layers of scaffolds in each tree. Tree height is generally kept lower than many other architectures discussed in this section. For example, tree height for open vase peaches in different Florida orchards has been set around 2.5 m, which creates a pedestrian orchard (ladders are not required) for most of the manual field operations.

10.2.3.2 Kym green bush (KGB)

KGB canopies, used primarily in cherries, also create freestanding trees constituting multiple vertical leaders (Fig. 10.3). These canopies are shorter compared with many other canopy structures discussed in this section, providing a pedestrian orchard for most of the field activities (this structure is also referred to as a pedestrian orchard). The shorter canopies allow mechanization systems (e.g. chemical application system) to go over the canopies, which allows improved efficiency. The trees are generally grown at 4.5–5.5 m row spacing and 1.5–2.5 m tree spacing (Long, 2010). Descriptions of several other types of architectures specific to sweet cherries can be found in Long *et al.* (2015).

10.2.3.3 Biaxial

Biaxial canopy is a special case of a multi-leader system, which constitutes two leaders in each tree canopy. A common system for biaxial orchards is to train both leaders of the canopy along the crop row, creating a vertical canopy structure. However, principally, an angle structure (Y or V) (Section 10.2.6) can be created by training each leader to one side of the angled canopy. This type of training system is used with various crops, including apples, cherries and pears.

10.2.3.4 Upright fruiting offshoot (UFO)

UFO (Fig. 10.4) is a trellised cherry orchard architecture developed by WSU professor Mathew Whiting (Whiting, 2008). In this system, the main trunk is grown about 30–50 cm above the ground before it is bent horizontally to form a base structure of the canopy. Limbs are then grown laterally

Fig. 10.3. A typical canopy architecture of a tree trained in a KGB structure in a commercial orchard in Tasmania (photo provided by Matthew Whiting, WSU Center for Precision and Automated Agricultural Systems, Prosser, Washington).

off the trunk, creating fruiting structures spaced regularly on a vertical plane. The limbs can be trained both vertically and in an angle (creating a V or Y architecture). In a typical vertical UFO, trees are grown up to a height of 3.5 m with a row spacing of 2.0–3.0 m and tree spacing of 1.0–2.0 m. This architecture provides one of the more friendly environments for mechanized/automated cherry harvesting, as individual fruiting offshoots can be shaken efficiently. Minimizing damage during fruit catching depends on various factors, but an angled canopy provides a better opportunity to catch fruit closer to their original position in the tree.

10.2.4 Conventional versus fruiting wall canopies

Depending on the complexity of free canopies, they can be divided into conventional and fruiting wall architectures. Tree canopy structures are present in a continuum from 'big-old-bushy' trees to formally trained narrow two-dimensional (2D) canopies. However, for the simplicity of discussion, they are divided here into conventional architectures (Fig. 10.5) and modern fruiting wall architectures (Fig. 10.6).

Over the past few centuries, numerous different types of training and pruning systems evolved around the world leading to many different

Fig. 10.4. Upright fruiting offshoots (UFO) cherry architecture planted in a Washington State University research orchard in Prosser, Washington (provided by Matthew Whiting, WSU Center for Precision and Automated Agricultural Systems, Prosser, Washington).

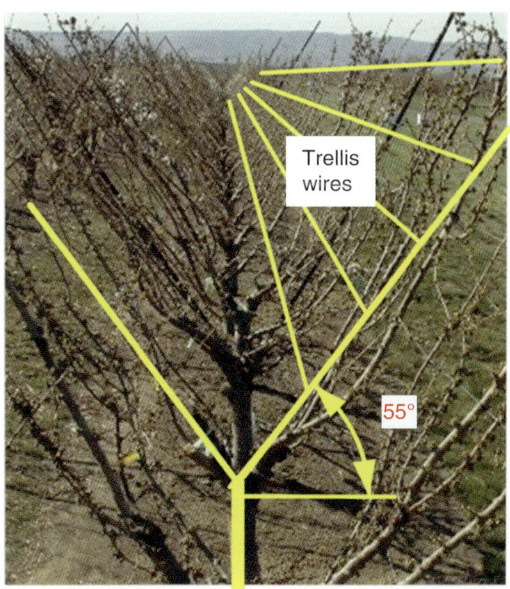

Fig. 10.5. A trellis-trained cherry tree architecture (from Zhou *et al*., 2014).

Fig. 10.6. A V-shaped (V) fruiting wall apple orchard of Auvil Fruit Co., Orondo, Washington (from De Kleine, 2014).

types of tree architectures (Fideghelli *et al.*, 2003). Conventionally, trees were grown with a comparatively low level of training and pruning and generally without a trellis system to support the structure. These conventional fruit tree canopies could be quite tall, wide and deep, creating 4D architectures, with three geometric dimensions and variability between individual trees as the fourth dimension.

In recent decades, however, the shape and size of fruit trees have been increasingly controlled using semi-dwarf and dwarf rootstocks (at least for some crops, such as apples and cherries) and more intensive training and pruning, leading to tree architectures that are easier to manage and offer increased labor productivity for operations such as thinning and harvesting. The newer tree canopies are kept comparatively short and narrow, creating a wall of fruiting surface referred to as 'fruiting wall canopies'. Trees trained in narrower canopy structures, such as super spindle or UFO trees, are used to create fruiting wall canopies. Modern fruiting wall canopies are developed and maintained in vertical as well as angled (V or Y) canopies (Section 10.2.6). These 2D canopies (which may still have 30–60 cm of canopy depth, but can be considered 2D canopies for practical purposes) provide uniform and sufficient light distribution to achieve higher levels of fruit quality and yield. Equally importantly, these canopies are designed to make the majority of branches and fruits visible and accessible to machines, which is highly desirable for mechanization and automation of various orchard operations. For this reason, these modern canopy architectures are also called SNAP architectures. As discussed in Sections 10.2.1 and 10.2.2, a planting rate of 4000 trees/ha is commonly found in these orchards with some orchards having as many as 12,000–15,000 trees/ha. As the trees are planted close together, these orchards are called 'high density' and sometimes even 'super density' orchards.

10.2.5 Formal versus random canopies

Based on how the fruiting branches are trained in a fruiting wall architecture, canopies can be divided into 'formal' and 'random'. In the formal canopies, only a finite number of branches or limbs are grown laterally from the vertical stem/trunk (or leaders) and these are tied to horizontal trellis wires. Six to seven trellis wires are quite common in modern fruiting wall apple orchards, creating a 2D tree canopy with six to seven distinct layers. Most of the fruit are grown on these horizontal limbs, thus creating horizontal columns of fruit, though some fruit can still be found on the vertical stems. In a random canopy architecture, branches or limbs are allowed to grow in all directions from the vertical stem or leaders and are not trained to trellis wires in any particular fashion, though trees are supported by the trellis system (e.g. a super spindle apple orchard). To maintain a 2D fruiting wall structure, branches are kept short through dormant pruning as well as occasional summer/fall pruning.

Both random and formal architectures offer the same type of accessibility to automated or robotic harvesting systems as long as the canopies are sufficiently narrow. However, it is important to note that a fruit standing freely and away from branches and trunks is always easier to pick than a fruit resting on another fruit or a branch. A formal architecture provides more opportunities compared with a random architecture to avoid fruiting in the vertical trunk/leader and to keep most of the fruit hanging down a fruiting limb. For a shake-and-catch harvesting system, a formal architecture provides an opportunity to shake only targeted fruiting limbs and catch the fruit just under those limbs, which has the potential to keep fruit quality at a desirable level for the fresh market (He *et al.*, 2016).

10.2.6 Angled versus vertical canopies

Based on the orientation of tree canopies, they can be divided into angled and vertical architectures. Angled canopies could be structured in a V-shape or Y-shape, generally referred to as V-canopy or Y-canopy. The angle of fruiting surface can vary widely (e.g. from 40 degrees to 80 degrees to the ground) depending on a number of factors, including type of crop, row spacing, geographic location and specific management practices. Some of the UFO cherry orchards in Washington State have been planted at a 60-degree angle to the ground surface. A vertical canopy is a special case of the angled canopy with the fruiting surface angle of 90 degrees to the ground.

Many different types of crop, including apples, cherries, pears, and plums, are grown in angled as well as vertical canopy architectures. When trees are planted in an angled canopy (e.g. V-architecture), they are planted in slightly wider row spacing compared with a vertical architecture of the same crop but generally half the inter-plant spacing. For example, super spindle apple trees in an angled canopy are generally planted with 3.0–4.0 m row spacing with inter-plant spacing of ~0.45 m.

Angled and vertical canopies offer their own advantages and disadvantages for mechanized or automated harvesting. Vertical architectures generally present canopies such that most of the fruit are in the outer surface when we look at the canopy from both sides. If a shake-and-catch harvesting system is to be used, a vertical canopy allows a catching mechanism to go from both sides of the canopy, thus potentially improving the collection efficiency. However, if appropriate pruning and thinning is used, fruit can be grown in the outer surface of an angled canopy as well, which then offers the same level of accessibility for a robotic harvesting system. Angled canopies, particularly when the orientation angle is low (flatter canopies), also provide an opportunity to design a catching surface that can be placed parallel to the fruiting surface, which can minimize the drop height if a shake-and-catch mechanism is to be used for harvesting. Flatter canopies also allow most of the fruit to hang down from the limbs, which can be beneficial for robotic as well as shake-and-catch harvesting.

10.3 Overview of Mechanical Harvesting

Researchers in the past have investigated different technologies for tree fruit mechanical harvesting. Despite wide differences in the specific mechanisms used from one system to another, almost all the investigated systems were designed to perform either shake-and-catch harvesting (mechanical mass harvesting) or pick-and-place harvesting (robotic individual picking). In principle, shake-and-catch harvesting applies vibratory excitation to the canopy, trunk or a branch of a tree to create a detaching force on fruit, and uses some type of catching device to collect the detached fruit. Mechanical tree fruit harvesters based on this approach have been commercially used in harvesting fruit destined for the processing market, due to their high productivity. However, there has been limited success in fresh market fruit harvesting, due to high levels of harvest-induced damage from shake-and-catch operations. When it comes to pick-and-place fruit harvesting, the general process includes: locate the target fruit, approach and detach the fruit, and then place the picked fruit in a designated container. In general, robotic harvesting technology has achieved very limited success, primarily due to inadequate accuracy, speed and robustness. As both of these technologies have some important limitations, research and development by both the public and private sectors have continued in order to fill the existing gaps so that a satisfactory solution to mechanical harvesting of fresh market tree fruit crops can be developed.

In the following subsections a brief review of the history of mechanical harvesting research will be presented followed by some fundamental principles for consideration during research and development. While labor-assist aids, such as mechanized platforms that raise the worker to the fruit location, are important tools for improving productivity, the focus here is on mass and selective harvesting technologies that save human labor.

10.3.1 Historical background of mechanical harvesting and handling

10.3.1.1 Mass harvesting systems

The tree fruit industry has sought for many decades to reduce harvesting and handling costs and alleviate labor pressure through mechanization. A sufficient labor supply was a concern as early as 1917 when the US government created guest worker programs for migrant workers to enter the USA for temporary agricultural work (Peterson, 1992). Likewise, harvesting labor expenditures have consistently accounted for a significant portion of total production costs. For example, in the 1956–1957 Florida citrus season, picking labor accounted for 63% (US$25,000,000) of the total cost of fruit production (Jutras and Coppock, 1958). Extremely low citrus prices in 1947–1948, when reported prices dropped to as low as five cents per box (Phillips and Grierson, 1957), were further motivation to reduce production costs. Accordingly, significant research and development of mechanical harvesting technologies for mass fruit removal started in the late 1950s (Lamouria *et al.*, 1958) and early 1960s (Coppock, 1961). Many ingenious fruit removal techniques – too many to review in detail – have been considered. For thorough reviews of the subject the reader is referred to articles by Peterson (2005), Sanders (2005), and Li *et al.* (2011).

Much of the earliest research was focused on mass harvesting of fruit for the process market in California and Florida. Adrian and Fridley (1961) developed an inertia tree shaker that was tested in prune, peach, and olive groves in California. Coppock (1961) applied a similar design in Florida citrus groves. Multiple projects used direct contact between the machine and fruit in order to separate the fruit, such as rotating spindles (Coppock, 1961) or rods (Peterson, 1982) that press on the fruit. However, most mass fruit removal methods have implemented some type of a mechanical vibration generator. In brief, a mechanical device applies force to the tree whereby the resulting reactionary force causes separation of the fruit. Some of the general fruit removal methods considered for oranges included detaching the fruit by shaking the tree (Coppock and Hedden, 1968), shaking the branch (Sumner and Hedden, 1975), and oscillating forced air (Whitney and Patterson, 1972). Similar mass removal techniques that used shaking have been studied for deciduous tree fruit, such as apples (Upadhyaya *et al.*, 1981; Peterson *et al.*, 1994), sweet cherries (Markwardt *et al.*, 1964; Halderson, 1966), and peaches (Webb *et al.*, 1973; Peterson and Monroe, 1977).

As would be expected, mechanical harvesting research was initially empirical in nature. The growing collection of experimental data in the 1960s led to the development of theoretical models describing the relationship between machine and crop. Diener *et al.* (1965) studied the vibration characteristics of trellis-trained apple trees and classified fruit movement into five modes of oscillation. Rumsey (1967) completed a theoretical analysis of steady-state forced vibration of the fruit-stem system in oranges. Perhaps one of the most influential analyses in mechanical harvesting research was the linearized, three degrees-of-freedom (DOF) model of the fruit-stem system developed by Cooke and Rand (1969). This model was used to accurately predict the frequencies for fruit separation with or without the stems. Optimization of machine dynamics based on fruit detachment models (Gupta *et al.*, 2016) remains an active field of research today.

For political reasons summarized by Peterson (1992), publicly funded research of mechanized fruit harvesting declined in the USA in the late 1970s. Up until the 1980s, researchers adapted their mechanical harvesters to the existing design of freestanding trees. However, it was recognized that crop cultural/canopy modifications were required to increase the commercial viability of mechanized fruit harvesting (Peterson, 1985). Researchers, most of whom were located outside the USA, began to consider a systems approach where the tree design was modified so as to increase compatibility with the machine. Examples include the Tatura Trellis training system (Gould *et al.*, 1986) and the Lincoln Canopy System (Dunn and Stolp, 1980; Domigan *et al.*, 1988). Because modern apple orchard systems have led to improved fruit quality and yield and are thought to be more compatible with mechanized harvesting systems, the current industry trend is increased plantings of formal orchard architectures with planar tree canopies (Sections 10.2.5 and 10.2.6).

10.3.1.2 Robotic harvesting

Aided by continued advances in processor and sensing technology, in the 1980s scientists and engineers began research and development of robotic technologies for selective fruit harvesting (Grand D'Esnon, 1985; Harrell and Levi, 1988). The typical approach was to use vision sensors to identify and localize the fruit and a mechanical manipulation system to detach the fruit from the tree. The earliest fruit-picking robots typically incorporated articulated or spherical manipulators with three to four DOF. Grand D'Esnon *et al.* (1987) developed a prototype (MAGALI) harvester with a spherical manipulator that could execute one prismatic movement and two rotations. Developed in France, the system used a suction end-effector to detach apples. In 1986, the University of Florida developed a spherical, three-DOF manipulator actuated with servo-hydraulic drives for citrus harvesting (Harrell and Levi, 1988). The harvesting efficiency for the early prototype systems was approximately 75%. This relatively low performance level was attributed to poor fruit identification and the inability to negotiate natural obstacles inside the tree canopy (Sarig, 1993), due to the manipulator's limited degrees of freedom.

In different laboratory and orchard experiments, robotic harvesters studied in the past have achieved approximately 50–75% fruit detection accuracy, a slightly lower percentage of harvested fruit, and a harvesting speed of one fruit every 3–10 s (Grand D'Esnon *et al.*, 1987; Harrell *et al.*, 1989; Rabatel *et al.*, 1995; Muscato *et al.*, 2005; Li *et al.*, 2011). Detection and harvest accuracies varied substantially with different lighting conditions and different types of orchards, and a substantial proportion of harvested fruit was downgraded because fruit were detached with bruising and/or without the stem.

It is important to note that researchers in the 1980s deliberately designed harvesting manipulators with less coupling between the degrees of freedom so that the control system of the mechanical components could be simplified, thereby enabling dedication of most processor capacity to the vision system (Muscato *et al.*, 2005). Because of improvements in processor speed, modern control systems can solve the inverse kinematics problem much faster than their predecessors. Kinematically redundant harvesting manipulators with more than six DOF have been reported (Baur *et al.*, 2012). For an excellent review of the past 30 years of research on harvesting robots, see Bac *et al.* (2014).

10.3.2 Theories and principles

The highly unstructured agricultural orchard poses unique engineering challenges compared with the orderly manufacturing lines of industrial factories where automation technologies have augmented or altogether replaced manual labor. The basic functional requirements of a harvesting machine are to efficiently detach and collect fruit from the tree in such a manner that the product meets market acceptance requirements. When designing mechanized harvesting technology, the following principles should also be considered.

10.3.2.1 Fruit quality

Tree fruit are sensitive products that can be damaged easily. The harvested fruit must meet market acceptance criteria for size and external defects, such as bruising and punctures. Mechanized harvesting has been more difficult to implement for fresh market fruit because of the lower tolerance for bruising and external defects compared with the process market.

10.3.2.2 Yield benefits

The harvesting method should not negatively impact overall yield. For example, one issue reported with shake-and-catch systems is a negative impact on the following year's crop due to removal of immature green fruit (Li *et al.*, 2005). Damage to the tree is also undesirable. Likewise, the removal of fruit spurs along with the harvested fruit should be avoided, as it reduces the number of fruiting positions for the next crop.

10.3.2.3 Machine/orchard compatibility

Developments in the 20th century in technologies for the mechanical harvesting of tree fruit would have been more robust if the shape and structure of the target plants had been initially adapted to the requirements of the proposed harvesting machine. However, this adjustment would require the total transformation of orchards, which was considered an unacceptable risk because of the expenses involved in making a change for an unproven technology (Muscato *et al.*, 2005). For today's orchards, the harvesting machine must be able traverse orchard rows, and the kinematics of the machine should be compatible with tree size, tree spacing, tree shape, and spatial fruit distribution.

10.3.2.4 Handling and storage

The harvested fruit must be collected and stored before it can go to market. Mass mechanical harvesting systems commercially implemented, such as harvesters for tart cherries and prunes, use collection surfaces to catch falling fruit. If a proposed system allows the fruit to fall to the ground, human labor may still be required to pick up the fruit from the ground and place it in a storage container. Robotic systems selectively harvesting individual fruit also require an efficient method of transportation and storage.

10.3.2.5 Cost

Detailed economic assessments, such as the one completed by Gallardo and Brady (2015) for mechanized platforms, are required to properly determine commercial viability of harvesting technology. Mechanical performance measured through the contexts of picking times, overall throughput, and product damage is a major consideration during evaluation of the system's overall economic feasibility. Depending on the terrain and labor support available, a relatively robust and simple design could potentially be another system requirement. For example, because agricultural technicians responsible for maintaining equipment may not possess the same level of robot-specific training as industrial robotics technicians, maintenance requirements and reliability impact cost evaluations.

10.3.3 Latest developments

Mechanized systems have been commercially implemented for the mass harvesting of process market fruit. In addition to olives and nuts, which are not prone to mechanical damage, two of the most prominent success stories are tart cherries in Michigan and prunes in California. Historically very labor intensive to harvest, the majority of both crops are now harvested mechanically (Peterson, 1992) using shake-and-catch systems.

Commercial shake-and-catch systems have also been used to harvest oranges in Florida for the juice market. In 2008–2009, mechanical harvesting of Florida oranges peaked at 35,600 of the 496,518 total acres, or approximately 8% (USDA-NASS, 2010). However, mechanical harvesting of Florida oranges has been on a downward trend over the past several years. One reason for this trend is growers' perception that the visible scarring caused by mass mechanical harvesting, like removed leaves, broken branches, and bark scuffs, adds to the physiological stress of trees already suffering from citrus greening (Moseley *et al.*, 2012), a widespread bacterial disease presently impacting the industry.

Scientists and engineers continue to actively research and develop mechanical harvesting systems for fresh market crops (De Kleine and Karkee, 2015a). However, mass mechanized harvesting has not yet reached commercial viability for fresh market tree fruit, primarily because of excessive fruit damage. The most common causes of damage reported are fruit-to-fruit and fruit-to-branch contact during shaking, fruit-to-branch contact during falling, and fruit-to-fruit contact on the catching surface (Peterson, 2005). Minimizing damage caused by fruit-to-fruit/branch (Zhou *et al.*, 2016a) contact as well as damage resulting from fruit impact with the catching surface (Zhou *et al.*, 2016b) are subjects of ongoing research.

An early review of agricultural robotics research (Sarig, 1993) noted that the community was generally optimistic that harvesting systems utilizing robotics technology might be commercially available by the end of the 20th century. Over 20 years later, there are still no known commercial implementations of robotic harvesting systems for any tree fruit, whether for the process market or fresh market (Bac *et al.*, 2014). The absence of available systems is not for lack of effort, as numerous robotic harvesting projects have been undertaken over the past 30 years. Rather, the unstructured orchard environment has proven very challenging for robotics technology. Some of these challenges are variable outdoor environmental conditions, complex tree structures, fruit clusters and occlusion, inconsistency in fruit shape and size, and fruit sensitivity to damage.

As the operating environment is complex and challenging, most of the prototype robots evaluated in the past generally lacked the desired level of robustness and productivity. Robust fruit detachment is the central component of the robotic harvesting technology and remains as the most challenging issue that researchers must address to increase the likelihood of commercial adoption. Most reported studies used modified industrial robots to detach a fruit from the tree, which could be one of the major attributors to the resulting performance deficiency as such systems may require more complicated grabbing and detaching motions for fruit picking than most industrial picking applications. Other important areas for further innovation would be to develop techniques to lower cycle times, and improve fruit detection and localization.

Academic institutions have conducted most agricultural robotics research over the past three decades. However, recognizing the industry need and magnitude of the labor problem, private commercial ventures have recently entered the field. Some of these companies include Abundant Robotics (Menlo Park, California), FFRobotics (Israel), and Energid Technologies (Cambridge, Massachusetts). The additional resources brought by private firms should help to realize additional technological advances in robotics systems for the harvesting of fresh market tree fruit.

10.4 Mechanical Harvesting Systems and Components

Contrary to the structured and well defined industrial environment where objects to be handled are generally sturdy with well defined shape, size, color and location, machines for fruit harvesting need to have the unique capacity to interact with complex and delicate plants and produce in unstructured outdoor environments. The machines need to be designed to handle these materials gently and robustly so that no plant or produce is damaged while harvesting. However, biological objects to be dealt with in fruit harvesting are complex, with variable and uncertain location, shape, size and color parameters. This section presents two types of harvesting systems and their components as well as different materials and methods that can be used to handle delicate plants and produce without damaging them.

10.4.1 Mass harvesting system components

As discussed in Section 10.3, mechanical or automated harvesting of tree fruit crops can be approached in one of two ways: mass harvesting; and individual fruit (robotic) harvesting. Components of individual fruit harvesting will be discussed in Section 10.4.2. Mass harvesting techniques (sometimes also called shake-and-catch harvesting) generally apply some type of shaking energy to trees to excite more than one fruit (often dozens or even hundreds of fruit) at a time (Fig. 10.7). When excitation energy exceeds the fruit's retention strength at stem–fruit or stem–branch junction, the fruit is detached from the branch. Detached fruit are then collected by specially designed catching mechanisms that try to minimize bruising, cuts and other types of harvest-induced fruit damage while maximizing the percentage of fruit caught by the mechanism. In the following subsections, different types of shaking methods and catching mechanisms are described.

10.4.1.1 Shaking parameters

Harvesting efficiency of a shaking mechanism depends on how efficiently the shaking energy is transmitted to different locations in the tree canopy. Different characteristics of a shaking operation can impact the efficiency of energy transmission and therefore efficiency of fruit removal. A shaking operation during fruit harvesting can be characterized primarily by shaking pattern, shaking frequency, shaking amplitude (or stroke), shaking location, shake type and duration.

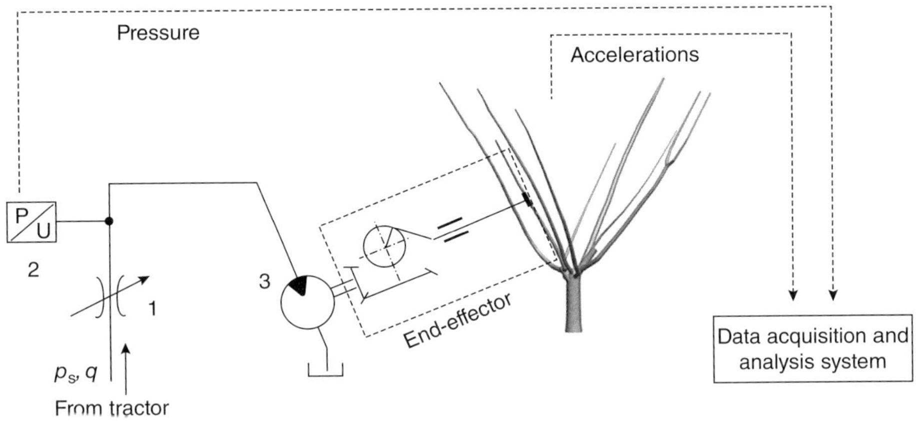

Fig. 10.7. Block diagram of a typical hydro-mechanical shaking mechanism (from He *et al.*, 2013). Data acquisition and sensing components are used for system performance evaluation. 1: A flow control valve; 2: A pressure sensor; 3: A hydraulic motor.

'Shaking pattern' defines the three-dimensional displacement of the shaking end-effector during its engagement with canopy components (e.g. a branch or trunk). For canopy shaking, multiple individual shaking units can use the same or different patterns. A linear horizontal signal is the most common shaking pattern and is intended to shake canopy components back and forth in the orthogonal direction of the vertical plane of the canopy. A vertical linear shaking could be another potential pattern. In addition, various other shaking patterns such as the ones depicted by Fig. 10.8 can be used or studied to improve fruit detachment efficiency and reduce harvest-induced fruit damage for different types of crops grown in varying crop canopies.

'Shaking frequency' defines how long it takes to complete a cycle of the shaking pattern, whereas 'shaking amplitude' or stroke defines how large the displacement is for a given pattern. For a linear shaking pattern, amplitude represents how far the end-effector moves in both directions from its neutral position. The shaking frequency f is calculated as $f = 1/T$, where T= period in seconds.

The amount of input energy necessary to detach a given fruit depends on many factors, including crop types, varieties, and canopy architecture. An optimal level of shaking energy can be transmitted to the tree by adjusting shaking duration in addition to frequency and amplitude. An intermittent shaking (turning the shaker on and off several times) can sometimes increase the detachment efficiency, as intermittent signals introduce an infinitely large number of shaking frequencies into the canopy (He *et al.*, 2013).

'Shaking location' in a tree canopy is another important parameter for shake-and-catch harvesting, as it makes a large impact on how efficiently the energy is transmitted from the shaking mechanism to the desired fruit locations. Naturally, the closer the fruit are to the shaking location, the

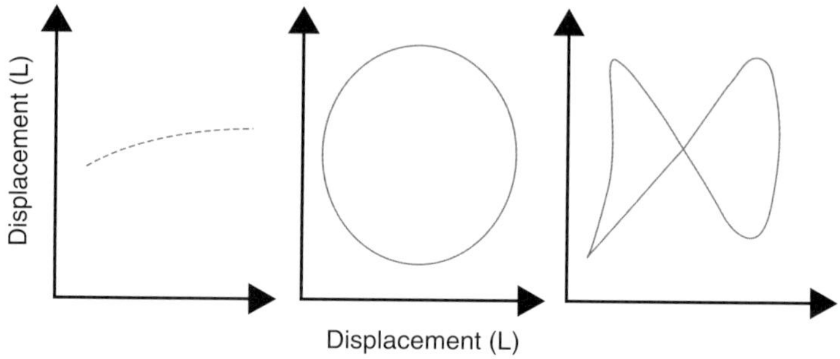

Fig. 10.8. A few possible patterns for shaking fruit trees and branches with a shake-and-catch harvester (from De Kleine and Karkee, 2015a).

higher the transmitted energy will be to fruit locations, leading to higher likelihood of fruit removal. Both distribution of fruit (Zhou *et al.*, 2014) and branch geometry need to be considered in selecting one or more shaking points that would achieve the desired level of fruit removal efficiency (Zhou *et al.*, 2014).

10.4.1.2 Canopy shaking mechanisms

Excitation energy can be transferred to fruit through application of a shaking signal, such as canopy shaking, trunk shaking and targeted branch-level shaking, at different levels in the tree canopy. Canopy shaking is the widest indiscriminant application of shaking energy to tree canopies that may occur through contact between shaking parts of the machine and various canopy components, including the trunk, branches and fruit. Some of the early efforts in developing mass harvesting of fruit investigated different types of machines to shake the entire tree canopy using finger-rods (Sumner, 1973) (Fig. 10.9) or horizontal bars. In the finger-type machine, dozens of soft fingers are inserted into the canopy and are then vibrated in different directions using signals of varying frequencies and patterns. Alternatively, a number of horizontal bars are pressed against the canopy and a certain level of vibratory signal (different level of frequency and stroke or amplitude) is applied to these bars. The energy from these devices (input units) is transferred into the canopy through the shaker and various components of the tree canopy. The energy is then transmitted to fruit locations through the vibration of canopy components, primarily the branches. As the process is non-selective, some fruit may come in direct contact with the shaking devices.

10.4.1.3 Trunk shaking mechanisms

Canopy shaking is a good method when the tree or bushes are big and transmitting a sufficient level of energy to individual fruit would not be efficient by shaking a trunk or a number of small stems. However, canopy shaking may cause some damage to fruit and other canopy components due to direct contact with the shaking mechanism. For crops with suitable

Fig. 10.9. Canopy-shaking citrus harvester Oxbo 3220 (Oxbo International Corp, from http://www.oxbocorp.com/Portals/0/Oxbo/Brochures/citrus2013.pdf).

canopy and trunk size, shaking energy can efficiently be transmitted to fruit locations using trunk shakers, which can also be simpler and more cost effective compared with a canopy shaker. For example, dwarf trees in central or multi-leader orchards (Sections 10.2.1 and 10.2.3) can provide an appropriately sized trunk or leaders for application of the amount of energy required to maximize fruit separation while keeping fruit and tree damage at a minimum level. To harvest fruit, a relatively powerful shaking end-effector is engaged with the tree trunk or leaders at a location that would optimize the energy transmission (generally 0.5–1.0 m above the ground surface), and a specific shaking signal (defined by frequency and amplitude) is applied for a given duration or until no more fruit are removed from the branches.

10.4.1.4 Branch shaking mechanisms

Trunk shaking generally requires a high level of energy to be applied to the trunk so that even the farthest fruit in the canopy can get a sufficient level of excitation to overcome the fruit retention force. When enough shaking energy is generated at the tip on long thin branches, shorter and thicker branches might get more energy than necessary to detach a fruit. A higher level of excitation in these locations can cause unnecessarily higher levels of velocity and acceleration of the detached fruit, causing fruit to jump outside the catching surface and, therefore, be more likely to be damaged during catching. To optimize the level of excitation for fruit at all locations in the canopy, it is appropriate to apply the correct amount of shaking energy to individual branches. In conventional fruit crop orchards (Section 10.2.4) with large and bushy trees (e.g. central leader), shaking individual tree branches might be challenging. However, in modern fruiting wall orchards with dwarf trees, it is possible to select individual branches and apply shaking energy sufficient to detach fruit in only the branch under consideration (Fig. 10.10). Through careful selection of frequency and amplitude of the shaking signal, variety-specific optimal shaking may occur, thereby leading to high levels of removal efficiency with fruit damage levels maintained at practically acceptable levels (He *et al.*, 2013). More interestingly, shaking energy could potentially be optimized to detach only the ripe fruit and leave the remaining

Fig. 10.10. Washington State University researchers conducting branch shaking tests in a commercial orchard using a handheld shaking device.

for another round of harvesting, which is important for many crops, such as some apple varieties.

10.4.1.5 Catching mechanisms

Fruit catching is an integral part of the mass harvesting technique. Two of the major considerations in designing a catching mechanism are: (i) to collect as many harvested fruit as possible' and (ii) to maintain fruit damage at a practically acceptable level. Location (relative distance between fruit in the canopy and catching surface), orientation, shape, and size of the catching mechanism play important roles in achieving these goals. A few important features of an effective catch mechanism include: (i) rapid singulation of fruit to minimize fruit-to-fruit contact during catching; (ii) reduction of flight distance before fruit land on a catching surface; and (iii) sufficient cushioning to reduce the magnitude of impact during fruit-to-catching surface and, potentially, during fruit-to-fruit contact. Many of these parameters are constrained by the specific shaking method used in a harvesting machine. When a canopy shaking mechanism (e.g. the citrus harvester shown in Fig. 10.9) is used for harvesting, a catching mechanism generally consists of a series of flaps installed at the bottom of the machine. When branch-level shaking is performed, the catching mechanism can be developed such that fruit can be caught with a relatively small catching mechanism, with a cushioning and singulation mechanism installed on the catching surface.

10.4.2 Robotic harvesting systems components

A robotic system designed for individual fruit picking (sometimes also referred to as the pick-and-place method) will rely on some sort of sensing, control, manipulation and end-effector technologies to pick individual fruit and place them in a container or conveying mechanism. Sensing and manipulation mechanisms can be designed so that a robotic system can selectively harvest individual fruit based on maturity parameters, such as color and size. The following sub-sections describe some of the major components of a robotic harvesting system.

10.4.2.1 Sensing and machine vision

Robotic harvesting systems commonly use machine vision and/or other types of 3D sensing methods for fruit detection and localization. One or more color cameras and other sensors can be installed in the robotic system at different locations and orientations so that the crop canopy can be viewed from one or more directions. Fruit detection is primarily based on color, shape, edge, and texture parameters (Hannan and Burks, 2004; Tabb *et al.*, 2006; Bulanon and Kataoka, 2010; Silwal *et al.*, 2014). A charge coupled device (CCD) camera can be used to capture images that will be analyzed using various image-processing techniques for background segmentation, scene classification, and object detection. An object(s) of interest could simply be a fruit (if no obstacle is expected to be interfering with the harvesting robot) or other canopy components like the trunk and branches could be potential obstacles that need to be avoided. Some of the techniques used for 3D localization of fruit include a laser sensor (Jiménez *et al.*, 2000), a stereo-vision system (Tarrío *et al.*, 2006), or a 3D camera (Silwal *et al.*, 2017). Laser sensors provide the most accurate 3D information, but they are relatively slow and expensive. Stereo vision provides information for both detection and localization of fruit, but its localization accuracy is limited. Recently, 3D camera technology has shown improved localization accuracy, but the technology has its own limitations such as low resolution and the need for registering with color camera images for fruit detection.

During field use, sensing and machine vision systems can suffer from various factors such as variable lighting conditions, fruit clustering, and occlusion of fruit by leaves, other fruit, and branches. With the modern fruiting wall canopy architectures (Section 10.2.4), fruit occlusion has been reduced. In addition, practical techniques such as hierarchical sensing and harvesting methods (Silwal *et al.*, 2016b) can be used to minimize problems caused by occlusion and clustering. In this method, only clearly visible fruit are harvested at the first iteration. As some fruit are harvested, the remaining fruit become better exposed to the machine, which makes detection and harvesting easier in successive iterations.

10.4.2.2 End-effector

Another important component of the robotic harvesting system is an end-effector, which is the only part of the robot that actually contacts the fruit. Various types of end-effector have been designed and evaluated in the past for robotic apple harvesting (Bulanon and Kataoka, 2010). Mechanical and pneumatic fingers, mechanical cups, scissors and vacuum-based methods are some of the end-effector techniques evaluated around the world. There are various pros and cons for each technique. End-effectors with a few fingers can offer higher degrees of freedom to achieve different types of motion for effective fruit removal with no or minimal damage. However, these end-effectors tend to be slower compared with vacuum-based or scissor end-effectors. Vacuum end-effectors can be fast but handle fruit less gently, thus potentially causing more fruit damage. Scissor end-effectors could be faster than fingers, but require a sensing system to locate the fruit stem, which is not a trivial task. In general, end-effector technology remains one of the more challenging areas of research for successful robotic harvesting of fresh market tree fruit crops.

10.4.2.3 Manipulation

After visual detection and localization of the fruit in the canopy, it is the task of the manipulator to move the end-effector to and from the desired locations for harvesting each targeted fruit. Although manipulators are considered universal devices, economy and intended application often dictate and influence their design (Craig, 1990). Apart from this, other factors such as payload capacity, size, speed, number of joints, and geometric arrangement impact important aspects of manipulator design, such as the reachable workspace and mechanical stiffness. There are several different varieties of manipulator, such as Cartesian, cylindrical, polar, SCARA (Selective Compliance Assembly Robot Arm), and revolution manipulators (Pires, 2007). Mechanical linkages in the manipulator are connected using different possible joints, such as revolute, cylindrical, prismatic, spherical, and planar joints. The manipulators seen in agricultural research and applications are a combination of one of the several joint parameters and rigid linkages at varying degrees of freedom.

Selection of the appropriate manipulator for specific tasks is a factor that affects the overall performance of a robotic system (Edan and Miles, 1994). Most projects incorporate kinematic designs based on qualitative assessments of the workspace. The design parameters described in the above paragraph are closely related and optimal selection is a difficult task. Often, engineers and roboticists rely on computer simulation to optimize parameters. It is important to note that the number of degrees of freedom, which indicates the complexity of the manipulator design, has direct influence on the complexity of the control system, speed, and optimization of the design. As an example, to simplify the machine, the earliest tree fruit-picking robots typically incorporated articulated or spherical manipulators with just three to four DOF (Grand D'Esnon *et al.*, 1987; Harrell and Levi, 1988)

10.4.2.4 Path planning and inverse kinematics

A general approach to cause a manipulator to move from one point to another (or from source to destination point) is to move each joint following a specific function of time. The path that describes the journey to its destination is a set of intermediate locations through which the manipulator/end-effector has to pass. Each and every point in the path is converted into a set of joint movements that drives individual actuator accordingly. For smooth operations, it is necessary that each joint starts and ends its motion at the same time. Often the calculation of motion function is a problem of trajectory generation (Craig, 1990). Agricultural workspace, which is often described as a highly unstructured environment, requires complex motion planning especially for difficult tasks such as obstacle avoidance.

Calculation of all possible joint positions and velocities to achieve a desired end-effector position and orientation, such as the location of a fruit in the canopy, is a fundamental problem in kinematics and use of manipulators in general. In general, the position and orientation of manipulator and end-effector are not computed directly (Pires, 2007). Instead, inverse kinematics is used to compute the end-effector's desired location. It is computed using the number of DOFs, which are the total independent variables (i.e. individual joint positions), and the kinematics of the robot to obtain joint angles of the manipulator. The task of obtaining the inverse kinematic solution for a general manipulator is more complex than forward kinematics. Often, the solution requires solving non-linear equations, and in many cases a close-form solution may not exist. Even with the increased computational power that has significantly decreased numerical computation time, the determination of feasible solutions to the inverse kinematics problem is not always significant. In a given manipulator design, the existence of the solution defines the reachable workspace of the manipulator (Craig, 1990).

10.4.3 Materials and methods for fruit collection and handling

Various factors, including fruit type (fruit parameters such as firmness and shape), fruit weight, drop height, catching-surface materials, orientation of the catching surface, and catching-surface shape and size play important roles in improving catching efficiency and reducing fruit damage. Catching-surface material should be selected such that it increases the impact duration (reduced peak acceleration) when a fruit hits the catching surface, while minimizing fruit bouncing. A smaller bouncing distance is important to minimize the likelihood of contact with other fruit as they fall into the catching surface. In other words, a good cushion material will be capable of absorbing most of the kinetic energy that a fruit has when it hits the catching surface so that it does not get damaged and/or bounce back. Any fruit can sustain a certain amount of stress before its cell wall is damaged and the fruit becomes bruised. If the catching surface can keep the compressive stress on the fruit surface within that threshold, no bruising occurs.

10.4.3.1 Foam materials

Foam materials provide good cushion during impact, which makes these materials attractive for fruit catching and handling. When a foam material is impacted, compressive stress and strain follow a trend with three specific zones (Avalle *et al.*, 2001; Zhou *et al.*, 2016b) (Fig. 10.11). Depending on various parameters of the target fruit and harvesting system (e.g. shape, size, density, firmness, and drop height), different types of foam material may be appropriate for disseminating sufficient impact energy without damaging the fruit. The best energy-absorption capacity of the foam material occurs when the impact process falls in the plateau region of the material. For handling a specific fruit crop, drop height and fruit weight play important roles defining which surface material can provide sufficient cushion. As the drop height and/or fruit weight is increased beyond a certain threshold, the kinetic energy of the fruit before impact may reach a level that can push the foam material to its densification zone, particularly when it is softer (lower firmness) (Fig. 10.11). If cushion material reaches the densification zone during impact, compressive stress is built up rapidly, thus increasing the likelihood of fruit damage.

10.4.3.2 Non-Newtonian fluid

Non-Newtonian shear-thickening materials have unique rheological characteristics (Brown and Jaeger, 2014) that can provide improved cushion

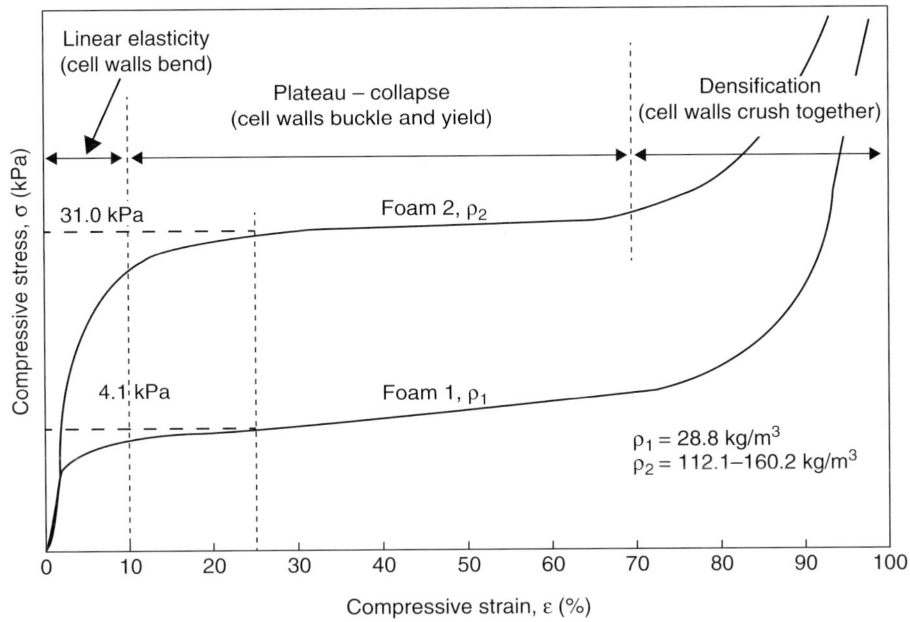

Fig. 10.11. The stress–strain diagram of two cushion materials used in Zhou *et al.* (2016b).

and deceleration during impact, thus reducing the likelihood of fruit damage during. A fruit drop test was conducted by De Kleine and Karkee (2015b) to measure the performance of a non-Newtonian shear-thickening fluid (corn starch and water mix) in catching peaches, pears and apples. The material performed better than conventional foam materials in disseminating energy and minimizing fruit damage during impact. For several varieties of apples tested, no bruising was found when the apples were dropped from heights up to 1.2 m on a plastic bag filled with the corn starch and water mixture. This finding may lead to mass harvesting systems for tree fruit crops with practically acceptable fruit damage levels. However, further study is necessary to develop materials or methods to make non-Newtonian shear-thickening materials practically useful in fruit harvesting machines.

10.4.3.3 Air suspension

Another potentially revolutionizing method for fruit catching could be the use of pressurized air to create a suspension zone above a catching surface. Such a cushion reduces impact duration and peak acceleration for both fruit-to-catching-surface and fruit-to-fruit contact. No other catching mechanism previously discussed can reduce the impact duration during fruit-to-fruit contact. A prototype air-suspension catching device evaluated by Ma *et al.* (2016) (Fig. 10.12) showed a substantial reduction in fruit damage during harvesting compared with a catching surface without air suspension.

Fig. 10.12. An air-suspension catching device.

10.4.3.4 Fruit transportation and bin filling

Mechanical conveyance, water conveyance and vacuum transportation are the three major ways fruit can been conveyed in the field from the fruit collection point to the bin or container. Various orchard platforms and harvest aid machines have already implemented some of these techniques for commercial adoption. Water conveyance is a convenient way to transport fruit with minimal impact and damage. Packing houses use water conveyance widely to move fruit through various grading, sorting and packaging units. Some researchers have also studied the use of water for fruit transportation in orchards. However, issues with cross-contamination of produce as well as the logistics of handling large volumes of water in the field environment still need to be addressed for this method to be practically feasible. Vacuum transportation is another method that has been practically adopted in orchard platforms (Fig. 10.13). As fruit are accelerated during vacuum transport, a deceleration mechanism is used before a container is filled with those fruit; deceleration can be a challenge with fruit that are particularly susceptible to bruising (Luo *et al.*, 2012).

All of these conveyance or transportation systems feed fruit to a bin/container-filling mechanism so that fruit can be gently downloaded on the container. Elephant-ear bin filling has been widely used because it can evenly distribute fruit into a container (Schupp *et al.*, 2011) (Fig. 10.14). These bin-filling mechanisms are designed to maintain a low height above the fruit-fill height in the container so that fruit-to-fruit impact is negligible.

Fig. 10.13. A harvest assist machine with vacuum transportation system (from De Kleine, 2014).

Fig. 10.14. Elephant-ear distributor in a bin-filling system (from Schupp *et al.*, 2011).

A simple laser sensor could be used to detect the depth to the fruit surface, which then can be used as an input to a control system that can raise the filling mechanism dynamically as the container gets filled.

10.4.4 Performance evaluation of harvesting systems

Performance of a fruit harvesting system is evaluated based on the efficiency of the system in removing, collecting and maintaining fruit quality. Some of the important performance measures are defined below.

10.4.4.1 Fruit removal efficiency

Fruit removal efficiency is defined as the ratio of the number of fruit removed from the defined harvest zone to the total number of the fruit within the zone, which is given by:

$$\eta_r = \frac{N_r}{N_t} * 100\% \tag{10.1}$$

where η_r = removal efficiency, N_r = number of fruit removed by the harvester, and N_t = total number of fruit within the investigated zone.

10.4.4.2 Fruit collection efficiency

Fruit collection efficiency is defined by the ratio of the number of collected fruit and number of removed fruit, as given by:

$$\eta_c = \frac{N_c}{N_r} * 100\% \tag{10.2}$$

where η_c = collection efficiency, N_c = number of collected fruit using the catching mechanism, and N_r = number of fruit removed by the harvester.

10.4.4.3 Fruit damage percentage

Fruit damage percentage is given by:

$$\lambda_d = \frac{N_d}{N_c} * 100\% \tag{10.3}$$

where λ_d = damage percentage, N_c = number of fruit collected by the catching surface, and N_d = number of fruit damaged during harvest.

Fruit damage percentage can be further divided into different levels of damage based on specific government or local regulations. For example, the US Department of Agriculture (USDA) provides standards to evaluate fruit damage for apples, cherries and other crops, and to classify into various groups such as ‘Extra Fancy’, ‘Fancy’ and ‘Culls’. The total number of fruit that can be sent out to the market is defined as market-quality fruit, which excludes culls from the total collected fruit.

The overall harvest efficiency of the harvesting system can then be defined as the ratio of the number of market-quality fruit and the total number of fruit in the canopy, represented as:

$$\eta = \frac{N_m}{N_t} * 100\% \quad (10.4)$$

where η = overall harvest efficiency, N_m = number of market-quality fruit, and N_t = total number of fruit in each canopy.

Using equations 10.1 to 10.3, equation 10.4 can be represented as:

$$\eta = \lambda_r * \lambda_c * (1 - \lambda_d) \quad (10.5)$$

Depending on fruit size and harvesting method, these quality measures can also be estimated using bulk weight of the fruit rather than individual counts (e.g. cherries).

10.4.4.4 Fruit removal condition

The condition of stem attachment to fruit has also been one critical parameter defining the performance of a harvester. In general, stem attached to fruit such as cherries and apples is considered the best outcome. There have been concerns in the past with potential degradation of the fruit as well as market acceptability when no stem is attached. However, some recent studies have indicated that stem detachment from different varieties of apple does not lead to any adverse effect on fruit quality and storability. Similarly, there have been findings that stem-off cherries are liked equally in various regions of the USA and Europe. In that regard, stem detachment status may not be an important parameter to evaluate a harvester. Finally, some harvesting machines may damage trees, which needs to be considered during design, development and operation.

10.4.4.5 Throughput and cycle time

For mass harvesting systems, ‘throughput’ can be defined as the number of fruit harvested per unit time. Commercial harvesters are also evaluated using machine productivity, which is defined as the area harvested by the machine per hour of harvest time. For a robotic harvester, ‘cycle time’ is an important performance measure, which is defined as the average time taken by the robot to harvest one fruit. Cycle time will include the time taken by the sensing and machine vision system to detect and localize the fruit, time spent in inverse kinematic computations, time taken by the manipulator to approach the fruit, time taken by the end-effector to detach the fruit, and the time taken by the manipulator and end-effector to move the fruit to a collection area and drop it for storage. From the standpoint of practical adoption, harvesting cost per piece of fruit is an important measure, but such a measure can be estimated only when both cycle time or throughput and cost of the machine are known.

Fig. 10.15. Different types of harvest-induced fruit damage in apples (from De Kleine and Karkee, 2015a).

10.4.5 Comparison of fruit harvesting techniques

The shake-and-catch harvesting technique is relatively simple and can be highly labor efficient and cost effective compared with manual or robotic harvesting. Mass harvesting of fruit trees with shake-and-catch techniques has been quite successful for crops destined for the processing market, such as olives and juicing citrus, and for crops with hard shells, such as nuts. However, when it comes to harvesting fresh market crops such as sweet cherries and apples, the technology has been limited by the unacceptable level of fruit bruising, cuts and other types of harvest-induced damage (Fig. 10.15). For some varieties of crop, detachment of fruit with shaking can also be limited (He *et al.*, 2016). Recent studies at WSU looked at different strategies for reducing the level of fruit damage while increasing fruit detachment efficiency. The adopted technique has shown good promise for some varieties of apples and cherries grown in a fruiting wall architecture. WSU's efforts on cherry harvesting will be presented in Section 10.5.1 as a case study for a shake-and-catch harvesting system.

Implementation of the robotic harvesting technique, on the other hand, has suffered from the complexity of the machine, speed and cost per unit of harvested fruit. The system hierarchy consists of several sub-systems within vision and manipulation. To improve robustness, redundancies are embedded into the design within each sub-process at the cost of computational time. The design of a successful harvesting system requires a proper balance between speed, cost, and robustness (Silwal *et al.*, 2016a) and sometimes these factors are contradictory to each other. Further discussions on the pros and cons of both shake-and-catch and robotic harvesting methods can be found in Sections 10.5, 10.6 and 10.7.

10.5 Case Study: Shake-and-catch Cherry Harvesting

10.5.1 Introduction

Manual sweet cherry harvesting is a highly labor-intensive operation, as the fruit are small and distributed throughout large canopies. Even the modern dwarf trees in fruiting wall canopies are 12–14 ft (3.7–4.3 m) tall, requiring the use of ladders or platforms to pick fruit. A large seasonal labor force is used in sweet cherry production areas around the world, including eastern Washington State. While improvement in machines that have been used by some growers producing tart cherries is essential for their wider adoption, development of practically adoptable harvesters for sweet cherry crops is considered critically important for the long-term sustainability of this industry in the light of decreasing availability and increasing cost of human labor. Researchers and engineers around the world have long been investigating methods to harvest cherries with shake-and-catch and other methods. Primarily, two philosophies of mechanization are in play in developing such machines: (i) assisting labor to be more productive and safe; and (ii) developing machines that would completely mechanize/automate the harvesting operation.

Development of a sweet cherry harvesting system resulting in minimal fruit damage requires understanding of fundamental knowledge on: (i) what is the achievable productivity and fruit quality from a shake-and-catch process, and how to control major parameters for achieving such a result; and (ii) what tree architectures could optimize horticultural attributes and also be machine-friendly for automated mass harvesting. Given the multifaceted nature of the above identified problems, the WSU CPAAS team used a truly trans-disciplinary approach to gain a comprehensive understanding from physical and biological aspects and used the gained knowledge to create possible solutions for removing or minimizing the specified challenges. In this section, a case study of two types of harvesting machines investigated at WSU over the past several years will be presented.

10.5.2 Optimizing shaking parameters for cherry harvesting

Studies have found that fruit removal efficiency for sweet cherry crop is heavily dependent on crop cultivar and canopy architecture (Du *et al.*, 2013). Varieties such as 'Van' and 'Skeena' have lower pedicel–fruit retention force (PFRF) at harvest and are therefore easier to detach compared with varieties such as 'Bing' and 'Chelan' (Smith and Whiting, 2010). In terms of canopy architecture, UFO trees trained in Y-trellis canopies can offer leaders or trunks of appropriate size for efficient transmission of shaking energy to fruit (Long, 2010; Du *et al.*, 2012; Zhou *et al.*, 2014). In contrast, more conventional architectures such as central leader trees would offer highly variable branch size and location, making it difficult for optimal shaking. One set of harvesting system parameters does not fit all variations in crop cultivars and architectures. It is, therefore, important to optimize various shaking parameters for specific crop cultivars and architectures so that the desired level of harvesting efficiency and fruit quality can be achieved.

10.5.2.1 Shaking frequency

In sweet cherry harvesting, shaking frequency has been found to be the factor influencing fruit removal efficiency and quality the most. An impact harvester such as the one developed by Peterson and Wolford (2003) was suitable for shaking larger trees but was found less effective compared with a continuous sinusoidal shaking at a certain frequency (Chen *et al.*, 2012) when it was used in modern UFO orchards. Chen *et al.* (2012) and Du *et al.* (2012, 2013) studied energy transmission to branches with a linear shaking mechanism (see Section 10.4.1 for different types of shaking methods) of varying frequencies up to 50 Hz. Shaking with sinusoidal signal achieved comparatively more uniform and consistent transfer of energy to all measured locations in the tree.

It was found that branch size and its location in the canopy affect the resonance frequency of individual branches. In these studies, 6–10 Hz, 12–15 Hz and 24–26 Hz shaking were found to have resonant effects on most of the branches in UFO trees of 'Skeena' variety, which may be the most effective frequencies for detaching cherries. A study by Zhou *et al.* (2014) also found 14 Hz to be an effective frequency for fruit removal while optimizing fruit quality. However, another study (He *et al.*, 2013) used 18 Hz shaking to increase the removal efficiency for 'Bing' and 'Skeena' varieties.

10.5.2.2 Shaking location

Depending on the crop architecture, the number and location of shaking points per branch/leader can vary widely for efficient fruit removal. It was found that two shaking locations (roughly at one-third and two-thirds of the branch height) were sufficient for harvesting 'Chelan' cherries trained in a UFO Y-trellis system (Section 10.4.1) (Zhou *et al.*, 2014).

10.5.2.3 Shaking duration

Shaking duration can vary between different cherry varieties. Cherries with smaller PFRF, such as 'Van', were shown to be detached within a few seconds of shaking. Often, multiple intermittent shaking of 5 s each (totaling 20 s shaking duration) was necessary to achieve higher level of removal efficiency with varieties such as 'Bing' and 'Skeena' (He *et al.*, 2013).

10.5.3 Optimizing catching parameters for cherry harvesting

As discussed in Section 10.4.1, various factors, including drop height, catching-surface materials, orientation of the catching surface, and catching-surface shape and size, play important roles in improving catching efficiency and reducing fruit damage (Fig. 10.16). In one of the studies at WSU CPAAS, Zhou *et al.* (2016a) developed a laboratory set-up to evaluate different foam materials in terms of their performance in capturing sweet cherries. A fruit guide tube was installed on a vertical shaft to drop fruit vertically. The drop height could be varied by moving the guide tube up and down. A set of fruit was then dropped on a catching surface set at varying tilt angles to the horizontal surface. The catching surface was padded with foam materials of varying firmness, density, and thickness. Interaction between the fruit and catching surface was recorded using a high-speed camera able to record 1500 images every second, showing full details on how the deformation and bouncing process occurs during fruit catching. Changes in impact force were also recorded during the catching process. The study found that foam material with lower firmness (4.1 kPa at 25% deflection) provided better cushion for sweet cherries compared with one with higher firmness (31.0 kPa at 25% deflection) when the drop height was < 1.2 m. In angled fruiting wall orchards (V- or Y-canopy, see Sections 10.2.4–10.2.6), the catching surface and machine can be designed such that the drop height is limited to around 1.0 m. In such a case, the softer cushion material evaluated in this study could be a reasonable choice.

This study also evaluated the effect of varying tilt angle of the catching surface on fruit quality (Fig. 10.17). It was found that a higher tilt angle has the potential to reduce fruit damage, as it reduces the compressive stress on the fruit. However, it might cause fruit to bounce further,

Fig. 10.16. Examples of different types of cherry fruit damage (from Zhou *et al.*, 2014).

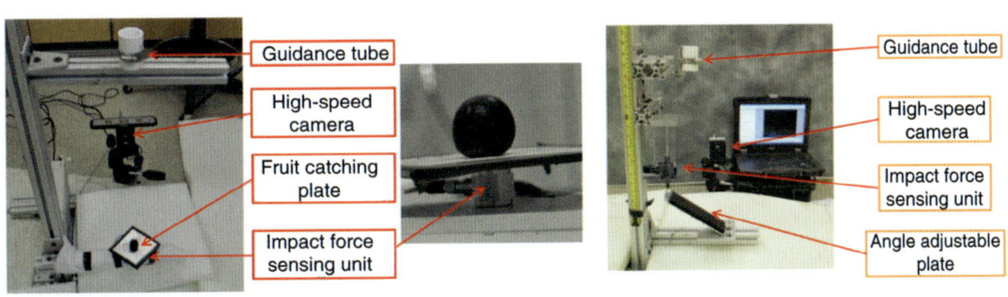

Fig. 10.17. Experimental set-up to evaluate effect of cushion materials and tilt angle of catching surface on sweet cherry fruit quality.

which has the potential to increase fruit-to-fruit contact. There can also be a practical limitation in terms of how steep a catching surface angle can be, depending on the type of orchard architecture.

10.5.4 Cherry harvesting systems

10.5.4.1 Hand-held tools for sweet cherry harvesting

Manual cherry harvesting involves frequent movement and placement of ladders, picking fruit while climbing up and down the ladder and carrying a heavy load (often more than 10 kg) of fruit in a bucket placed in the front of the picker, and getting off the ladder and walking back and forth to a bin or container to empty the bucket. This operation is not only labor intensive, it is also highly inefficient and involves a high risk of fall, and of repeated motion and other types of ergonomic injuries. Hand-held harvesting tools can be designed to improve the productivity of labor while avoiding or minimizing the use of ladders and avoiding or minimizing the need to walk to the fruit container. The tools also reduce the intensity of manual work in the field (i.e. no need to climb up and down the ladder with heavy loads) and improve worker health and safety. Hand-held devices can be an important tool for orchard operations, particularly for conventional canopies where machine harvesting (discussed in the following sub-section) could be challenging.

Over the past several years, WSU has investigated various types of hand-held shaking and catching mechanisms. These mechanisms were designed primarily as tools that a pair of laborers can use as a team to improve their productivity. One worker could operate the hand-held machine-assisted shaker, while another worker could operate the catching and fruit transportation mechanism. Workers switched roles multiple times over their shift to optimize productivity and minimize strain. Fruit catching surfaces were padded with foam material with a hole in the middle that led to the fruit transportation mechanism. Transportation mechanisms basically included a tube padded with some type of foam material and fruit were transported down through the tube using gravity. For example, Zhou *et al.* (2014) used a hand-held shaker and reached up to 97% removal efficiency for 'Chelan' variety when UFO cherry leaders/trunks were shaken at two locations. Fruit damage rate was around 23%, which is substantially higher than that of 13% with hand-picked fruit. As expected, fruit catching with hand-held devices is easier in well trained fruiting wall orchards. Further research is essential to improve the robustness and commercial adoptability of hand-help cherry harvesting tools.

10.5.4.2 Sweet cherry harvesting machines

For modern fruiting wall canopies, larger machines are more promising for mechanized or even complete automated harvesting with a potential for substantially reducing labor demand, and associated cost and health risks. One such harvesting machine (Fig. 10.18) was developed and evaluated by Peterson and Wolford (2003). The two parts of this mirrored harvester were both front-wheel steered with a driver seat at the rear end of the machine. Two operators worked from two opposite sides of the tree canopy in coordinating shaking and catching operations. A rapid displacement actuator (RDA), which was a hydraulically operated end-effector, rapidly displaced fruit-loaded leaders or trunks. The end-effector had a rubber padding to minimize tree damage. A cushioned catching surface was used in both machines to capture and convey fruit to a container or bin. Leaves and other dirt were blown away by a fan. Catcher pans were used to close the catching surface below the trees. When evaluated in a research orchard at WSU, this machine achieved a removal efficiency of 82% and fruit damage rate of 25% with 'Skeena' cherries trained in UFO Y-trellis architecture (Chen *et al.*, 2012). When a sinusoidal shaking was used rather than the impact shaking, they found that the removal efficiency could go up to 90%.

Fig. 10.18. A mirrored shake-and-catch fruit harvesting machine being evaluated in an experimental sweet cherry orchard, Washington State University, Prosser, Washington (from Larbi and Karkee, 2014). S_T: Inter-tree distance; S_R: Inter-row distance.

Based on the findings of Chen *et al.* (2012) in terms of improved efficiency and lower tree damage with sinusoidal shaking, the harvesting machine was modified and evaluated by Larbi *et al.* (2015b). The modification included the replacement of RDA with a linear continuous shaker, which was actuated using a crank and slider mechanism. The modified machine also avoided the use of an operator seat and was operated using remote control units, which improved worker productivity (Larbi *et al.*, 2015b) and potentially worker safety. The modified machine achieved a fruit removal efficiency of 87% in UFO Y-trellis cherries with up to 90% fresh market quality fruit. This work showed a substantial improvement in fruit removal efficiency and fruit quality of a cherry harvesting machine. However, fruit collection efficiency was found to be only 79%. As in many other fruit harvesting systems, most of the fruit drop to ground was through the gaps in the catcher pan around tree trunks. Further improvement on catcher pan design and extending it to the other side of the tree trunk might be helpful.

10.5.5 Future direction for cherry harvesting

Recent studies with various shake-and-catch systems and parameters showed a fruit removal efficiency of up to 97% (Chen *et al.*, 2012; Du *et al.*, 2013; Zhou *et al.*, 2014) when one to two shaking points were used in a fruiting wall orchard. These studies also reported up to 90% fresh market quality fruit. It was found that undetached fruit are generally in low-caliber, longer branches, particularly weak and pendant branches. This finding suggests that further improvement in horticultural practices such as training, pruning and thinning would be important to grow fruit in appropriately sized branches so that harvesting efficiency, collection efficiency and fruit quality can be further increased. As these relatively minor canopy modifications are practically implementable, shake-and-catch sweet cherry harvesting systems show a commercialization promise for some varieties such as ‘Skeena’. There are a few other aspects discussed below that might help further improve the efficiency and robustness of sweet cherry harvesting systems.

10.5.5.1 Biological aspects

Some cherry varieties have much lower PFRF and are good candidates for shake-and-catch harvesting, while others are relatively difficult to detach. Use of stem-loosening material (e.g. Ethephon) has also shown the potential to increase the efficiency of shake-and-catch cherry harvesting systems for some varieties (e.g. ‘Bing’), and thus increase the potential for its practical adoptability (Smith and Whiting, 2010). Studies on the effect of these chemicals on fruit quality and shelf-life are still not sufficient. There could also be genetic and marker-assisted breeding studies to find and introduce genes responsible for reduced levels of PFRF, which may lead to a variety with desired consumer characteristics and with good potential for mechanized harvesting.

10.5.5.2 Automating shake-and-catch harvesting

Operator performance is key in achieving high harvesting efficiency and productivity. Because trees are in their full foliage stage during harvesting, it is difficult for operators to see branches; therefore, they take significant time to locate branches and engage a shaking end-effector. Additionally, an operator in the field very close to these large machines is exposed to safety hazards (Larbi *et al.*, 2015a).

Fully automated shake-and-catch harvesting that detects and shakes only a targeted canopy region and catches the fruit directly beneath the area of fruit separation could offer the potential to achieve high productivity at a relatively low cost while keeping fruit damage within an acceptable level. Studies in the past (Amatya, 2015; Amatya and Karkee, 2016; Amatya *et al.*, 2016) have shown that automated detection of shaking branches under full foliage is possible, which shows promise for this kind of automated shake-and-catch harvesting.

10.6 Robotic Apple Harvesting System

Aimed at developing a road-map for fully automated mechanical harvesting systems practically usable in harvesting fresh market apples, a joint task force consisting of an industry advisory group and a technology development group from Washington State has stated that an ideal solution should be capable of harvesting more than 95% of the fruit with less than 5% of harvest-induced cullage in SNAP fruiting wall orchard systems using less than 20% of the current level of human labor. It was expected that a system capable of achieving a harvest speed of one apple per second would be economically competitive with current harvest methods.

To address this challenge, a trans-disciplinary team at WSU CPAAS has been working on understanding fundamental knowledge on sensing and end-effector technologies and integrating a prototype robot for harvesting apples grown in fruiting wall orchards (Section 10.2.4). The following sub-sections describe system design, proof-of-concept evaluation in the laboratory and a field evaluation study of the prototype. This description is followed by a brief discussion on the challenges, lessons learned, and potential lines of research for further development and commercial adoption.

10.6.1 Working environment and design specifications

Canopy architecture has greatly limited performance of harvesting robots for high-value crops such as (but not limited to) apples, citrus, cucumbers, and peppers. It has been shown in the past that unstructured conventional canopy structures (Section 10.2.4) have greatly affected cycle time and overall performance of the harvesting system (Grand D'Esnon *et al.*, 1987; Van Henten *et al.*, 2002). A pragmatic approach to this challenging problem is to modify and improve the canopy structure to simplify the task of detection and manipulation. Modification of the working environment relates to a common practice while implementing robotics into an industrial system (Bac *et al.*, 2014). All research activities in this work, therefore, are conducted in commercial apple orchards with modern fruiting wall canopy architectures (Sections 10.2.4–10.2.6; Fig. 10.6).

Often, design specifications for harvesting systems are arbitrarily set, as a detailed economics analysis of robotic apple harvesting is lacking in the literature. While an economics assessment of harvest-assist platforms has recently been completed (Gallardo and Brady, 2015), a similar assessment of robotics technology is needed to assist with the selection of specific design criteria. As per Silwal *et al.* (2017), some fundamental tasks and requirements for an apple harvesting robot include: (i) detect, localize and optimally sequence identified fruit for harvesting; (ii) detach each fruit without bruising and safely guide it to a storage container whilst avoiding obstacles and hitting/dropping fruits; (iii) be operable under night-time and natural daylight conditions and different weather; (iv) be robust to harvest multiple varieties of apple with varying shape, size, and color, grown in different orchard architectures; and (v) use a simple and cost-effective mechanical design that can be repaired in the field.

An apple-harvesting robot for selective fruit harvesting requires integration of visual sensors and mechanical manipulators with an attached end-effector. These components are required, respectively, to detect, localize, and physically detach the fruit from the tree. In our design, at the beginning of each harvesting cycle, a Red, Green, Blue, and Depth (RGB-D) camera system detects and localizes every apple in images acquired within its field of view. After optimally sequencing the harvesting priority of each identified apple, fruit coordinates are passed to the mechanical system. Because all fruits in the scene are identified at the beginning of a cycle, no additional visual sensing between successive fruit picks is required. In this chapter, such a vision system is referred to as a global camera set-up. Then, a serial link manipulator with seven DOF and an under-actuated grasping end-effector approaches, grasps, and detaches each fruit from the tree. Fruit picking is continued until all prioritized apples are removed from the scene, marking the end of a harvest cycle. The generalized method of the harvesting cycle described above is common to both laboratory experiments and field evaluations.

10.6.2 Developing components for the robotic apple harvester

10.6.2.1 Machine vision system

The machine vision system consists of a color camera (Prosilica GC1290C, AVT Technologies) mounted on top of a time-of-flight (ToF) based 3D camera (Camcube 3.0, PMD Technologies) (Fig. 10.19a). This configuration was used to acquire color images to detect apples and then estimate their 3D coordinates using data from the ToF camera. A fruit detection algorithm used Circular Hough Transformation (CHT) iteratively to identify clearly visible as well as individual apples in clusters and blob analysis (BA) to identify partially visible apples. The camera set-up was installed in an over-the-row

Fig. 10.19. (a) 2D and 3D camera used in vision system; (b) over-the-row sensor platform used to acquire images in a commercial apple orchard; and (c) identified apples marked by circles and picking sequence generated with Traveling Salesman Problem (TSP) (Silwal *et al.*, 2016a)

platform (Fig. 10.19b) that provided a controlled lighting environment, prevented fruit motion from wind during imaging, and protected instruments from precipitation during field experiments. The method achieved more than 90% accuracy in detecting apples.

To improve the accuracy of machine vision system and therefore harvesting efficiency of the robot, an iterative approach to image acquisition and harvesting of only clearly visible fruit was used (Silwal *et al.*, 2016b). The apples selected for priority harvesting were then sequenced in logical order for harvesting using a Traveling Salesman (TS) algorithm to maximize the efficiency of the robotic arm's movement during harvesting (Fig. 10.19c). The total distance required to move the manipulator from a known position to each fruit location was minimized using the Nearest Neighbor (NN) algorithm. Additional details on the vision algorithm, camera calibration, hand–eye coordination and localization can be found in Silwal *et al.* (2014, 2016a, b, 2017).

10.6.2.2 Study of human hand picking

In order to improve manipulation performance and reduce damage, it is important to understand the dynamics of human hand picking and utilize the knowledge in designing robotic end-effectors. Studies by Nguyen *et al.* (2012) and Tong *et al.* (2014) have also indicated that the optimum picking technique often requires sensory knowledge of the fruit's orientation. To understand the dynamics of human hand picking, multiple hand-picking methods were used and compared for their effectiveness in picking apples (Davidson *et al.*, 2016a). Four different picking methods, viz. horizontal pulling, horizontal pulling with twisting, vertical pulling, and vertical pulling with twisting, were experimented on five major apple varieties ('Jazz', 'Envy', 'Pacific Rose', 'Fuji' and 'Pink Lady') and tree cultivation systems. The peak distal normal forces at the point of fruit separation and the angle of fruit rotation at separation point were measured with force sensors (flexible force sensor, Tekscan, South Boston, Massachusetts) installed on the fingers (Fig. 10.20a) and an inertial measurement unit (IMU, a nine-axis attitude sensor, Sunkee, Amazon Inc., Seattle, Washington) installed on the hand (Fig. 10.20b). Total displacement from the onset of motion to separation is also shown in Fig. 10.20c.

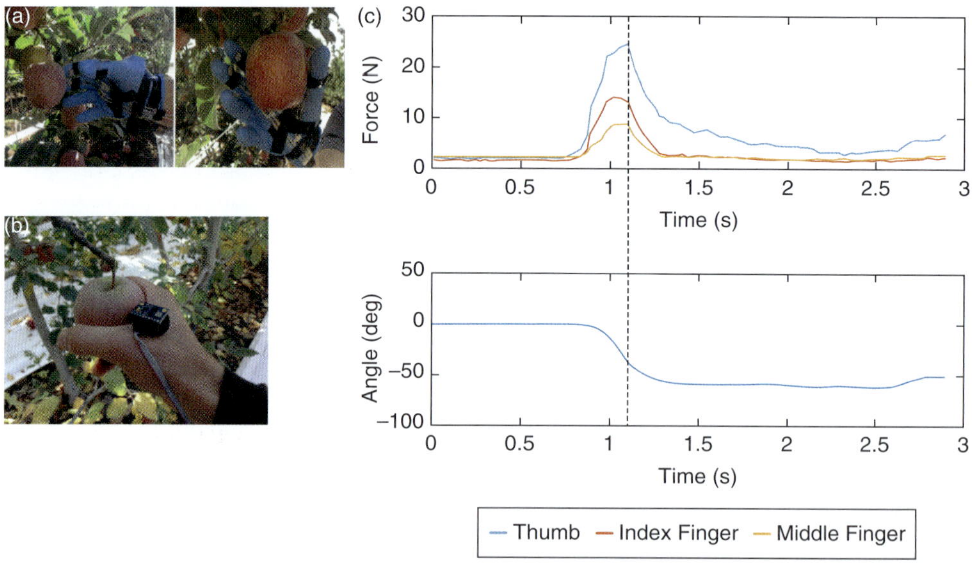

Fig. 10.20. (a) Hand approaching the fruit during a horizontal grasp; (b) the location of the nine-axis attitude IMU is shown with the glove and force sensors removed; and (c) a plot of the normal forces (top) and rotation angle (bottom) during a picking sample. The dashed line is where the stem abscission joint is severed and the fruit is detached (Davidson and Mo, 2015).

Experimental results included normal contact forces on fingers, angle of rotation around the axis of the forearm, and rates of stem detachment. Dynamic analysis showed that peak normal forces occur at the point of fruit detachment. The average angle of fruit rotation at the detachment point varied from 32 to 54 degrees, with contact forces as high as 40 N during picking. The outcomes of this study served as a prelude to the design and optimization of an under-sensed end-effector to pick apples without bruising.

10.6.2.3 Manipulator and end-effector design

A custom manipulator and end-effector were fabricated for the robotic apple harvester. The manipulator is a six-DOF serial link design incorporating the modular Dynamixel Pro actuators (Robotic Inc, Irvine, California). The Computer Aided Design (CAD) model of the integrated end-effector and manipulator are shown in Fig. 10.21a. The end-effector has three tendon-driven fingers that produce a spherical power grasp; each finger has two links and two flexure joints. The end-effector also includes a gripper that applies pressure against the apple's stem while the fruit is grasped. To minimize manipulator payload, all end-effector components were fabricated with additive manufacturing. More detailed information about the design and fabrication can be found in Davidson *et al.* (2016b).

10.6.2.4 Path planning

In modern canopy architectures, apples are distributed in a planar fruit wall. As most of the fruit in these orchards are visible and accessible to sensors and end-effectors, the workspace was assumed to be obstacle free. During a single picking sequence, the manipulator guides the end-effector to an approach position which is 10 cm away from the fruit, using point-to-point motion with trapezoidal velocity profiles. Using differential kinematics, the end-effector approaches the fruit along its normal vector with a certain translational velocity. After 1.5 s, assuming that quasi-static equilibrium has been reached and the apple is securely grasped, the manipulator simultaneously rotates the apple 30 degrees around the end-effector normal and moves it horizontally 13 cm away from the tree canopy. The end-effector is then opened to release the fruit, marking the end of the detachment process.

(a) (b)

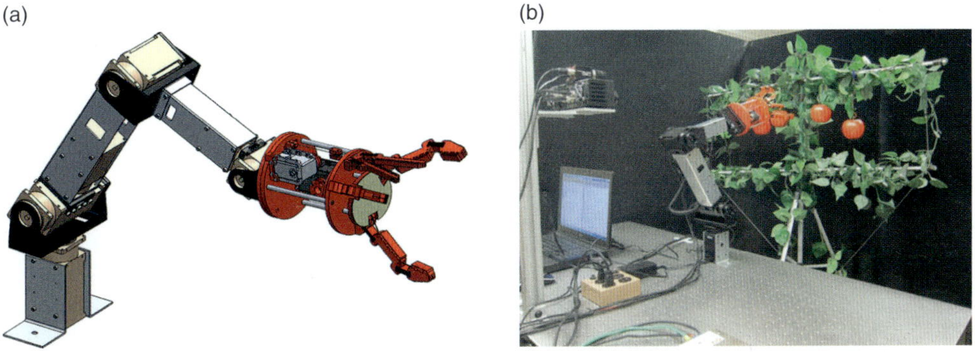

Fig. 10.21. (a) CAD model of robot arm and end-effector; and (b) mockup (from Davidson *et al.*, 2016b).

The inverse kinematics solver developed for the system uses a numerical algorithm developed by Wang and Chen (1991). If any of the joint solutions violate a joint limit or present outside the manipulator's workspace, the respective fruit is excluded from the harvesting cycle.

10.6.3 Laboratory experiments with the robotic harvester

The performance of the robotic system was first studied in a mock-up environment at a laboratory. The experimental set-up (Fig. 10.21b) included a replica apple tree with some foliage, branches, and suspended fruit. The replica tree was created with a planar design so as to represent a section of fruiting wall canopy architecture (Section 10.2.4). A black curtain was suspended behind the tree to facilitate image segmentation similar to the environment created inside an over-the-row system (Silwal *et al.*, 2014) described in Section 10.6.2. During laboratory studies, the vision system was mounted 0.5 m behind and 0.5 m above the base of the manipulator. The distance from the tree to the vision system varied from approximately 0.9 to 1.1 m, which is the representative distance expected during the field trials.

In the experiment, the time elapsed in different steps of the harvesting process including machine vision, approach, and picking was recorded to evaluate the performance of the harvester. Out of a total of 100 fruit the system attempted to pick, 95% (95/100) were successfully detached. Approximately 56% (56/100) of the fruit had their stems gripped during the picking sequence. On average, fruit localization time was found to be 1.18 ± 0.25 s per fruit, path planning was 0.06 ± 0.01 s, and picking time was 6.82 ± 0.26 s per fruit. This study indicated that the global vision system and open-loop, look-and-move fruit-picking technique may achieve acceptable harvesting efficiencies. It was learned that not applying stem pressure increased the likelihood of a stem pull or spur detachment. There were several limitations in the laboratory set-up that restricted comprehensive performance assessments. Because of the relatively few apples present during each picking cycle, it was difficult to explore the trade-off between fruit position error tolerance and interference from adjacent fruit and/or stems.

10.6.4 Field evaluation of the robotic apple harvester

The integrated robotic system was mounted on the cargo box of a John Deere 'Gator' electric utility vehicle (John Deere, Moline, Illinois) (Fig. 10.22a). A few essential modifications, described below, were made to the vision system and robot arm based on the experience from the laboratory study.

Though the initial machine vision system (Section 10.6.2) utilized an over-the-row platform to create uniform lighting conditions, images during the field trials of the robotic harvester were acquired in broad daylight conditions without supportive structures. In such natural daylight conditions, images suffered from irregular exposures that altered color saturation and contrast. In extreme cases, highly over/under-exposed

Fig. 10.22. (a) A robotic platform with a manipulator, an end-effector and a vision system in a commercial orchard in Prosser, Washington and (b) output of an exposure fusion algorithm (Silwal *et al.*, 2016a).

images negatively impact fruit identification. To overcome this limitation, an exposure fusion algorithm (Mertens *et al.*, 2007) was used. Five images at different exposure values were acquired for exposure fusion. Implementation of this method created a uniformly exposed image of apple tree canopies while minimizing shadows and overly saturated regions (Fig. 10.22b). On the mechanical side, a prismatic base was installed to increase the depth of the manipulator's reach.

This robotic apple picker achieved a cycle time of 6 s per fruit in a formally trained high-density fruiting wall tree canopy. In the first round of evaluation in a commercial orchard, the integrated robotic system picked 85% of the fruit that it attempted. The majority of failures were observed in cases where the apples were located very close to the branch and trellis wire. In such instances, the end-effector grabbed the obstruction along with the fruit and failed to detach it. Additional sources of failure came from the error in hand–eye calibration and limited workspace. About 68% of the fruit were harvested with stem attached with them. Based on the qualitative observation in the field, no fruit was damaged or bruised.

10.6.5 Overall discussion, potentials and challenges

The overall goal of this research was to evaluate a low-cost robotic system for apple harvesting in fruiting wall commercial orchards. The total price for all equipment excluding the electric utility vehicle was within US$25,000. The bottom-up approach described above assesses the performance in a modern orchard system. Best-case conditions that were within the practical realm of canopy management and with the agreement of the orchard manager were used in this experiment. These included ideal fruit distribution (five to eight fruit per trellis wire between two trees) (Silwal *et al.*, 2016a) and a uniform background behind the canopy to facilitate image processing. The system performance including the speed, however, does not yet meet the productivity and fruit quality levels that are desired for a commercially adoptable harvester.

10.6.5.1 Vision

A global sensing set-up was found to be robust in detecting and localizing the apple centers in 3D space. Additional imaging was not required between successive fruit picks. In the outdoor environment with natural lighting conditions, exposure fusion showed potential for enhanced vision performance by removing hard shadows and saturated regions of the images acquired. The hierarchical approach of iterative imaging and harvesting strategically reduced canopy complexity. This approach can achieve very high fruit identification/ harvesting accuracy in fruiting wall apple orchards. Machine vision in an outdoor environment is always a challenging problem. More robust and effective vision algorithms such as deep learning-based image segmentation and object recognition could improve the accuracy of fruit identification.

10.6.5.2 Manipulation

Even under the noticeable localization error, the end-effector was able to grasp a fruit. It was possible because of the passive compliance incorporated in the end-effector design that enhanced grasping robustness. It was learned that, even with planar canopy structure, complex manipulation and path planning may be required to improve the robustness of the system as well as its application to wider types of orchard architectures.

10.6.5.3 Overall system

It was also found that the overall system performance is a function of speed and robustness. Often one factor must be sacrificed to enhance the other, which makes them contrary to each other. Acquisition and operation cost of the system makes the optimization more complex. In this study, speed was prioritized over robustness while keeping the system development cost as low as possible. Additional sensing might be necessary for improved harvesting efficiencies and overall robustness. Studies of the force profile during the hand-picking process (Davidson *et al.*, 2016a) showed that it is possible to detect the instance a fruit is detached from the branch. Integration of force sensors in the end-effector could provide feedback to determine if a fruit was missed or not detached.

There are other practical considerations, including weather proofing and fruit preparation for storage, which are essential for commercial viability. A supportive structure such as the over-the-row platform could provide protection against natural elements. A practical design of such a mechanical structure or system should also consider compatibility with multiple orchard architectures.

10.7 Status, Challenges and Opportunities for Fruit Harvesting

The biggest opportunity for tree fruit harvesting exists in the fresh market. As discussed before, every piece of fresh market produce found in the grocery store was picked by the human hand. Despite decades of research, there are still no mass or selective mechanical harvesters in commercial use. In general, fruit damage from mass mechanical harvesting has been unacceptably high for fresh market acceptance. Likewise, the transfer of industrial robotics technology directly to field-based, biologically driven environments has resulted in limited success. The limitations of robotic systems have been well documented (Bac *et al.*, 2014) and include insufficient cycle time, challenges with fruit detection, and limitations with robust manipulation for fruit detachment.

In the agricultural robotics community, there presently seem to be two prevailing directions of research. One of these has studied relatively sophisticated functionality, including kinematic redundancy (Baur *et al.*, 2012), grasp planning (Eizicovits and Berman, 2014), and obstacle detection (Bac *et al.*, 2013) and avoidance (Nguyen *et al.*, 2013). The second line of research is attempting to simplify the harvesting task. Researchers in California have studied linear fruit reachability (Vougioukas *et al.*, 2016) in high-density orchard systems with the goal of reducing the harvesting robot's degrees of freedom and simplifying its control. While simplifying the harvesting task will most likely require modifications to the crop environment, it has the potential to result in robust, cost-effective harvesting systems. Other potential research areas for consideration that will be discussed in this section include model-based design, multipurpose robotic systems, and human–machine collaboration.

10.7.1 Model-based design

The usual development cycle in agricultural robotics has tended to follow a design–build–field-evaluate–redesign model. To measure performance, field studies in production orchards are typically required. This type of development model is expensive and inefficient. A promising direction of research is model-based design whereby models of orchard systems can be used in virtual harvesting simulations to optimize the robotic system before hardware selection and fabrication. Arikapudi *et al.* (2015) recently presented work on the digitization of pear trees with the goal of developing open-source design tools for simulation of virtual harvesting. Such efforts are critical for early design optimization as well as reducing the cost/time of the development cycle, especially in agricultural applications where the duration available for field studies is highly constrained.

Looking to the future, systematic approaches for agricultural automation that combine machine design and horticultural practices remain an opportunity for improvement. As discussed by both Sanders (2005) and Peterson (2005), it is widely recognized that to maximize the potential of robotic fruit harvesting, compatible horticultural systems are required. In these systems, plant characteristics, like canopy growth, tree spacing, and fruit position, as well as cultural practices are developed in conjunction with machine designs. Though canopy architectures have come a long way towards machine-friendly structures as discussed in Section 10.2, further improvements in training, pruning and thinning of apple trees to minimize occlusion and reduce doubles and triples (multiple fruits growing together) while ensuring that fruit are hanging down into an open space without leaning against a branch, trunk or any other structure could further improve fruit visibility and accessibility. Closer collaboration between growers, horticulturalists, and engineers is vital to achieve these desirable parameters in the designs of future orchards.

10.7.2 Multipurpose robotic systems

Modular robotic systems able to perform multiple tasks in the orchard are increasingly attractive. Modularity and the ability to swap different end-effector configurations could enable the same robot to be used for chemical spraying/thinning, mechanical pruning, and autonomous harvesting. Capital costs required to purchase an expensive robot could be offset by savings from reduced labor expenditures across several labor-intensive tasks in the orchard. Results from a recent project to develop a multipurpose robot for greenhouse applications (Belforte *et al.*, 2006) demonstrates the viability of such a system.

10.7.3 Human–machine collaboration

The final research direction suggested for improved harvesting performance is human–machine collaboration and haptic interfaces whereby the sensory abilities of the human are used to augment the capabilities of the machine. Previous robotic harvesting research tended to focus on completely autonomous systems with no user input. A semi-autonomous system that integrates human–machine collaboration could potentially improve overall harvest success rates. For example, if the machine vision system were unable to identify the fruit because of occlusion or clustering, the system operator could use a haptic interface and camera feed to complete the picking sequence manually, with the robot replicating the human operator's physical motions.

10.8 Summary

Effectively removing fruit at greater speed/throughput and minimizing harvest-induced fruit damage are two of the major challenges for developing mechanized or automated harvesting systems practically usable for fresh market tree fruit production. These challenges require scientists and engineers to go beyond just adopting the technologies from existing applications and develop innovative solutions. One area for innovation could be an overall system integration approach, including interaction and integration of human–machine–tree systems, which also requires further improvement in horticultural systems. If we can create a technology allowing machine operators to cooperate with machines/robots and help them a little, the complexity of the robotic task can be substantially reduced, which could help make the robotic harvesting solution more adaptable and affordable.

In recent years, many academic institutions, companies and venture capitalists have been attracted to developing robotic solutions for fruit harvesting. Among others, there are large research projects ongoing in the USA (including WSU's projects), the Netherlands, China, and New Zealand funded through government research programs. In terms of private companies, a robotic apple picker is being developed by Abundant Robotics, Inc., a California-based company, with support from the Washington Tree Fruit Research Commission. The researchers creatively developed a vacuum-based picker, which could help avoid damaging or dislodging the target fruit, adjacent fruit or part of the tree in ways that a grasping machine might achieve at very high picking speed. Other companies from around the world, such as FFRobotics (Israel), are also putting a lot of effort into developing robotic fruit harvesters. Based on current progress in technology development in both academia and industry, it is reasonable to expect that fully automated tree fruit harvesting systems could become commercially available for US growers in 3–5 years. However, a detailed economic assessment of fruit harvesting technology is still lacking, which poses some uncertainty in terms of selecting specific design criteria and performance specifications for future research and development.

References

Adrian, P.A. and Fridley, R.B. (1961) Excessive vibration eliminated by new tree shaker mounted on fruit catching frame. *California Agriculture* 15(8), 12–13.

Amatya, S. (2015) Detection of cherry tree branches and localization of shaking positions for automated sweet cherry harvesting. PhD Dissertation, Washington State University, Pullman, Washington.

Amatya, S. and Karkee, M. (2016) Integration of visible branch sections and cherry clusters for detecting cherry tree branches in dense foliage canopies. *Biosystems Engineering* 119, 72–81.

Amatya, S., Karkee, M., Gongal, A., Zhang, Q. and Whiting, M.D. (2016) Detection of cherry tree branches in planner architecture for automated sweet-cherry harvesting. *Biosystems Engineering* 146, 3–15.

Anonymous (2016) Pruning and training. Available at: http://www.oliveoilsource.com/page/pruning-and-training (accessed 15 September 2016).

Arikapudi, R., Vougioukas, S. and Saracoglu, T. (2015) Orchard tree digitization for structural-geometrical modeling. In: Stafford, J.V. (ed.) *Precision Agriculture '15: Proceedings of the 10th European Conference on Precision Agriculture (ECPA), Volcani Center, Israel.* Wageningen Academic Publishers, Wageningen, The Netherlands, pp. 329–336.

Avalle, M., Belingardi, G. and Montanini, R. (2001) Characterization of polymeric structural foams under compressive impact loading by means of energy-absorption diagram. *International Journal of Impact Engineering* 25(5), 455–472.

Bac, C.W., Hemming, J. and Van Henten, E.J. (2013) Robust pixel-based classification of obstacles for robotic harvesting of sweet-pepper. *Computers and Electronics in Agriculture* 96, 148–162.

Bac, C.W., Van Henten, E.J., Hemming, J. and Edan, Y. (2014) Harvesting robots for high-value crops: state-of-the-art review and challenges ahead. *Journal of Field Robotics* 31(6), 888–911.

Baur, J., Pfaff, J., Ulbrich, H. and Villgrattner, T. (2012) Design and development of a redundant modular multipurpose agricultural manipulator. In: *IEEE/ASME International Conference on Advanced Intelligent Mechatronics (AIM)*, Kaohsiung, Taiwan. IEEE, Piscataway, New Jersey, pp. 823–830.

Belforte, G., Deboli, R., Gay, P., Piccarolo, P. and Aimonino, R.D. (2006) Robot design and testing for greenhouse applications. *Biosystems Engineering* 95(3), 309–321.

Brown, E. and Jaeger, H.M. (2014) Shear thickening in concentrated suspensions: phenomenology, mechanisms and relations to jamming. *Reports on Progress in Physics* 77(4), 046602.

Bulanon, D.M. and Kataoka, T. (2010) Fruit detection system and an end effector for robotic harvesting of Fuji apples. *Agricultural Engineering International: CIGR Journal* 12(1), 203–210.

Chen, D., Du, X., Zhang, Q., Whiting, M.D., Scharf, P.A. and Wang, S. (2012) Performance evaluation of mechanical cherry harvesters for fresh market grade fruits. *Applied Engineering in Agriculture* 28, 483–489.

Cooke, J.R. and Rand, R.H. (1969) Vibratory fruit harvesting: a linear theory of fruit-stem dynamics. *Journal of Agricultural Engineering Research* 14(3), 195–209.

Coppock, G.E. (1961) Picking citrus fruit by mechanical means. *Proceedings of the Florida State Horticultural Society* 1362, 247–251.

Coppock, G.E. and Hedden, S.L. (1968) Design and development of a tree-shaker harvest system for citrus fruit. *Transactions of the ASAE* 11(3), 339–342.

Craig, J.J. (1990) *Introduction to Robotics: Mechanics and Control.* Addison-Wesley, Reading, Massachusetts.

Davidson, J.R. and Mo, C. (2015) Mechanical design and initial performance testing of an apple-picking end-effector. Paper no. IMECE2015-50482. *ASME 2015 International Mechanical Engineering Congress and Exposition (IMECE)*, Houston, Texas, 13–19 November, 2015. Vol. 4A: Dynamics, Vibration, and Control, pp. 4–11.

Davidson, J., Silwal, A., Karkee, M., Mo, C. and Zhang, Q. (2016a) Hand-picking dynamic analysis for undersensed robotic apple harvesting. *Transactions of the ASABE* 59(4), 745–758.

Davidson, J.R., Silwal, A., Hohimer, C.J., Karkee, M., Mo, C. and Zhang, Q. (2016b) Proof-of-concept of a robotic apple harvester. Paper presented at IEEE/RSJ International Conference on Intelligent Robots and Systems (IROS), 9–14 October 2016, Daejeon, Korea, pp. 634–639 (accepted). doi: 10.1109/IROS.2016.7759119.

De Kleine, M. (2014) Semi-automated end-effector concepts for localized removal and catching of fresh-market apples in fruiting wall orchards. PhD Dissertation, Washington State University, Pullman, Washington.

De Kleine, M.E. and Karkee, M. (2015a) A semi-automated harvesting prototype for shaking fruit tree limbs. *Transactions of the ASABE* 58(6), 1461–1470.

De Kleine, M.E. and Karkee, M. (2015b) Evaluating a non-Newtonian shear-thickening surface during fruit impacts. *Transactions of the ASABE* 58(3), 907–915.

Diener, R.G., Mohsenin, N.N. and Jenks, B.L. (1965) Vibration characteristics of trellis-trained apple trees with reference to fruit detachment. *Transactions of the ASAE* 8(1), 20–24.

Domigan, I.R., Diener, R.G., Elliott, K.C., Blizzard, S.H., Nesselroad, P.E., Singha, S. and Ingle, M. (1988) A fresh fruit harvester for apples trained on horizontal trellises. *Journal of Agricultural Engineering Research* 41(4), 239–249.

Du, X., Chen, D., Zhang, Q., Scharf, P.A. and Whiting, M.D. (2012) Dynamic responses of sweet cherry trees under vibratory excitations. *Biosystems Engineering* 111(3), 305–314.

Du, X., Chen, D., Zhang, Q., Scharf, P.A. and Whiting, M.D. (2013) Response of UFO (Upright Fruiting Offshoots) on cherry trees to mechanical harvest by dynamic vibratory excitation. *Transactions of the ASABE* 56(2), 345–354.

Dunn, J.W. and Stolp, M. (1980) Mechanical harvesting apples and raspberries grown on the Lincoln Canopy system. *Acta Horticulturae* 114, 261–268.

Edan, Y. and Miles, G.E. (1994) Systems engineering of agricultural robot design. *IEEE Transactions on Systems, Man, and Cybernetics* 24(8), 1259–1265.

Eizicovits, D. and Berman, S. (2014) Efficient sensory-grounded grasp pose quality mapping for gripper design and online grasp planning. *Robotics and Autonomous Systems* 62(8), 1208–1219.

Elkins, R.B., Meyers, J.M., Duraj, V., Tejeda, D.J., Mitcham, E.J. *et al.* (2011) Comparison of platform versus ladders for harvest in northern California pear orchard. *Acta Horticulturae* 909, 241–249.

Fathallah, F.A. (2010) Musculoskeletal disorders in labor-intensive agriculture. *Applied Ergonomics* 41(6), 738–743.

Fideghelli, C., Sartori, A. and Grassi, F. (2003) Fruit tree size and architecture. *Acta Horticulturae* 622, 279–293.

Gallardo, R.K. and Brady, M.P. (2015) Adoption of labor-enhancing technologies by specialty crop producers: the case of the Washington apple industry. *Agricultural Finance Review* 75(4), 514–532.

Gallardo, R.K., Taylor, M. and Hinman, H. (2010) *2009 Cost Estimates of Establishing and Producing 'Gala' Apples in Washington.* Washington State University Extension Factsheet FS005E. Washington State University, Pullman, Washington.

Godin, C., Costes, E. and Sinoquet, H. (1999) A method for describing plant architecture which integrates topology and geometry. *Annals of Botany* 84, 343–357.

Gould, I.V., Young, G.S. and Godley, G.L. (1986) *Mechanized Fruit Harvesting from the Tatura Trellis.* ASAE Paper No. 386-10. American Society of Agricultural Engineers, St Joseph, Michigan.

Grand D'Esnon, A. (1985) *Robotic Harvesting of Apples.* ASAE Paper No. 185, 210–214, presented at the meeting of Agri-Mation 1, Chicago, Illinois, 25–28 February 1985. American Society of Agricultural Engineers, St Joseph, Michigan.

Grand D'Esnon, A., Rabatel, G., Pellenc, R., Journeau, A. and Aldon, M.J. (1987) *MAGALI: a Self-propelled Robot to Pick Apples.* ASAE Paper No. 87–1037. American Society of Agricultural Engineers, St Joseph, Michigan.

Gupta, S.K., Ehsani, R. and Kim, N. (2016) Optimization of a citrus canopy shaker harvesting system: mechanistic tree damage and fruit detachment models. *Transactions of the ASABE* 59(4), 761–776.

Halderson, J.L. (1966) Fundamental factors in mechanical cherry harvesting. *Transactions of the ASAE* 9(5), 681–684.

Hannan, M.W. and Burks, T.F. (2004) *Current Developments in Automated Citrus Harvesting.* ASABE Paper No. 043087. American Society of Agricultural and Biological Engineers, St Joseph, Michigan.

Harrell, R.C. and Levi, P. (1988) Vision controlled robots for automatic harvesting of citrus. *Agricultural Engineering* 8, 2–5.

Harrell, R.C., Slaughter, D.C. and Adsit, P.D. (1989) A fruit-tracking system for robotic harvesting. *Machine Vision and Applications* 2(2), 69–80.

He, L., Zhou, J., Du, X., Chen, D., Zhang, Q. and Karkee, M. (2013) Energy efficacy analysis of a mechanical shaker in sweet cherry harvesting. *Biosystems Engineering* 116(4), 309–315.

He, L., Fu, H., Sun, D., Karkee, M. and Zhang, Q. (2016) *A Shake and Catch Harvesting System for 'Jazz' Apples Trained in Vertical Fruiting Wall Architecture*. ASABE Paper No. 162461420. American Society of Agricultural and Biological Engineers, St Joseph, Michigan.

Jiménez, A.R., Ceres, R. and Pons, J.L. (2000) A vision system based on a laser range-finder applied to robotic fruit harvesting. *Machine Vision and Applications* 11(6), 321–329.

Jutras, P.J. and Coppock, G.E. (1958) Mechanization of citrus fruit picking. *Proceedings of the Florida State Horticultural Society* 71, 201–204.

Kader, A.A. (ed.) (2002) *Postharvest Technology of Horticultural Crops*, 3rd edn. University of California, Davis, California.

Lamouria, L.H., Hartmann, H.T., Harris, R.W. and Kaupke, C.R. (1958) *Mechanical fruit harvest – olives, peaches, and pears*. Updated (1961) as: Mechanical harvesting of olives, peaches and pears, in: *Transactions of the ASAE*, 4(1), 12–14. doi: 10.13031/2013.40994.

Larbi, P.A. and Karkee, M. (2014) Effects of orchard characteristics and operator performance on harvesting rate of a mechanical sweet cherry harvester. *GSTF Journal on Agricultural Engineering* 1(1), 1–11.

Larbi, P., Vong, C.N. and Karkee, M. (2015a) A study of operator performance for a mechanical sweet cherry harvester: comparison between manual and remote-controlled operation. *Journal of Agricultural Safety and Health* 21(3), 145–157.

Larbi, P., Amatya, S., Karkee, M., Zhang, Q. and Whiting, M. (2015b) Modification and field evaluation of an experimental mechanical sweet cherry harvester. *Applied Engineering in Agriculture* 31(3), 387–397.

Li, K., Syvertsen, J. and Burns, J. (2005) Mechanical harvesting of Florida citrus trees has little effect on leaf water relations or return bloom. *Proceedings of the Florida State Horticultural Society* 118, 22–24.

Li, P., Lee, S. and Hsu, H. (2011) Review of fruit harvesting method for potential use of automatic fruit harvesting systems. *Procedia Engineering* 23, 351–366.

Long, L.E. (2010) KGB training system for cherries. Available at: http://extension.oregonstate.edu/wasco/sites/default/files/kgb_training_system.pdf (accessed 25 September 2016).

Long, L., Lang, G., Musacchi, S. and Whiting, M. (2015) *Cherry Training Systems*. PNW 667. Oregon State University Extension Service, Cornvallis, Oregon.

Luo, R., Lewis, K.M., Zhang, Q. and Wang, S. (2012) *Assessment of Bruise Damage by Vacuum Apple Harvester Using an Impact Recording Device*. ASABE Paper No. 121338094. American Society of Agricultural and Biological Engineers, St Joseph, Michigan.

Ma, S., Karkee, M., Fu, H., Sun, D. and Zhang, Q. (2016) Air suspension-based catching mechanism for mechanical harvesting of apples. In: *Proceedings of 5th IFAC Conference on Sensing, Control and Automation Technologies for Agriculture*, Seattle, Washington. International Federation of Automatic Control, Laxenburg, Austria. doi:10/1016/j.ifacol.2016.10.065.

Markwardt, E.D., Guest, R.W., Cain, J.C. and LaBelle, R.L. (1964) Mechanical cherry harvesting. *Transactions of the ASAE* 7(1), 70–74.

Mertens, T., Kautz, J. and Van Reeth, F. (2007) Exposure fusion. In: *Proceedings of the 15th Pacific Conference on Computer Graphics and Applications*. IEEE, Piscataway, New Jersey, pp. 382–390. Reproduced (2009) in *Computer Graphics Forum* 29(1), 161–171.

Mika, A. (1991) Trends in fruit tree training and pruning systems in Europe. *I International Symposium on Training and Pruning of Fruit Trees* 322, 29–36.

Moseley, K.R., House, L.A. and Roka, F. (2012) Adoption of mechanical harvesting for sweet orange trees in Florida: addressing grower concerns on long-term impacts. *International Food and Agribusiness Management Review* 15(2), 83–98.

Muscato, G., Prestifilippo, M., Abbate, N. and Rizzuto, I. (2005) A prototype of an orange picking robot: past history, the new robot and experimental results. *Industrial Robot: An International Journal* 32(2), 128–138.

Nguyen, T.T., Van Eessen, D., De Baedemaeker, J. and Saeys, W. (2012) Optimum detaching movement for apples-harvesting robot. *International Conference of Agricultural Engineering (CIGR)* (Automation Technology: Precision Farming. International Conference of Agricultural Engineering – CIGR-AgEng 2012, Agriculture and Engineering for a Health Life, Valencia, Spain, 8–12 July 2012, pp. P1788). CIGR-EurAgEng, Valencia, Spain.

Nguyen, T., Kayacan, E., De Baedemaeker, J. and Saeys, W. (2013) Task and motion planning for apple harvesting robot. *4th IFAC Conference on Modelling and Control in Agriculture, Horticulture and Post-Harvest Industry*, Espoo, Finland, 28–30 August 2013. International Federation for Automatic Control, Laxenburg, Austria, pp. 247–252.

Perry, R. (2010) Different approaches to tall spindle establishment in apple. Available at: http://msue.anr.msu.edu/uploads/files/2010_NW_orchard_show/Perry_AppleTallSpindle.pdf (accessed 01 September 2016).

Peterson, D.L. (1982) Rod press fruit removal mechanism. *Transactions of the ASAE* 25(5), 1185–1188.

Peterson, D.L. (1985) Cultural modifications of deciduous tree fruits for mechanized production. *HortScience* 20(6), 1015–1018.

Peterson, D.L. (1992) Harvest mechanization for deciduous tree fruits and brambles. *HortTechnology* 2(1), 85–88.

Peterson, D.L. (2005) Harvest mechanization progress and prospects for fresh market quality deciduous tree fruits. *HortTechnology* 15(1), 72–75.

Peterson, D.L. and Monroe, G.E. (1977) Continuously moving shake-catch harvester for tree crops. *Transactions of the ASAE* 20(2), 202–205, 209.

Peterson, D.L. and Wolford, S.D. (2003) Fresh-market quality tree fruit harvester. Part II: apples. *Applied Engineering in Agriculture* 19(5), 545.

Peterson, D.L., Miller, S.S. and Whitney, J.D. (1994) Harvesting semidwarf freestanding apple trees with an over-the-row mechanical harvester. *Journal of the American Society of Horticulture Science* 119(6), 1114–1120.

Phillips, R.V. and Grierson, W. (1957) Cost advantages of bulk handling through the packing house. *Proceedings of the Florida State Horticultural Society* 70, 171–177.

Pires, J.N. (2007) Robot manipulators and control systems. In: Pires, J.N. *Industrial Robots Programming: Building Applications for the Factories of the Future*. Springer, New York, 35–107.

Rabatel, G., Bourely, A., Sevila, F. and Juste, F. (1995) Robotic harvesting of citrus: state-of-art and development of the French Spanish Eureka project. *Proceedings of International Conference on Harvest and Post-harvest Technologies for Fresh Fruits and Vegetables* 95, 232–239.

Robinson, T., Hoying, S., Sazo, M.M., DeMarree, A. and Dominguez, L. (2013) A vision for apple orchard systems of the future. *New York Fruit Quarterly* 21(3), 11–16.

Rumsey, J.W. (1967) Response of citrus fruit-stem system to fruit removing actions. MS thesis, University of Arizona, Tuscon, Arizona.

Sanders, K.F. (2005) Orange harvesting systems review. *Biosystems Engineering* 90(2), 115–125.

Sarig, Y. (1993) Robotics of fruit harvesting: a state-of-the-art review. *Journal of Agricultural Engineering Research* 54, 265–280.

Schupp, J., Baugher, T., Winzeler, E., Schupp, M. and Messner, W. (2011) Preliminary results with a vacuum assisted harvest system for apples. *Fruit Notes* 76, 1–5.

Sigler, D. (2010) Tall or super spindle? A matter of space. *Fruit Growers News*, September 2010, p.1.

Silwal, A., Gongal, A. and Karkee, M. (2014) Apple identification in field environment with over the row machine vision system. *Agricultural Engineering International: CIGR Journal* 16(4), 66–75.

Silwal, A., Davidson, J., Karkee, M., Mo, C., Zhang, Q. and Lewis, K. (2016a) *Effort Towards Robotic Apple Harvesting in Washington State.* ASABE Paper No. 162460869. American Society of Agricultural and Biological Engineers, St Joseph, Michigan.

Silwal, A., Karkee, M. and Zhang, Q. (2016b) A hierarchical approach to apple identification for robotic harvesting. *Transactions of the ASABE* 59(5), 1079–1086.

Silwal, A., Davidson, J.R., Karkee, M., Mo, C., Zhang, Q. and Lewis, K. (2017) Design, integration, and field evaluation of a robotic apple harvester. *Journal of Field Robotics*, 1–20. Early View (online) published 28 April 2017. doi: 10.1002/rob.21715.

Smith, E. and Whiting, M. (2010) Effect of ethephon on sweet cherry pedicel-fruit retention force and quality is cultivar dependent. *Plant Growth Regulation* 60(3), 213–223.

Sumner, H.R. (1973) Selective harvesting of Valencia oranges with a vertical canopy shaker. *Transactions of the ASAE* 16(6), 1024–1026.

Sumner, H.R. and Hedden, S.L. (1975) Harvesting oranges with a full-powered positioning limb shaker. *Proceedings of the Florida State Horticultural Society* 88, 117–120.

Tabb, A.L., Peterson, D.L. and Park, J. (2006) *Segmentation of Apple Fruit from Video via Background Modeling.* ASABE Paper No. 063060. American Society of Agricultural and Biological Engineers, St Joseph, Michigan.

Tarrío, P., Bernardos, A.M., Casar, J.R. and Besada, J.A. (2006) A harvesting robot for small fruit in bunches based on 3-D stereoscopic vision. In: *Proceedings of the 4th World Congress Conference on Computers in Agriculture and Natural Resources*, 24–26 July 2006 (Orlando, Florida). ASABE Publication No. 701P0606. American Society of Agricultural and Biological Engineers, St Joseph, Michigan, pp. 270–275.

Tong, J., Zhang, Q., Karkee, M., Jiang, H. and Zhou, J. (2014) *Understanding the Dynamics of Hand Picking Patterns of Fresh Market Apples.* ASABE Paper No. 141898024. ASABE and CSBE/SCGAB Annual International Meeting, Montreal, July 2014. American Society of Agricultural and Biological Engineers, St Joseph, Michigan.

Upadhyaya, S.K., Cooke, J.R., Pellerin, R.A. and Throop, J.A. (1981) Limb impact harvesting, part II: experimental approach. *Transactions of the ASAE* 24(4), 868–871, 878.

USDA ERS (2012) *United States Department of Agriculture Economic Research Service analysis of NASS Farm Labor Survey data (1990–2012).* US Department of Agriculture, Washington, DC.

USDA-NASS (2010) Florida citrus statistics. Available at: https://www.nass.usda.gov/Statistics_by_State/Florida/Publications/Citrus/Citrus_Forecast/2009-10/cit0610.pdf (accessed 25 August 2016).

Van Henten, E.J., Hemming, J., Van Tuijl, B.A.J., Kornet, J.G., Meuleman, J., Bontsema, J. and Van Os, E.A. (2002) An autonomous robot for harvesting cucumbers in greenhouses. *Autonomous Robots* 13(3), 241–258.

Vougioukas, S., Arikapudi, R. and Munic, J. (2016) Upper bound estimates of fruit reachability in orchard trees using linear motion. Paper 529, *Proceedings of the International Conference on Agricultural Engineering* (AgEng – CIGR 2016), Aarhus, Denmark.

Wang, L.C. and Chen, C.C. (1991) A combined optimization method for solving the inverse kinematics problems of mechanical manipulators. *IEEE Transactions on Robotics and Automation* 7(4), 489–499.

WSDA (2011) Washington Tree Fruit Acreage Report 2011. Washington State Department of Agriculture. Available at: http://www.nass.usda.gov/Statistics_by_State/Washington/Publications/Fruit/FruitTreeInventory2011.pdf (accessed 10 September 2016).

WSESD (2013) Agriculture Workforce Report. Washington State Employment Security Department. Available at: https://fortress.wa.gov/esd/employmentdata/docs/industry–reports/agricultural–workforce–report–2013.pdf (accessed 27 April 2016).

Webb, B.K., Hood, C.E., Jenkins, W.H. and Veal, C.D. (1973) Development of an over-the-row peach harvester. *Transactions of the ASAE* 16, 450–455.

Whiting, M. (2008) The UFO system: a novel architecture for high efficiency sweet cherry orchards. Conference paper 2011, ASHS Annual Conference. *Proceedings of the IXth International Symposium on Integrating Canopy Rootstock and Environmental Physiology in Orchard Systems.* American Society for Horticultural Science, Alexandria, Virginia.

Whitney, J.D. and Patterson, J.M. (1972) Development of citrus removal device using oscillating forced air. *Transactions of the ASAE* 15(5), 849–860.

Zhou, J., He, L., Zhang, Q. and Karkee, M. (2014) Effect of excitation position of a hand-held shaker on fruit removal efficiency and damage in mechanical harvesting of sweet cherry. *Biosystems Engineering* 125, 36–44.

Zhou, J., He, L., Karkee, M. and Zhang, Q. (2016a) Analysis of shaking-induced cherry fruit motion and damage. *Biosystems Engineering* 144, 105–114.

Zhou, J., He, L., Karkee, M. and Zhang, Q. (2016b) Effect of catching surface and tilt angle on bruise damage of sweet cherry due to mechanical impact. *Computers and Electronics in Agriculture* 121, 282–289.

11 Opportunity of Robotics in Precision Horticulture

Thomas Burks[1]*, Duke Bulanon[2] and Siddhartha Mehta[1]

[1]University of Florida, Gainesville, Florida, USA; [2]Northwest Nazarene University, Nampa, Idaho, USA

11.1 Introduction

The motivation towards adoption of mechanization and automation technologies for fruit production has been associated primarily with labor productivity, labor cost and availability, as well as other factors such as cultivar/varietal improvements, fruit quality and safety, disease and pest pressures, environmental concerns and regulations, and global market pressures. Although the vast majority of progress has been realized during the past 50 years, there seems to be an accelerated effort in developed countries in the past decade as two major factors come to bear. The first is rapidly escalating labor cost along with a shrinking labor force, while the second is a significant acceleration in agricultural automation technological development enabled by aerospace, defense and industrial efforts. The concept of appropriate automation becomes crucial, since global market pressures limit the cost of automation to competitive levels. Unlike the aerospace and defense industries, fruit producers must remain economically competitive with global suppliers. Consequently, the selection of appropriate technology is probably the most important aspect of automating any production practice. It is therefore necessary to consider the full spectrum of solutions when addressing a production problem, which include manual aids, traditional mechanization, mechatronically enhanced equipment, semi-autonomous robotics with human assistance and oversight, or ultimately fully autonomous systems.

In the past several decades, many tree fruit producers have promoted the development of mechanized/automated solutions for various production tasks, including harvesting. However, successful harvesting development has largely

* Corresponding author, email: tburks@ufl.edu

been limited to processed applications, where fruit damage during harvesting is minimally problematic, since the fruit will be typically processed within 24 h of harvest. Several fresh market horticultural commodity groups around the USA are facing growing global market pressures that threaten their long-term viability. For instance, Brazilian orange growers can produce, process, and ship juice to Florida markets more cheaply than can Florida growers. In the event that tariffs are eliminated, numerous horticultural commodities across the nation will not be able to compete in either domestic or international markets with their counterparts in Latin America and Asia. The combination of low commodity prices both domestically and abroad, high labor prices and low labor productivity present significant challenges for US agriculture.

The potential societal benefits from agricultural mechanization/automation are numerous. By sustaining crucial commodities, the economic infrastructure that supports these industries will be reinvigorated. Rural communities will have new opportunities for better jobs that have less drudgery than traditional manual field labor. Opportunities to improve worker health and safety by automating dangerous operations have significant potential.

The objective of this chapter is to present an overview of the major production task areas in tree fruit production that either have already been automated or are currently in research and development stages. This chapter will specifically address topics in autonomous navigation, cultural practices, robotic selective harvesting and novel technologies for crop monitoring.

11.2 Autonomous Robotic Vehicle Guidance

Autonomous navigation is a valuable tool in agricultural systems development since it has the potential to significantly improve steering accuracy and repeatability, while freeing the operator to attend to higher-level activities. In orchard and grove applications autonomous guidance has several potential uses, ranging from mowing, spraying, disease and pest scouting to planting and harvesting. Most efforts to date have focused on semi-autonomous operations where human supervisors would be on-board the vehicle during operation. However, future opportunities exist for fully autonomous systems, though personnel safety, cost, reliability and legal constraints may delay adoption of fully autonomous operations in groves.

Vehicle position, heading, steering effort and speed with respect to the desired path are the most important issues that must be considered. Global positioning systems (GPS) in combination with inertial navigation systems have been widely used as positioning and heading sensors in traditional field agriculture application. Both real-time kinematic GPS (RTKGPS) and real-time differential GPS (DGPS) have been tested with success based on the degree of accuracy required in the navigation system. There is a trade-off between accuracy and cost in the selection of DGPS and RTKGPS, with the latter being more accurate and expensive. RTKGPS has been giving very accurate results (Benson *et al.*, 2001; Nagasaka *et al.*, 2002; Noguchi *et al.*, 2002). Gyros have been widely used for inclination measurement (Mizushima *et al.*, 2002). Fiber optic gyro (FOG) has given the best performance (Nagasaka *et al.*, 2002). At present gyros and inclinometers are available together as inertial measurement units (IMU) for pitch, roll and yaw and linear velocity measurements. With the combination of RTKGPS and FOG, accuracy of ± 5 cm (2.0 in) has been achieved (Noguchi *et al.*, 2002). GPS cannot be used alone for positioning in some tree fruit applications as it gives errors when the vehicle moves under tree canopies.

In addition to sensing global positions, the vehicle must be able to detect local obstacles that may impede the path. Several sensing technologies have been explored for this task. Ultrasonic sensors can map tree canopies while traveling at speeds of 1.8 m/s (5.91 ft/s), but measurement accuracy is better at lower speeds (Iida and Burks, 2002). The development of machine vision guidance techniques has become a very attractive sensing alternative, especially when combined with other proximity-based sensors (Zhang *et al.*, 1999; Benson *et al.*, 2001). They have proven to be reliable in several row-crop applications, but have not performed well in sparsely populated crops. Their reliability reduces with low lighting, shadows, dust and fog. Benson *et al.* (2001) overcame this by using artificial lighting. Laser radar (ladar) has been used for ranging and obstacle avoidance. It has higher resolution than ultrasonic sensing, and requires fewer computations than vision. Its performance degrades with dust and rain, like vision, and it is costlier than ultrasound. It provides planar data of the path, but can generate three-dimensional (3D) images by rotating the laser source to give a 3D view. O'Conner *et al.* (1995) found that sensor data was noisy, and could be filtered using Kalman filters to obtain robust sensor fusion.

Steering control is a major factor for accurate guidance. Proportional integral derivative (PID) control has given satisfactory performance (Zhang *et al.*, 1999). Neural networks have the inherent disadvantage of learning only what the driver does, so they are not robust. Behavior-based control is a new development that has been successfully used in small mobile robots. A behavior-based system in combination with a real-time control system (RTCS) is expected to do well in vehicle guidance. Fuzzy control has recently been tried, with results comparable with PID (Benson *et al.*, 2001). Senoo *et al.* (1992) pointed out that the fuzzy controller could achieve better tracking performance than the PID controller. It has wider adaptability to all kinds of inputs. Qiu *et al.* (2001) verified that the fuzzy steering control provided a prompt and accurate steering-rate control on the tractor. Kodagoda *et al.* (2002) found fuzzy control to be better than PID for longitudinal control. PID was also found to have large chatter, high saturation. A combination of fuzzy and PID control holds significant promise (Benson *et al.*, 2001). Efficient guidance can be achieved using a fuzzy–PID control system with vision, ladar and IMU as sensors. Subramanian *et al.* (2006, 2009) developed a successful in-the-row auto-guidance system for citrus groves relying on data fused from machine vision, ladar and IMU, without DGPS. This control implementation is demonstrated in Fig. 11.1.

11.2.1 Case Study: Autonomous navigation in citrus groves

Subramanian *et al.* (2006) instrumented a John Deere electric 'Gator' (e-Gator) all-terrain utility vehicle, as shown in Fig. 11.1, and implemented a fuzzy enhanced Kalman filter to fuse ladar, vision and IMU path estimates to predict optimal course correction efforts (Subramanian et al., 2009). Through the use of a Pure Pursuit algorithm, steering and speed commands were generated to drive the position error to zero. This approach provided more robust path correction than previous un-fused sensory data selected on a priority basis in combination with a PID-based steering controller. In addition to in-row navigation, a vision- and ladar-based approach was developed for accommodating various uncertainties, such as missing trees and end-of-row conditions. Figure 11.2 shows the e-Gator moving down the row with vision and ladar constantly scanning the forward-looking terrain. By observing abrupt changes in ladar range to tree canopy and variation in path versus tree spectral characteristics, algorithms were developed that could detect the end-of-row conditions. A visual odometry-based dead reckoning approach was implemented to assist end-of-row turning, thus providing the capacity for fully autonomous navigation within the grove.

The vision algorithm used color-based segmentation of trees and ground, followed by morphological operations to clean the segmented image. The path center was determined as the center between the tree boundaries. The error was calculated as the difference between the center

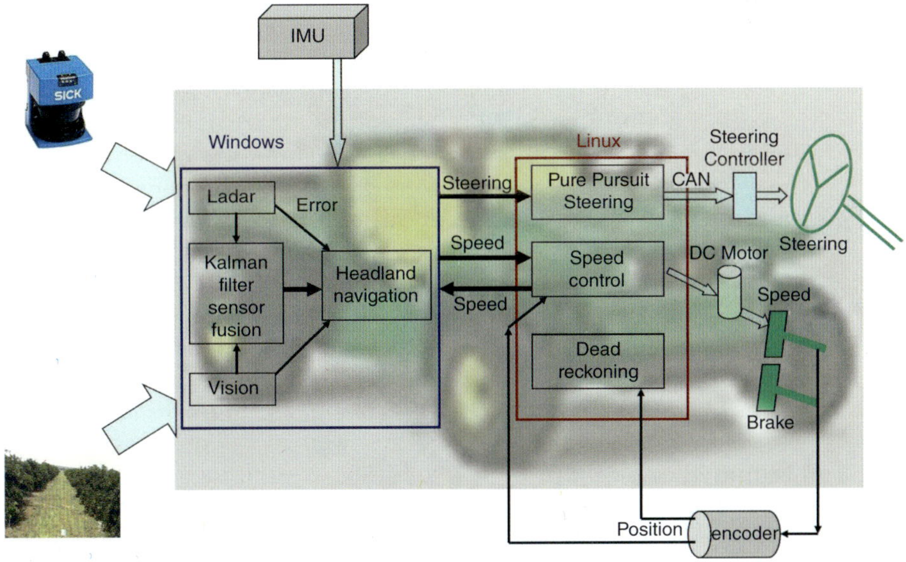

Fig. 11.1. Autonomous vehicle control architecture.

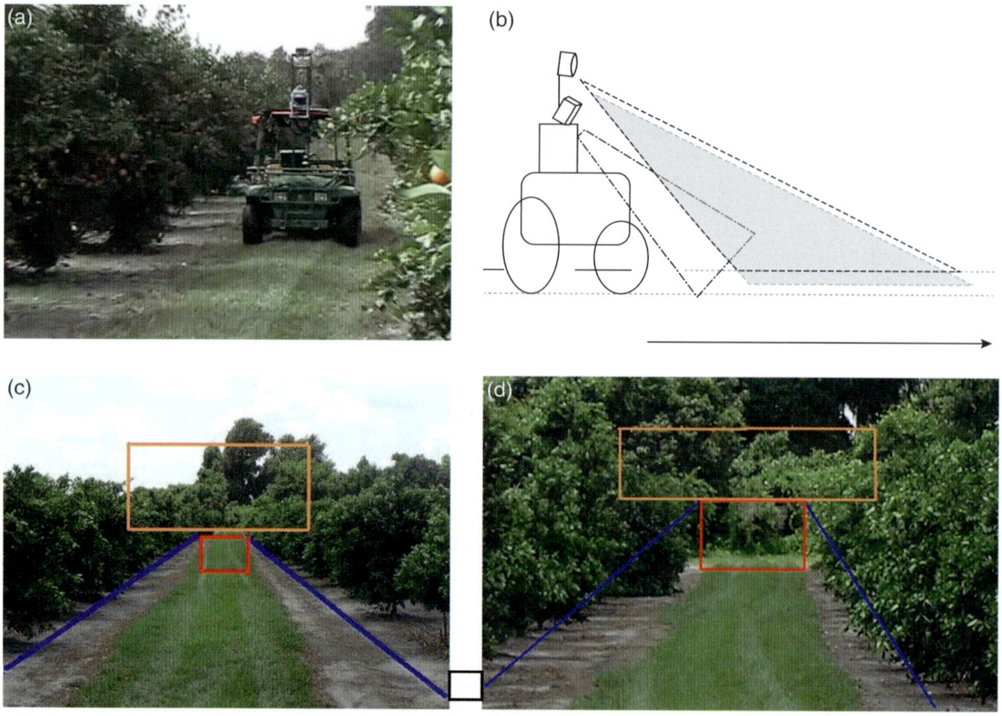

Fig. 11.2. (a) Autonomous vehicle in operation with in citrus grove; (b) ladar and vision field of view; (c) vision systems tracking path and canopy in-row; and (d) end-of-row detection.

of the image and the path center identified in the image. The error was converted to real-world distance using prior pixel-to-distance calibration. To calculate the required heading of the vehicle, a line was fitted for the path center in the image representing the entire alleyway. The angle between this line and the image center line was determined as the required heading for the vehicle to navigate the alleyway with low lateral error. The ladar algorithm used a distance threshold to differentiate objects from the ground. An object of a minimum width was identified as a tree. The midpoint between the trees identified on either side was determined as the path center. The errors measured using vision and ladar were adjusted for tilt using the tilt information from the IMU. The information from the sensors was used in the fusion algorithm, and the resulting error in lateral position was passed to a PID control system, which controlled the steering angle to reduce the error to zero.

11.2.2 Kalman filter design

A Kalman filter was used to fuse the information from the vision, ladar, IMU, and speed sensors. The models and implementation equations for the Kalman filter (Zarchan and Musoff, 2005) are described below.

11.2.3 State transition model

The state transition model predicts the coordinates of the state vector x at time k+1 based on information available at time k and is given by:

$$x(k+1) = Ax(k) + w(k) \tag{11.1}$$

where k = time instant; w = process noise, which was assumed to be Gaussian with zero mean.

$$x(k) = [d(k)\ \theta(k)\ \theta_R(k)v(k)]^T \tag{11.2}$$

where $d(k) \in R$ = lateral position error of the vehicle in the path, which is the difference between the desired lateral position and the actual lateral position; $\theta(k) \in R$ = current vehicle heading; $\theta_R(k) \in R$ = required vehicle heading; $v(k) \in R$ = vehicle linear velocity.

It is to be noted that only the change in heading was used for estimating the position error, which affects the overall guidance. The absolute heading was included in the filter to remove noise from the IMU measurements. Figure 11.3 shows the vehicle in the path with the state vector variables.

The state estimate error covariance matrix, $P \in R^{4\times4}$, gives a measure of the estimated accuracy of the state estimate $x(k$+1).

$$P(k+1) = AP(k)A^T + Q(k) \tag{11.3}$$

where $Q \in R^{4\times4}$ denotes the covariance matrix for process noise w.

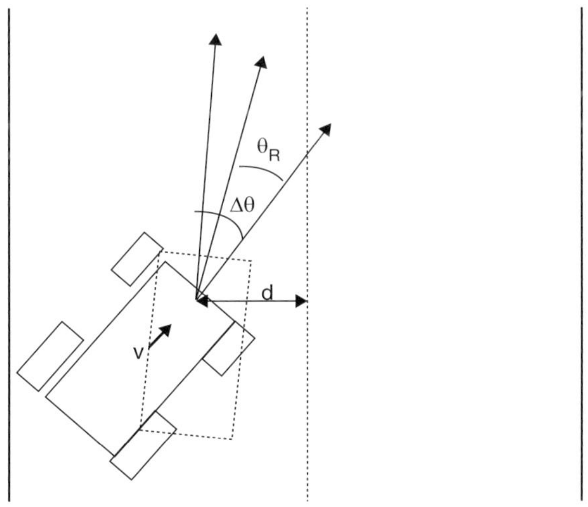

Fig. 11.3. Vehicle in the path with the state vector variables.

11.2.4 Measurement model

This model defines the relationship between the state vector, x, and the measurements processed by the filter, z:

$$z(k) = Hx(k) + u(k) \tag{11.4}$$

$$z(k) = [x_C(k) x_L(k) \theta_C(k) \theta_{\text{IMU}}(k) v(k)]^T \tag{11.5}$$

where $z(k) \in R^5$ denotes a state vector composed of measured values of:

- lateral position error of the vehicle in the path from vision algorithm, denoted by $x_C(k) \in R$;
- lateral position error of the vehicle in the path from ladar algorithm, denoted by $x_L(k) \in R$;
- required heading of the vehicle determined from vision algorithm, denoted by $\theta_C(k) \in R$;
- vehicle linear velocity measured using the speed sensor, denoted by $v(k) \in R$; and
- current heading of the vehicle measured using IMU, denoted by $\theta_{\text{IMU}}(k) \in R$.

11.2.5 Filter gain

The filter gain, G, is the factor used to minimize the posteriori error covariance, P:

$$G(k) = P(k+1)\, H^T (HP(k+1)\, H^T + R)^{-1} \tag{11.6}$$

where R is the measurement noise covariance matrix.

To determine R, the vehicle was set stationary in the middle of the path, each sensor (except the speed sensor) mounted on the vehicle was turned on independently, and the information from the sensors was collected for 30 s.

Figure 11.4 shows the operation of the Kalman filter. The filter estimates the process state x at some time k+1 and estimates the covariance P of the error in the estimate. The filter then obtains the feedback from the measurement z. Using the filter gain G and the measurement z, it updates the state x and the error covariance P. This process is repeated as new measurements come in and the error in estimation is continuously reduced.

11.2.6 Reliability factor of primary guidance sensors in the Kalman filter

Ladar is accurate at short range for the given mounting arrangement, and machine vision is good at providing the overall heading of the vehicle. By

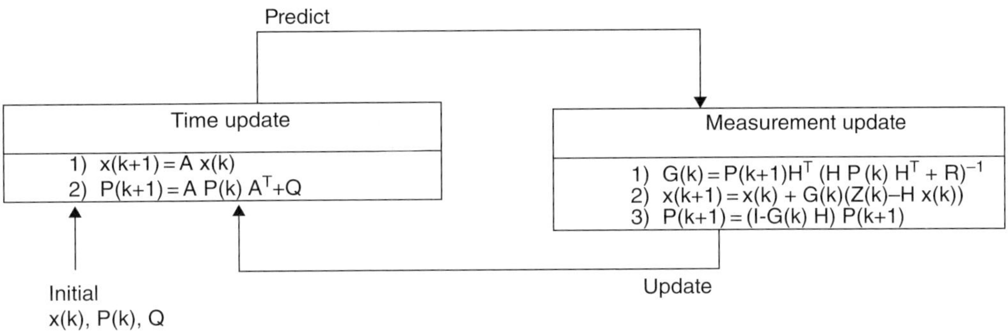

Fig. 11.4. Kalman filter operation.

experimentation in a citrus grove, the following observations were made: tree foliage is highly variable, trees can be absent on either or both sides of the path, and some trees can be small enough for ladar to not recognize them as trees. In such a variable path condition, it is not feasible to have constant noise values for the ladar and machine vision. For example, when the ladar does not recognize trees on either side, the guidance becomes more reliable if the process noise for the ladar is increased to a high value such that vision takes over as the only guidance sensor. When the vehicle gets to the end of a tree row, vision is no longer useful, so ladar can be made the sole guidance sensor to cross the last tree. A fuzzy logic system was used to decide the reliability of the sensor. The reliability was changed in the Kalman filter by changing the measurement noise covariance matrix R.

11.2.7 Fuzzy logic sensor supervisor

A fuzzy logic based supervisor was used to decide which sensor is more reliable at different locations in the grove alleyway. The fuzzy logic algorithm was implemented in software using C++ in the PC-based controller. The input to the fuzzy logic supervisor was the horizontal distance of the vehicle centerline from the trees on either side of the vehicle. Both vision and ladar input their corresponding distance values. Altogether, there are four input values: vision left tree distance, vision right tree distance, ladar left tree distance, and ladar right tree distance. These inputs were divided into three linguistic variables: reasonable, unreasonable, and zero (Fig. 11.5).

A triangle-based fuzzification method was used to fuzzify the input values. The input membership function relates the input x to the linguistic variables reasonable, unreasonable, and zero (Fig. 11.5b). The meaning of the linguistic variables is literally whether the distance from the tree row is reasonable or not.

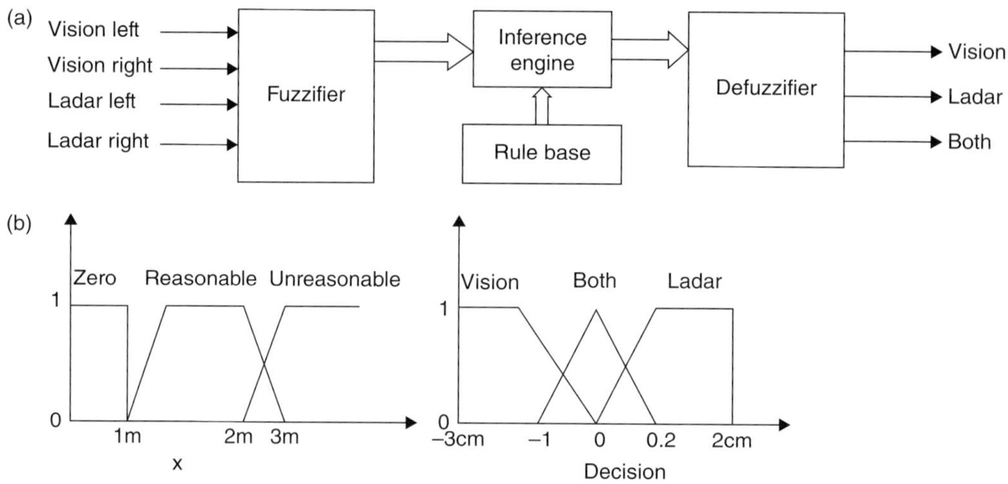

Fig. 11.5. (a) Fuzzy logic structure and (b) membership functions for sensor supervisor.

11.3 Novel Technologies for Robotic Crop Status Monitoring

During the last decade numerous studies have sought to develop various forms of crop status monitoring using technologies like machine vision in the visible and near-infred (NIR) spectral regions, ladar, ultrasonics and more recently biosensors. These sensor technologies are being used to monitor crop factors such as yield estimation, canopy size, canopy volume, leaf density, disease pressures and pest pressures. These data are then being used in various forms of precision agriculture to control fertilization cost, chemical usage, and even irrigation demands. This section will briefly introduce several of the applications under development. Once coupled with autonomous robotic platforms, these technologies can become the basis for a new era in robotic precision agriculture. In the following sections, numerous technologies will be introduced that could be used in conjunction with autonomous robots.

11.3.1 Laser used in precision sprayer

Campoy *et al.* (2010) established a tree canopy modeling and precision spraying system by using laser sensors and 3D maps. Chen *et al.* (2011) developed a lidar-guided sprayer to synchronize spray outputs with canopy structures. A density algorithm was developed by using the depth value from the laser scanner. Variable rate application (VRA) was realized through the use of a pulse width modulation (PWM) solenoid valve.

11.3.2 Laser used in yield estimation

Swanson *et al.* (2010) developed a multi-modal system using ladar (laser detection and ranging) for yield prediction in citrus trees. In this study, a ladar and stereo color cameras were mounted on vehicles to capture canopy features, such as canopy volume, canopy density and orange counts. They provided two models to estimate the yield: linear regression model and kernel regression model.

11.3.3 Laser/lidar used in tree canopy volume

Tumbo *et al.* (2002) also did an investigation of laser measurement on citrus canopy volume. The information given by laser measurements could be used to calculate a laser canopy volume index (LCVI). This information was acquired by using Schwartz Eloctro-Optics (SEO) laser scanners (Wangler *et al.*, 1992). Wei and Salyani (2005) extended this laser scanning system to calculate foliage density. They defined foliage volume as the space contained within the laser incident points and the tree row plane, and canopy volume as the space enclosed between outer (smoothed) canopy boundary and the tree row plane. Then they defined foliage density (D_f) as the ratio of foliage volume to tree canopy volume.

Swanson *et al.* (2010) built a linear and kernel regression model to estimate the citrus yield based on canopy volume, density and fruit counts. The laser scanner was mounted on a vehicle. The vehicle went along the tree row. The 3D point clouds were collected using multiple scans. Then 3D point clouds were mapped into a 3D grid composed of voxels. The volumes of all of the voxels were summed together for each slice to form the total canopy volume.

11.3.4 Machine vision used in yield estimation

Annamalai and Lee (2003) developed an image processing algorithm to locate citrus fruits in a canopy image. By studying the pixel distribution of hue and saturation color plane, the thresholds to separate different classes (citrus fruits, leaves, and background) were estimated. For the algorithm: (i) color image was transformed to binarized image through the threshold in hue, saturation and intensity (HSI) color image; (ii) then erosion and dilation were applied to remove noise; (iii) after that, the gap within a fruit was removed by applying dilation and erosion; and (iv) finally the number of fruits was counted by using blob analysis. The result showed that the correlation coefficient was 0.76 for the regression analysis between manual method and machine vision algorithm. Annamalai *et al.* (2004) made an improvement on the previous algorithm by adding a luminance component to the threshold to make it less dependent on the brightness level.

Swanson *et al.* (2010) provided methods to calculate tree canopy volume, density, immature orange counts and mature orange counts.

They used intensity profiles to detect green oranges. For mature oranges, they firstly converted the image into Luminance plus color a and b (Lab) color space. Then, the oranges were separated from background by calculating the minimum distance from each pixel to the sample region means. Finally, morphological operators and watershed transform were used to filter and segment fruit clusters.

Han and Burks (2010) employed a Plucker coordinates system to reconstruct a 3D image scene of a citrus canopy using monocular vision. A single video camera was mounted on a robot manipulator to capture multi-sequential views of the citrus canopy. The centers of tree leaves were detected as the leaf features. These feature points were used for feature matching in the speeded-up robust features (SURF) method. After feature matching, two methods (8-points algorithm and Plucker coordinate system) were used to reconstruct the 3D citrus canopy. The Plucker coordinates system showed a better result than the 8-points algorithm.

11.3.5 Machine vision used in detecting citrus greening on leaves

Pydipati *et al.* (2006) developed a disease detection approach for citrus leaves using color vision-based texture analysis. The image data of infected leaves from various diseases common in citrus were collected and texture features were extracted using the color co-occurrence method reported in Burks *et al.* (2000). Significant texture features were selected and used to train both statistical classifiers and neural network classifiers. Classification accuracies approaching 97% were achieved in laboratory conditions. Kim (2011) developed a machine vision-based method for detecting citrus greening on leaves using color texture features under controlled lighting in order to discriminate between citrus greening and leaf nutrient deficiency conditions that are commonly confused with citrus greening. This approach used low-level magnification to enhance features and was conducted in a laboratory setting, achieving classification accuracies above 95%.

Mishra *et al.* (2007) developed a spectral method for the detection of huanglongbing (HLB). Canopy reflectance spectral data was collected with an ASD FieldSpec spectroradiometer. Using techniques of spectral wavelength discriminability, spectral derivative analysis and spectral ratio analysis, the research aimed at identifying optimal wavebands for accurate detection of HLB in citrus. It was concluded that the visible region (400–700 nm) has good (0.89–0.85) discrimination. Results from the finite difference second derivative method revealed that wavelengths of 480 nm, 590 nm, 754 nm, 1041 nm, and 2071 nm have the potential to differentiate HLB. Using the spectral ratio analysis, it was found that the reflectance of HLB-infected trees at 530–564 nm was higher than that of healthy trees. A second sensitive point was observed at 710–715 nm. Sankaran *et al.* (2010) used mid-infrared (MIR) spectroscopy to analyze and detect HLB-infected citrus leaves. The healthy, nutrient-deficient, and HLB-infected leaves were ground under liquid nitrogen and analyzed using a portable MIR instrument. The preprocessed data were analyzed with principal component analysis and the samples were classified using quadratic discriminant analysis (QDA) and k-nearest neighbor (kNN)-based algorithms. The statistical models, QDA and kNN yielded high overall classification accuracies of > 80%, with a diseased (HLB) class classification accuracy of > 90%.

11.4 Cultural Practices Mechanization and Automation

Tree fruit production is very labor intensive, and as a result of increasing labor costs, shrinking labor pools and global market pressure from developing countries its economic viability has reached a critical stage. Although harvesting is typically recognized as the most labor-intensive operation, cultural practices such as weed and grass control (mowing), control of tree size and shape (hedging and pruning), and control of fruit yield and size (thinning) have become a popular target for automation. This section will discuss the state of mechanization and automation in cultural operations in tree crops. Many of these cultural practices have potential for automation and robotic applications.

11.4.1 Hedging and pruning automation in orchard production

Pruning of fruit trees: (i) adjusts tree shape and the ratio of framework to fruit-bearing area of the canopy; (ii) alters the top/root ratio; and (iii) changes the food storage status of the tree (Tucker *et al.*, 1994). Proper control of crop growth is essential for the maintenance of a healthy and productive orchard. In addition, pruning improves sunlight access for the tree, which provides the energy for photosynthesis. Light becomes a limiting factor in crowded groves.

The mechanization of pruning began in California in the early 1960s through an effort to mechanically top lemon trees (Jutras and Kretchman, 1962). Mechanical toppers were used to eliminate hand pruning of vigorous shoots at the top of the tree. The machine consisted of a modified sickle-bar mower blade mounted on towers, which were adjustable for height. In later years, topping machines used a series of circular saws mounted on a horizontal boom (Sansavini, 1978). Mechanical pruning is based on a predetermined cutting plan: horizontal top cutting (topping) and vertical walls, or oblique hedging (house top). Consequently, hedging is normally surface pruning. However, for certain crops such as citrus, deep cuttings on alternating sides of the tree on an every-other-year basis is often employed.

11.4.2 Fruit thinning by hand, string mechanisms and electromechanical methods

Hand thinning is a necessary but costly management practice in peach production. Organic apple production also may require hand thinning to adjust crop load. Mechanical devices to aid in thinning have been developed but have not proven efficient or capable of completely replacing hand thinning. The introduction of narrow canopy training systems and novel peach tree growing approaches will create new opportunities to examine mechanical methods for thinning peach and apple trees (Schupp *et al.*, 2008). A spiked-drum shaker was used to thin pillar peach trees at 52–55 days after full blossom. The spiked drum was a vibrating direct-drive double spiked-drum shaker designed for harvesting citrus. The shaker was mounted on a tractor-towed trailer and consisted of two rotating drums each measuring 8 ft (2.4 m) in diameter and 5 ft (1.5 m) in height. Each drum was composed of six whorls of nylon rods spaced 12 in (0.3 m) apart on a central axis. Each whorl was made up of 16 individual rods and the whorls were radially spaced at equal angles around the axis of the drum. Results of the drum shaker trial, conducted at a commercial orchard, showed that, although this type of mechanical thinning generated larger fruit, the level of crop reduction and disproportionate removal of fruit over the canopy were a concern. It also broke some small shoots and twigs, and caused bark damage when rods got entangled in the branches.

A rotating string thinner was designed by H. Gessler, a German grower, to remove apple blossoms in organic orchards (Schupp *et al.*, 2008). The string thinner consisted of a tractor-mounted frame with a vertical spindle 3.0 m tall in the center of the frame. Attached to the spindle were 36 steel plates securing a total of 648 plastic cords, each 0.5 m long. The speed of the rotating spindle was adjusted by a hydraulic motor. The height and angle of the frame were adjustable to conform to the vertical inclination of the tree canopy, and the intensity of thinning was adjusted by changing the number of strings and the rotation speed. In the 2007 commercial orchard trial, the string thinner effectively reduced flower density in the upper canopy part as compared with hand thinning. It was also observed that the string thinner had a much greater blossom removal on branches parallel to the drive row. Access to interior canopy and blossoms was limited. The researchers suggested refinements in both machinery and canopy design to obtain maximum efficacy.

A study was conducted to improve mechanical thinning by exciting individual branches at a precise frequency and duration to achieve a superior distribution of fruit remaining on the tree (Rosa *et al.*, 2008). A unique and precise electromagnetic limb shaker that required no branch clamping was developed and evaluated under field conditions and tested on nectarines, peaches and prunes. Results from tests conducted using this electromechanical fruit thinning showed that more fruits were removed from the top part (33%), followed by the middle part (28%). The lower part of the tree had the lowest removal percentage (14%). A low removal rate at the top portion is preferred, because high-quality fruits are usually located at the top. Although results showed some potential, this type of fruit thinning will be more effective if it has the ability to identify individual branches.

11.4.3 Robotic pruning/thinning

An alternative to mechanical pruning/thinning uses robotic approaches that can emulate the manual pruning. Ideally, a robotic system would provide similar or better quality pruning at a much faster rate and can work for extended periods of time compared with what a human pruner could accomplish. These robotic systems are equipped with machine vision for pruning point detection, and robotic manipulators with special end-effectors designed to prune or thin.

Vision Robotics Corporation (VRC, 2011) developed a robotic vineyard pruner. According to VRC, the robotic pruner automatically spur-prunes the grape vines with a quality comparable to hand pruning. The robot's stereo camera pre-scans the vine and the robot creates a 3D model of the vine which includes the vine, canes, and the buds. The pruning plan is determined from the model and used to guide the robot. Tho robotic arms are equipped with hydraulic pruning shears. Initial results showed that the robot could prune a typical acre in about 3.5 h.

11.4.4 Precision spraying applications

Berenstein *et al.* (2010) developed algorithms for a machine vision-based grape cluster and foliage detection autonomous selective vineyard sprayer. One algorithm called Foliage Detection Algorithm (FDA) was used to detect foliage to apply pesticide on leaves. Three algorithms called Grape Detection Algorithms (GDA1, GDA2, and GDA3) were used to identify grape clusters to guide application of hormones to the grapes. They found that 90% of the grape cluster could be identified and pesticides usage was reduced by 30%.

11.4.5 Yield monitoring

Yield monitoring systems have been developing over the past couple of decades applying new techniques to document and respond to the spatial variability of crops (Whitney *et al.*, 1999). Yield maps, with results of yield monitoring systems, have provided growers with essential information for spatial analysis and evaluation of crop production management at a within-field level. Yield monitoring systems can be divided into two main categories by their mass flow sensing techniques, either of which can be used in robotic harvesting applications: (i) indirect method based on machine vision techniques; and (ii) direct method based on weighing techniques. Vision-based yield monitoring techniques can be attractive not only for measuring mass flow of production but also for monitoring fruit quality, screening for problems with diseases, maturities, etc. However, according to Lee and Slaughter (2004), the challenges of recognizing occluded fruits using two-dimensional (2D) vision and execution rates between 14.6 s and 19.7 s are the two key weakness that need to improve in order to apply these systems to mass or selective harvesting machines.

Chinchuluun *et al.* (2006) built an automatic machine vision system for citrus fruit yield estimation using charge coupled device (CCD) cameras, ultrasonic sensors and DGPS to develop image-processing algorithms for fruit detection. Rather than measuring crops on platform beds or transporting belts, this system detected fruits on the tree before harvesting, and could not only count fruits on trees but also estimate size. The coefficient of determination (R^2) was found to be up to 0.83 for the number of fruit from manual counting versus the number of fruit counted by the vision-based counting algorithm. However, the coefficient between the number of fruit counted by the algorithm and actual harvested fruit was only 0.64.

On the other hand, efforts using weighing-based yield monitoring techniques have been successful in several applications. Weighing-based techniques can be classified by intermittent and continuous weighing methods. Measuring the weight of a truck loaded with fruits using weighing sensors is one common method of intermittent weighing. In spite of its price competitiveness and high measuring accuracy, it has been replaced by the continuous weighing system due to the latter's time-saving efficiency and the necessity to re-establish material flow (Dawson *et al.*, 1976).

Whitney *et al.* (1999) integrated the geographical information system (GIS) and GPS with a fruit road-siding truck in order to investigate weight-based yield mapping. They integrated a load-cell weighing system at each corner of the truck lift bed and pressure transducers on the lift cylinder in order to record weight data for citrus. Each time the truck operator loaded a fruit tub, he would push a button to enable the data logging system to record load cell and pressure transducer readings, while at the same time acquiring positioning data from the GPS receiver. Although their yield monitoring system was only a prototype, it successfully created color-coded yield maps which were easily interpreted by growers. Whitney *et al.* (2001) implemented a RTKGPS with the yield monitor and compared position accuracy with two other commercial DGPS systems.

Upadhyaya *et al.* (2006) developed an electronic weighing device with an impact plate and a conveyor-speed sensing system to measure mass flow of tomatoes. Tomatoes impacted the plate as they dropped off the harvester boom conveyor, where the impact force and conveyor speed data were recorded continuously on a data logger. This weighing system was integrated into a commercial tomato harvester and tested during the 2004 and 2005 harvesting seasons. A weigh wagon was used to verify the measurements of the impact-type electric weighing system. The results of tests suggested very good potential for impact-type weighing systems, with R^2 exceeding 0.96.

11.5 Robotic Tree Fruit Harvesting Background

Robotic solutions for fresh market tree fruit harvesting have been studied by numerous researchers around the world during the past several decades. However, very few developments have been adopted and put into practice. The reasons for this lack of success are due to technical, economic, horticultural and producer-acceptance issues. In industrial automation applications, a robot's environment is designed for optimal performance, eliminating as many variables as possible through careful systems planning. In agricultural settings, environmental and horticultural control can be a significant hurdle to successful automation. Not only must the plant system be designed for successful automation, but also the cultural and horticultural practices employed by the producers must often be changed to provide a plant growth environment in which robotic systems can be successful. According to Sarig (1993),

> The major problems that must be solved with a robotic picking system include recognizing and locating the fruit, and detaching it according to prescribed criteria, without damaging either the fruit or the tree. In addition, the robotic system needs to be economically sound to warrant its use as an alternative method to hand picking.

A successful robotic harvesting system must be able to satisfy the following constraints: (i) picking rate of fruits should be faster than or equal to manual picking; (ii) fruit quality should be equal to or better than manual picking; and (iii) the system should be economically justifiable.

Economic analysis of robotic citrus harvesting was carried out by Harrell *et al.* (1988), who identified 19 factors that affect harvesting costs and concluded that the cost of robotic citrus harvesting was still greater than that of hand harvesting. They found that robotic harvest cost was primarily affected by harvest inefficiency, followed by harvester purchase price, average picking cycle time and harvester repair expense. They concluded that robotic harvesting technology research and development should continue and should concentrate on the following areas: (i) harvest inefficiency; (ii) purchase price; (iii) harvester reliability; and (iv) modifications in work environment that would improve performance of robotic harvesters. Furthermore, it was found that the robotic harvest cost was most sensitive to harvest inefficiency. Therefore, it was recommended that the primary design objective would be to minimize harvesting inefficiency. They concluded that a harvesting efficiency of 93–99% would be required before robotic harvesting reached breakeven point with manual labor at current harvesting costs.

Robots tend to perform well in structured environments, where the position and orientation of the target is known or targets can be set up in the desired position and orientation. Harvesting fruit crops robotically in unstructured environments creates a new set of challenges, since many of the aspects relied upon by industrial robots do not exist. Challenging design conditions might include, for example, non-uniform lighting ranging from direct sunlight to overcast and twilight conditions, variable temperature and humidity, wet and dry conditions, variable fruit sizes and maturity, non-uniform plant size and fruit position, fruit occlusions and limb obstacles, mobile power supplies, and a dirty, harsh environment. Add to these the need for low-cost equipment solutions and there is a very difficult engineering design requirement, which explains the low development success rate.

The objective of the next two sections is to present an overview of the major horticultural and engineering aspects of robotic harvesting systems for tree crops. In order to provide the reader with sufficient breadth of information, this section is primarily a survey that tries to identify the key issues that robotic system developers and horticultural scientists should consider to optimize plant–machine system performance.

11.5.1 Horticultural aspects of robotic harvesting

Modifications and improvements in cultural practices for mechanization are continually being made through research and experience (Sims, 1969). In order to have a successful automated/mechanized system, the cultural practices must be designed for the machine and the variety (Davis, 1969). A systems development approach must be followed to ensure that the cultural practices are suited for the crop variety and machinery systems being considered (Sims, 1969). The major aspects related to cultural practices that affect fruit and vegetable mechanical harvesting include field conditions, plant population and spacing, and plant shape and size. Efficient harvesting mechanization cannot be achieved by machine design alone. Establishing favorable field conditions for the harvesting system under development has to be considered before the harvesting system can be effectively developed (Wolf and Alper, 1983).

Peterson *et al.* (1999) developed a robotic bulk harvesting system for apples. They trained the apple trees using a Y-trellis system and found them to be compatible with the mechanical robotic harvesting. Fruit was trained to grow on the side and lower branches to improve fruit detection and removal. They further suggested that pruning could enhance the harvesting process by removing unproductive branches that block effective harvesting. Further research was suggested to determine the variety and rootstock combinations most compatible with the training and harvesting system. The concept of designing a grove for optimal economic gain requires an optimal combination of varieties, rootstocks, grove layout, production practices, and harvesting methodologies.

11.5.2 Plant population and spacing

Harvesting equipment can operate at maximum productivity when the workspace has been organized to minimize inefficient obstacles, standardize fruit presentation, provide sufficient alleyways, and maximize fruit density on uniform growth planes.

Certain tree species and even certain varieties within species have an optimal subsistence area for best fruit production, which provides a proper ratio between the number of leaves needed to produce carbohydrates and other organic compounds, and the number of developing fruits (Monselise and Goldschmidt, 1982). The woody mass – roots, trunk, scaffolds and branches – support the tree canopy, but contribute minimally toward fruit development once nutrient uptake and moisture demand are met. However, they continue to use the tree's resources to maintain themselves, presenting obstructions to robotic harvesting. Ben-Tal (1983) suggested that maximum yield per unit area would be achieved by a large number of relatively small trees, suggesting that smaller robotic systems may actually provide a better economic return.

Scalability of robotic systems is an important economic factor, which impacts the design of the plant growth system. The productivity of large multiple-arm systems versus smaller more agile human-like robots is an important economic question. Large equipment systems require wide row spacing, while smaller systems can work in a more confined grove configuration. Optimally, the fruit should be grown in a hedgerow configuration where the plants produce a maximum number of fruits over the surface area (Ben-Tal, 1983). This suggests that the trees or plants be grown at a close spacing so that the growth plane is uniform with minimal scalloping of the hedge between plants.

11.5.3 Plant shape and size

The ideal configuration for efficient robotic harvesting would be a vertical or slightly inclined hedge wall, 10–12 ft (3.0–3.7 m) tall, which was relatively uniform, smooth and continuous from start to row-end. The fruit would be located on the canopy surface with minimal occlusion. In reality this would not be the case, but it provides some insight into what the robot would need in order to maintain fast harvest cycle times and maximum fruit removal. Deviations from the ideal will cost removal efficiency and cycle time performance.

Orchards should have uniform plant sizes and predictable shapes for efficient robotic harvesting (Cargill, 1983). Standardization of tree sizes significantly improves harvesting throughput and thus economic benefit. These standard sizes should consider tree height, tree thickness, tree shape, and tree spacing within and between rows, so that the robotic equipment can maintain continuous harvesting, with minimal idle harvest time when traveling between trees. A number of these features are designed into the grove at planting, while others must be maintained mechanically through cultural practices. A common modern approach for maintaining both tree size and shape is mechanical pruning. The trees can be pruned to the desired shape before fruit set and allowed to grow during the remainder of the year. In some limited cases, severe pruning is being tested. Under this practice, alternating sides of the tree are pruned each year and allowed to set fallow, while the other side of the tree produces the current year's crop. When the canopy returns the following year, the woody mass is covered by the new growth and a relatively uniform vertical wall is achieved. Impact of annual fruit yield has not been reported on this technique to date.

Experiments conducted on apples demonstrated that tree shape contributed toward the suitability of mechanical harvesting (Zocca, 1983). Modifications to cultural practices for growing and harvesting fruit are important for successful mechanical harvesting. A mechanized pruner was developed that not only reduced the labor required for pruning, but also properly shaped the hedgerow for maximum harvesting efficiency of erect cane fruits (Morris, 1983).

Ben-Tal (1983) points out several problems that can arise when an orchard is prepared through pruning for a specific kind of equipment, such as reduced yield, fruit quality, and the number of years of production. Additional issues such as canopy light exposure and maximum height of a tree for proper spraying, pruning, etc. should be considered. The question of plant geometry and its relationship to productivity needs to be thoroughly examined (Rohrbach, 1983).

11.5.4 Tree genetics for optimal harvesting

Plant breeders developing new varieties of fruit must consider whether the variety will be accepted at market and whether it will be durable under machine handling. Attractive appearance and long shelf-life are imperative in the fresh market. Varieties must be resistant to bruising, cracking, and rupturing during machine handling. The fruit must be relatively easy to remove from the plant and the peduncle must remain attached (Davis, 1969; Lapushner *et al.*, 1983).

In addition to fruit-related issues, there are a number of tree factors that can be improved genetically which can enhance robotic harvestability. Two major obstacles impede efficient robotic harvesting: (i) locating fruit occluded by the leaf canopy; and (ii) harvesting fruit located in the tree or plant interior. In both cases, a plant system that presented the majority of the fruit at the canopy surface would improve harvestability. There are two possible solutions. The first would suggest a thin leaf canopy, so that the detection systems could more easily view the plant interior; and the second suggests a dense canopy that might force more fruit to grow at the surface. The two strategies seem to be in conflict under normal tree behavior. Sparsely leafed trees tend to have more interior fruit, which reduces fruit accessibility, while with densely leafed trees it will be more difficult to sense the interior fruit. A tree that naturally fruited at the limb extremities with minimal interior fruit might resolve this problem.

Another primary concern is canopy uniformity. Factors affecting uniformity in emergence, stand, growth, and maturity must be clearly understood in order to develop viable plant systems for mechanical harvesting (Davis, 1969). Cultural practices have been discussed that could produce a hedgerow system. However, trees that require severe hedging to maintain their shape often develop woody structures near the surface, which could be an obstacle to robotically harvesting interior fruit. A tree that grew to an appropriate mature height and shape and then maintained its size with either minimal hedging or woody mass build-up would be ideal.

Several plant breeding projects have contributed favorably to mechanical harvesting. Peach (*Prunus persica*) breeders increased fruit harvest by releasing varieties with varying maturities, effectively doubling or tripling the length of the peach season in many production areas (Carew, 1969). Dwarfing rootstocks in combination with apple varieties have provided size control of apple trees. Plant improvement through breeding can modify crop characteristics and assist in the introduction of mechanical harvesting systems (Carew, 1969).

11.6 Design Aspects of Robotic Harvesting

Robotic system developers from the USA, Europe, Israel, and Japan conducted independent research and development on harvesting systems for apples and citrus from the mid-1980s to 2000, achieving harvesting efficiencies of up to 75%. These low levels of performance were attributed to poor fruit identification and the inability to negotiate natural obstacles inside the tree canopy (Sarig, 1993). Harvesting cycle times for citrus were estimated at 2 s/fruit for a two-arm machine (or 4 s/fruit for a single-arm machine). Cycle times for apples were expected to be higher than citrus, due to improved canopy access (Sarig, 1993). These levels of harvesting performance and the resulting economic return on investment prevented producer acceptance.

The focus of most robotic fruit harvesting projects has been to design a harvesting system that can mimic the precision of a human harvester while improving harvesting efficiency and labor productivity. The typical design of a robotic fruit harvester consists of a vision system for detecting the fruit, a manipulator that acts like a human arm, and an end-effector to pick the fruit. However, a complete robotic harvesting system is actually much more complex, as can be seen in Fig. 11.6.

The system architecture illustrates the various functional areas of a robotic system, which begins with the vehicle platform that must provide mobility within the orchard. Such a system must be equipped with adequate power resources not only to propel itself but also to power the various devices that it carries. This will normally require significant hydraulic, pneumatic and electric power capacity. In addition, appropriate environment protection must be provided for the human supervisor and controller components, since, at this stage of development, it is unlikely that any system will be fully autonomous. The platform may provide some level of autonomous guidance, especially within the alleyway, to relieve the supervisor of driving responsibility. The sensory system's task is to locate the vehicle position within the grove and then the fruit position relative to the robot. These sensory data will be used to move the robot through the grove and localize fruit position

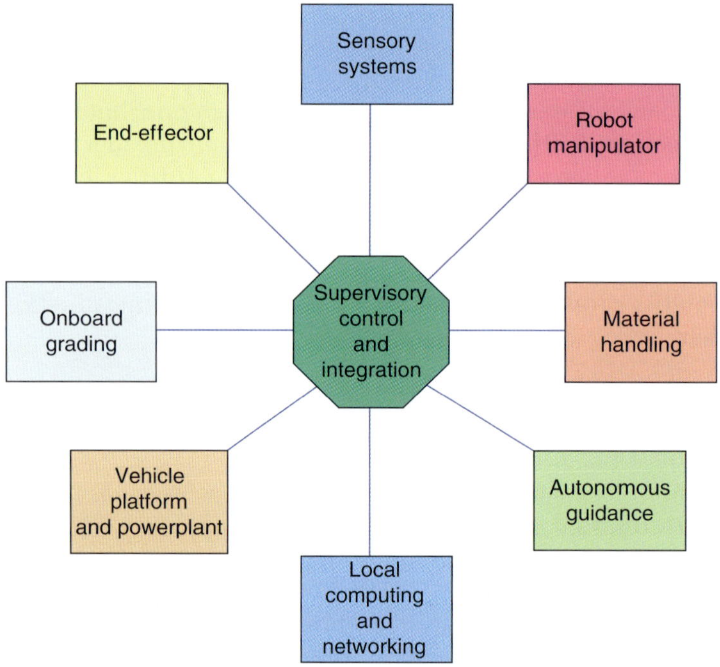

Fig. 11.6. Functional areas of the robotic system.

for harvesting. In some cases, fruit position maps may be created so that harvesting motions can be optimized for speed and energy consumption. The manipulator's task, which can take on many different configurations, is to move the harvesting end-effector into position to harvest the fruit. Once the robot platform is in harvesting position, the sensory suite will assist the manipulator through visual servo control to the final position where the fruit is within the grasp of the end-effector. The sensory suite may consists of machine vision cameras (stereo or monocular), laser, ultrasonic, or infrared range sensors, other proximity or tactile devices, global positioning systems, inertial measurement units and ladar. There are numerous end-effector approaches that have been implemented, each with their own unique characteristics.

Once the fruit has been harvested, the fruit is transported through internal conveyance to either an on-board containerization system, which fills and then offloads field boxes, or to a cross-conveyor, which offloads the fruit to a trailing transport vehicle. In either case, road-siding vehicles move the harvested fruit from the point of harvest to the road-side trucks, which will transport the fruit to the processor or packing house. The harvesting systems can be equipped with additional functionality using on-board grading systems that leave culls in the field, and yield monitoring systems that geo-reference fruit harvest rates for precision agriculture applications and traceability. Ultimately each robotic system must have an internal communications network based on ethernet or a Controller Area Network (CAN-Bus), which enables interfacing of all component controllers to the systems supervisor that will monitor individual component status and performance, and coordinate all system interactions.

11.6.1 Physical properties and fruit removal

A robotic harvester must be able to remove fruit quickly without damaging the fruit or the tree. An integral part of the harvester is the end-effector, i.e. a tool or device attached to the end of the manipulator that grabs and removes the fruit from the tree. Because of its direct interaction with the fruit and tree structure, it must be designed with the specific physical properties of the commodity to be harvested in mind.

There are several ways that a robot might damage the fruit or tree, such as: (i) end-effector applying excessive positive/negative pressure or force to the fruit during pick and place operations; (ii) inappropriate stem separation techniques for the type of fruit; (iii) fruit damage during retraction from the tree canopy or conveyance to bulk storage; or (iv) manipulator contact with the tree structure. Fruit damage may not be visually evident at harvest time. However, bruising, scratches, cuts, or punctures can result in decreased shelf life and increase food safety risks. Consequently, a properly designed end-effector must minimize or preferably eliminate fruit damage.

The fruit removal technique employed is typically the largest cause of fruit injury. In the case of oranges, the fruit must be harvested with the calyx intact and the stem removed flush with the calyx. If the peel is torn away from the calyx, the resulting fruit is unusable for the fresh fruit market due to contamination and reduced shelf life. This condition is referred to as 'plugging'. If a long stem remains on the fruit, the packer will either reject the fruit or require stem removal post harvest. The rind of oranges makes them one of the more durable fruit, in contrast with more delicate-skinned products, such as apples or peaches, but oranges are still susceptible to injury. Injury is more prevalent in less mature oranges (Juste *et al.*, 1988). Flood *et al.* (2006) extended the work of Juste by conducting rind resistance tests over a broader range of punch sizes, which was more representative of robotic harvesting end-effector contact areas. They determined contact pressure thresholds that would protect the fruit from puncture and bruising damage. Additional tests were conducted on various harvesting motions, resulting in the identification of optimal pitch and rotational modes of detachment that significantly reduced plugging (Flood, 2006).

When manually harvesting oranges, the fruit is detached using one of three methods depending on the variety and cultural practice. The laborer can use a set of clippers to detach the fruit, usually leaving as short a stem as possible. Secondly, the laborer can lift the fruit so that the stem axis is rotated 90 degrees and then pull down so that the force is perpendicular to the stem axis. Lastly, the laborer can add a twisting motion to the second method. Although the end-effector does not necessarily have to follow one of these methods, an understanding of manual procedures gives insight into some of the potential methods.

The first type of robotic harvesting end-effector developed was the cutting end-effector. Several cutting end-effector designs have been developed, as described in Ito (1990), Sarig (1993), Pool and Harrell (1991) and Bedford *et al.* (1998). This method is prevalent in several agricultural applications, since it produces the least amount of stress on the actual fruit. The basic premise is first to capture the fruit using a suction cup or gripper, and then to use a cutting device to sever the stem that is holding the fruit on to the tree. This can either be done blindly by swinging a blade around the outer edge or by detecting the stem's location and cutting it with a scissor device. The stem's location can be detected either through machine vision or through force/torque sensors. In the blind system, a blade would ideally pass around the encased fruit to sever the stem without damaging adjacent fruit or the tree. The blade must be large enough to encircle the fruit, and must maintain sharpness to achieve a clean cut. The scissor method reduces the chance of fruit damage but is substantially more complex, requiring a larger end-effector, more sensors, and more time. This approach is extremely difficult to implement successfully in clustered fruit.

The second type is the pull-and-cut end-effector. This method was proposed by Pool and Harrell (1991). In this method, the fruit is grasped either through suction or with a type of collection sock. The stem is severed as the end-effector retracts. This method disturbs the surrounding limb structure, making subsequent harvesting more difficult since the fruit is in motion, and still has some of the limitations of the cutting end-effectors previously mentioned.

The third type of end-effector design is the twisting method. This method was suggested by Juste *et al.* (1992) and Rabatel *et al.* (1995) to be the most promising of the three. This involves twisting the fruit, preferably perpendicular to its attachment axis, until the stem is severed. Twisting the fruit in this manner reduces the amount of disturbance to the tree and thus to the surrounding fruit. Twisting involves the least amount of force of the three methods and has the lowest plugging rate. Like the other two types, fruit size is a consideration here as well. Generally, the twisting action is achieved by use of a rotating suction cup. This cup must be of the right size to create a good seal while still providing enough force to keep the orange from slipping. One of the major advantages of this method is that there is a large flexibility in the angle of approach. Except at the stem, the cup can attach to any part of the fruit. However, experience has shown this approach to be relatively slow in some cases, due to numerous revolutions being required to achieve fruit separation.

Tutle (1985) suggested an approach that combined the twisting and pulling approach in US Patent 4,532,757. The end-effector design selected for a given application should be developed in conjunction with the manipulator, sensors, and control development to optimize the capabilities of the harvester. Flood *et al.* (2006) implemented and tested an approach similar to this with very low plugging rates once the appropriate harvesting sequence was identified.

11.6.2 Machine vision and sensing technologies

The first major task of a fruit-harvesting robot is to identify and locate the fruit. Once the fruit is located in the canopy, the robot can be directed toward the fruit for harvest. While humans can easily recognize fruit in the orchard, this is not an easy task for automatic harvesting. Fruits are objects that have variable shape, size, and color, and they are randomly positioned in a tree which also has variable size and different canopy density. In addition, these fruits are subjected to variable lighting conditions and other environmental elements like wind and moisture.

Schertz and Brown (1968) suggested the use of photometric information to determine the location of fruits on the tree, utilizing the light reflectance difference between the fruits and the leaves in the visible and infrared spectrum. With the advancement of computer and sensor technologies, the use of monochrome cameras fitted with a color filter or color video cameras has facilitated the discrimination of fruits from the canopy background, especially fruits that have contrasting color with their canopy, like the orange. In robotic fruit harvesting, machine vision has become one of the most popular sensing systems for fruit identification. A basic machine vision system includes a camera, optics, lighting, data acquisition system (USB, Firewire, Giga-E, or Frame Grabber), and an image processor, usually a personal computer. Vision systems are capable of determining either the 2D or three-dimensional (3D) position of the fruit, depending on the hardware and software implementation.

In pioneering research, Parrish and Goksel (1977) demonstrated the technical feasibility of using machine vision to guide a spherical robot for apple harvesting. In this research, a black-and-white camera was used to detect the apple fruits. A red filter was fitted in front of the camera to enhance the contrast between the fruit and the background. A few years later, Tutle (1985) developed a machine vision-based orange harvester, which used a photodiode array for image acquisition. Two filters were used with the photodiode; one filter was between 600–700 nm, which covers the chlorophyll absorption band, and the other filter permitted wavelengths between 750–850 nm, which is the water absorption band. Grand D'Esnon *et al.* (1987) used a color-based machine vision system for detecting apples. The image-processing algorithm was able to detect the red colored fruit, but problems were encountered in variable lighting conditions. At the University of Florida, Slaughter and Harrell (1989) developed an orange fruit-detection system with a 15-bit color camera using hue, saturation, and intensity to separate the fruit from the leaf canopy.

According to Sarig (1993), 'While major progress has been made with the identification of fruit on the tree and determination of its location, only 85% of the total fruits on the tree are claimed to be identified.' There are three major problem areas associated with the use of machine vision-based sensing: (i) partial and totally occluded fruit are difficult to detect accurately; (ii) light variability can result in low detection rates of actual fruit as well as high levels of false detections; and (iii) the computational time required to process images influences real-time control.

Fujiura (1997) developed robots with a 3D machine vision system for crop recognition. The vision system illuminated the crop using red and infrared laser diodes and used three position-sensitive devices to detect the reflected light. The sensors selected were suitable for agricultural robots that were required to measure the 3D shape and size of targets within a limited measuring range. Jiminez *et al.* (2000) developed a laser-based vision system for automatic fruit recognition to be applied to an orange-harvesting robot. The machine vision system was based on an infrared laser range-finder sensor that provided range and reflectance images and was designed to detect spherical objects in a non-structured environment. The sensor output included 3D position, radius, and surface reflectivity of each spherical target, and had good classification performance.

Plebe and Grasso (2001) presented a color-based algorithm for detecting oranges and determining the target centers. They also applied stereo imaging to these processed images to determine the range to the detected fruit. Their algorithm correctly identified 87% of the oranges, while 15% of the detected regions were incorrectly classified as oranges when they were not. Their approach had difficulty with both brightly and poorly lit oranges, brightly lit leaves, and certain types of occlusion. Bulanon *et al.* (2001) presented an algorithm that used a 240 × 240 pixel color image to detect apples. The apples were detected by thresholding the image using both the red color difference and luminance values. It was determined that the red color difference values were much more effective at detecting the apples than the luminance values. Bulanon *et al.* (2009a) demonstrated improved citrus fruit detection through multi-perspective viewing of a fixed boundary region of interest, achieving approximately 90% detection rates in orange canopies.

Numerous other sensors are commonly employed in robotic harvesting systems, such as ultrasonic range, laser range, capacitive proximity, light-emitting diode (LED) range, and so on. It is not likely that a single sensor will solve the complete sensing problem; rather, several sensors will need to be integrated together to form a complete sensory system.

11.6.3 Robotic manipulation and control

The manipulator is defined as a mechanical system, usually composed of a series of actuated links that function like a human arm capable of moving within one-, two- or three-dimensional space. In robotic fruit harvesting, the tool end of the manipulator is fitted with a fruit gripper, while the base is typically mounted on a mobile platform which positions it in the tree canopy. The manipulator's task is to move the gripper into position to pick the fruit and then place the fruit into a collection bin. The manipulator is composed of joints and links, similar to the human arm, with each joint having one degree of freedom (DOF). In general, a three-DOF robot can provide maneuverability to any point in 3D space without regard to orientation, while a six-DOF manipulator can move to any point within its work space volume with complete position and orientation capabilities. However, a six-DOF system is generally limited to a single pose and as such may not be able to avoid an obstacle in its work space. If 3D position and orientational tool frame accuracy is required, a redundant manipulator may be required. A redundant manipulator must have one more DOF than the degrees of positioning accuracy required. Therefore, if X, Y, Z Cartesian position and orientational degrees of pitch, roll and yaw are required, a minimal redundant manipulator would have seven DOF, with additional degrees being optional. However, numerous research and development efforts have been implemented that used less than seven DOF, often suffering from lack of maneuverability to avoid obstacles. Although additional degrees of freedom improve maneuverability and tool frame dexterity, they also increase manipulator cost and control complexity.

The manipulator's geometric configuration or architecture, forward and inverse kinematic algorithms and the manipulator dynamic equations of motion form the key design characteristics of a robotic manipulator. Unlike assembly lines in a factory, harvesting fruit trees (apples, oranges, etc.) is highly unstructured and the robot must have the workspace reach and end-effector dexterity necessary to reach fruit within a complex environment cluttered with limb and leaf canopy obstacles. There are several geometric configurations used in industrial applications that have been applied to fruit harvesting: Cartesian, cylindrical, spherical, articulated and redundant (Fig 11.7).

The first field prototype for harvesting apples was developed in France (Grand D'Esnon, 1985). The mechanical system consisted of a telescopic arm that moved up and down in a vertical framework. The arm was mounted on a barrel that could rotate horizontally. In 1986, a new prototype (MAGALI) was built (Grand D'Esnon *et al.*, 1987) that used a spherical manipulator servoed by a camera set at the center of the rotation axes. The manipulator used a pantographic prismatic movement (only rotational joints) along with two rotations.

In 1986, the University of Florida, along with other collaborators, initiated a program to develop a robotic system for citrus harvesting (Harrell *et al.*, 1988). The outcome of this research was a three-DOF manipulator actuated with servo-hydraulic drives. Joints 0 and 1 were revolute and Joint 2 was prismatic. This geometry was characteristic of a spherical coordinate robot. The feasibility of a robotic citrus harvester was ascertained by this research work.

The Franco-Spanish Eureka project (Rabatel *et al.*, 1995), started in 1991, was based on a feasibility study done at the University of Florida. The proposed robotic system had a dual harvesting arm configuration to achieve greatest economic return. However, the prototype consisted of only one harvesting arm. The arm had two modules: an elevating arm and a picking arm. The picking arm was of a pantographic structure rather than a linear structure. The elevating arm supported the picking arm and the associated camera. The elevating arm was equipped with a lateral DOF to avoid collision of the picking arm with the vegetation, while acting as a fruit conveyor as well.

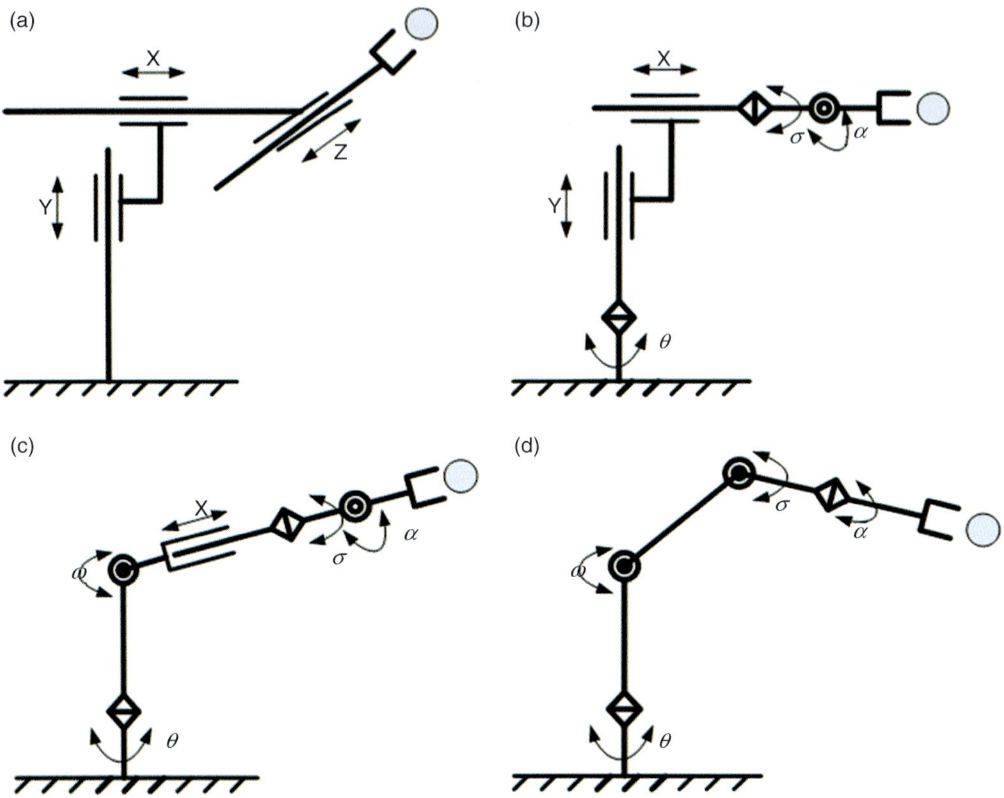

Fig. 11.7. Geometric configurations of industrial robots: (a) Cartesian, (b) cylindrical, (c) spherical and (d) articulated.

A mandarin orange harvesting robot for the orchard named 'Kubota', designed in Japan in 1989 by Hayashi, Ueda, and Suzuki, was reported by Sarig (1993) and Kondo and Ting (1998). The Kubota robot had an articulated arm with four DOF, but acted as a spherical coordinate robot due to the joint actuation schemes. The articulated arm of this robot utilized an end-effector with rotating stem-cutters that contained a color TV camera and a light source.

Ceres *et al.* (1998) presented a manipulator design for an aided fruit-harvesting robot (Agribot) that worked under human guidance. The articulated manipulator structure was designed based on a kinematic, dynamic and geometric study that took into account the fruit distribution on the tree. The parallelogram structure of the Agribot's picking arm had four DOF (all rotational), including the gripper. All of the joints were driven by electric motors. Fruit detection was done by a human operator using a laser telemeter and a joystick and fruit detachment was done through an end-effector with a suction cup that pulled the fruit into a V-shaped cutter.

Another development for a citrus-harvesting manipulator in Italy was reported by Cavalieri and Plebe (1996), Fortuna *et al.* (1996) and Muscato *et al.* (2005). The first research prototype had two spherically configured picking arms mounted at the tool point of a four-DOF positioning platform. The picking arms were driven by electric motors while the platform was driven by hydraulic actuators. Muscato *et al.* (2005) also presented a second prototype with two arms mounted on a 45-degree inclined platform carried on a caterpillar tractor. Both arms were of the Cartesian type and were driven by electric motors, but the upper arm had a telescopic link in place of a prismatic link as in the lower arm, due to space constraints. The research also presented a variety of end-effector developments, a three-finger pneumatic device that used a cutter to remove fruit from stem, a grasping device with a helix movement that brought the stalk into the cutter, and another pneumatic end-effector with jaws to capture the fruit, a sliding tray to hold the fruit and clippers to cut the stalk. Control was achieved through feedback from a camera and proximity sensor located on the end-effector.

Several manipulator architectures have been attempted for fruit harvesting. Of these, the articulated joint (six DOF) seems to work the best, since it closely resembles a human arm. In order to avoid obstacles and to harvest interior canopy fruit, the optimal configuration for a robotic harvester may require more degrees of freedom than a standard six-DOF articulated manipulator. Agricultural robotic arm developments in the past were simplified in terms of the arm mobility as well as their construction. This could be attributed to an intended reduction in development time and cost as observed by Sivaraman (2006). To address the issues reported from past arm developments as well as to realize an economically viable solution, significant task specific synthesis and performance evaluation are needed for the harvesting manipulator. Sivaraman (2006) conducted a synthesis of seven-DOF manipulators for the robotic citrus harvesting task, identifying candidate manipulator configurations using modern design tools such as RobotecPro and MATLAB Robotics Toolbox. These tools were used to evaluate workspace singularities, tool dexterity, actuator torque and acceleration requirements for various harvesting trajectories.

Figure 11.8 depicts the development process where in Fig. 11.8a, seven-DOF candidate manipulator configurations were identified; Fig. 11.8b depicts the workspace and singularity analysis that was done in MATLAB robotics toolbox that enabled link parameter selection based on workspace requirements; Fig. 11.8c depicts the tool space dexterity analysis completed in RobotectPro where the dexterity ellipse demonstrates excellent tool dexterity; and finally, Fig. 11.8d shows a harvesting workspace scenario where various harvesting trajectories were generated for each crossing point on the canopy map. These were used to determine the required actuator torques, velocities and accelerations necessary to achieve desired harvesting cycle rates.

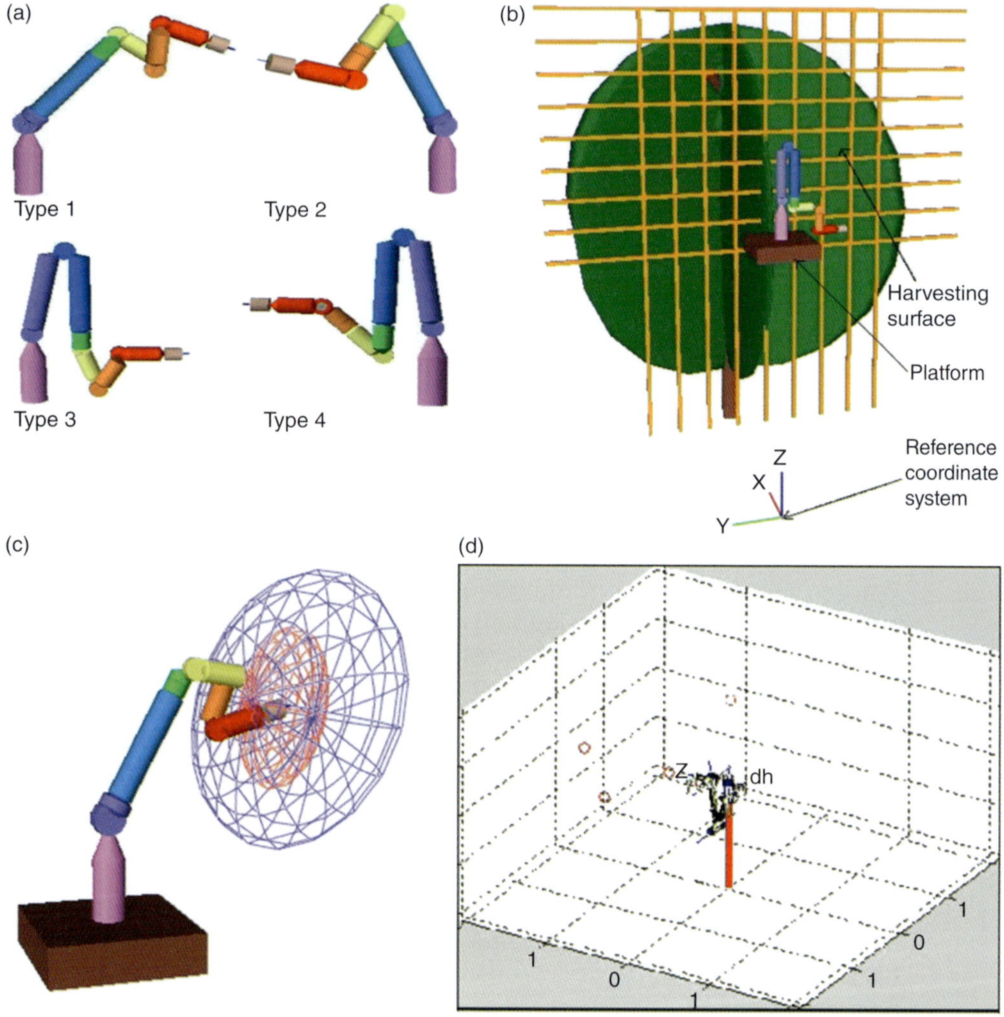

Fig. 11.8. (a) Seven-DOF articulated redundant manipulator configurations; (b) Workspace and singularity analysis model; (c) Dexterity analysis model; and (d) Harvesting trajectories model (Sivaraman, 2006).

11.7 Case Study: Robotic citrus harvester system development

The University of Florida and the Florida Department of Citrus began a collaborative investigation into the potential of using robotics for citrus harvesting in the early part of this century, which led to the development of a field test bed and eventually field trials from 2004 to 2009. This effort began with a thorough synthesis of the successes and failures of prior research efforts. Several observations were made: (i) fruit detection rates were inadequate to achieve commercial feasibility; (ii) harvesting efficiency as it relates to both fruit detection and fruit removal are too low; (iii) fruit located within canopy interior is very difficult to harvest; and (iv) current end-effector technologies are inadequate for the general harvesting problem, especially in fruit where the stem must be clipped. In the past three decades, there have been numerous technological advances that have improved the potential for agricultural robotics. The cost and speed of computers have vastly improved, redundant manipulators have been developed, the cost and performance of solid-state sensors such as the color CCD camera became available in the market, computer algorithms have improved to match computing speeds, and numerous other advances have been made in actuator design, mobile power, hyper-redundant manipulators, prosthetics and so forth. All of this provides encouragement that agricultural robotic harvesting may be on the horizon.

It has been proposed that, prior to any successful robotic development being realized, the technological barriers that have prevented previous efforts from being successful must first be overcome. Consequently, a robotic harvesting development test bed was conceived (Fig. 11.9). The purpose of the test bed was to create a development environment in which end-effectors, manipulators, control approaches, and sensory technology could be developed and performance could be validated. The test bed consisted of the fruit-sensing system (machine vision and ranging sensor), a macro-positioning system, a seven-DOF harvesting manipulator, an end-effector equipped with force torque sensor, and rack-mounted dual processors for control development, with all of these installed in a retrofitted panel van with a pull-behind trailer. In addition, the van was equipped with air conditioning, a clean electrical generator for the control room, and a see-through viewing window. Meanwhile, the trailer was equipped with its own electric generator, a hydraulic power supply for the two-DOF macro-positioning platform, pneumatic power supply (end-effector) and back-up fuel supply tank. The viewing window provided the operator/researcher with a first-hand view of the actual harvesting process.

11.7.1 Test bed robotic manipulator

The manipulator used in this study was a Robotics Research Corp. model 1207 seven-DOF manipulator. Figure 11.10 shows the manipulator and its kinematic model. It had a maximum reach of 50 in (127 cm) and its end-effector's velocity could reach up to 20 in (50.8 cm) per second. The maximum payload capacity was approximately 20 lb (9 kg). Similar to the human arm, it had a shoulder, elbow, and wrist where the fruit-gripping end-effector was attached. Although this manipulator was not specifically designed for fruit harvesting, it provided a high-precision manipulator that could be used for developing the other technologies. This seven-DOF manipulator was mounted on a two-DOF platform that allowed the manipulator to move vertically and horizontally into the canopy. The added

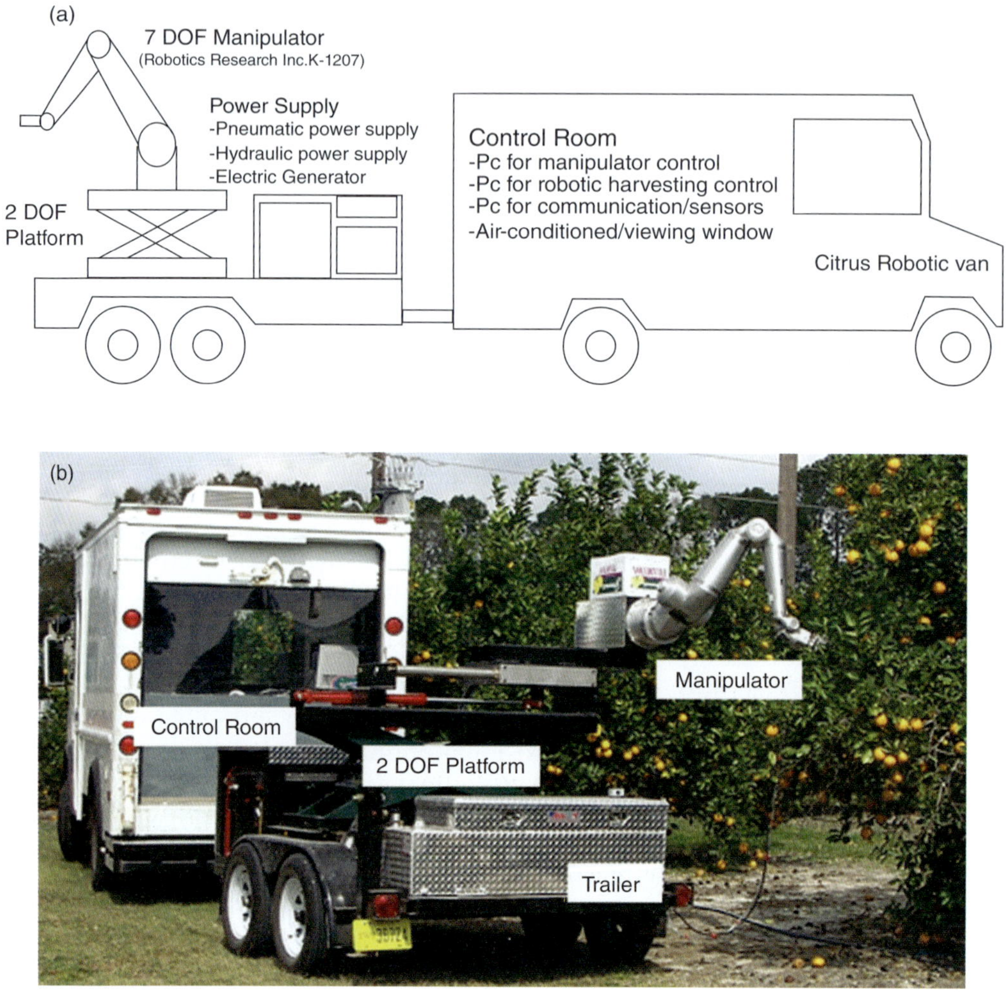

Fig. 11.9. Robotic harvesting development test bed: (a) set-up of test bed; (b) photo of robotic test bed in the orchard.

DOF increased the working space of the manipulator, allowing the robot to harvest fruits within the tree canopy.

11.7.2 Vision sensory system

The vision system detects the fruit and localizes its position. It then guides the end-effector towards the fruit using visual servo control strategies. The vision system is composed of a color CCD camera for image acquisition and a rack-mount PC for image processing. The CCD camera is connected to the PC using a frame grabber that digitizes the image to a

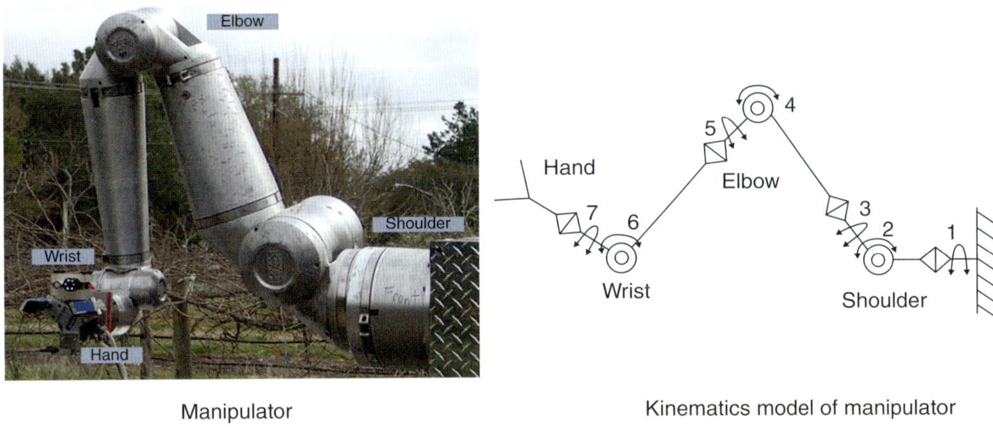

Fig. 11.10. Articulated manipulator for citrus harvesting (Robotics Research Corp.).

640 × 480 24-bitmap image running at 30 frames per second. The images are acquired under natural lighting conditions.

There are numerous fruit recognition approaches that can be adopted based on various features such as shape, size, spectral properties, or texture. In this study, the primary image processing steps are: (i) segmentation by spectral characteristics using an adaptive thresholding approach; (ii) blob analysis; (iii) circle detection; and (iv) centroid detection, which estimates fruit center position within the image. Segmentation separated the fruit pixel from the background pixel by using the color difference values with an adaptive thresholding approach. The color difference model removes intensity from the original color value, thus minimizing the effects of illumination variation within the canopy scene. Blob analysis differentiates large segmented regions from small segmented regions which can represent image noise or fruits on the back side of the tree that are out of reach of the robot. Individual fruit selection within a fruit cluster and partial fruit occlusion is a significant challenge. An algorithmic approach that combines edge detection with circle detection provides a method of recognizing the individual fruits in the cluster so that the top fruit can be harvested first. Figure 11.11 shows an acquired canopy image sample and the successive processed images. Both the red and blue regions are fruit pixels. Blob analysis further differentiated the fruits into large (blue) and small (red). Further processing was conducted on the large regions, while small regions were removed. The two clustered fruits were successfully separated by circle detection and their centroids were determined. These centroidal positions would then become the basis for fruit position localization, and the estimated destination for the visual servo control.

Since 2D images lack range information, depth to target was measured in two ways: (i) ultrasonic sensor gave a rough estimate of range to canopy; and (ii) using triangulation methods base on pseudo stereo imaging. Once

Fig. 11.11. Sample image processing for the automatic detection of orange fruits: (a) sample RGB image of orange canopy; (b) segmentation of fruit from background; (c) filtering operation to remove noise; and (d) location of centroids of fruits.

the manipulator reaches an estimated distance to target, the vision system grabs an image of the target. Then the manipulator jogs to a known offset position, and takes a second image frame. It then calculates the distance of the fruit by triangulation similar to stereo vision. The hand then approaches the fruit using this calculated distance.

11.7.3 Harvesting end-effector

The end-effector is pneumatically actuated with three custom-designed fingers for gripping the fruit. When a fruit is harvested, the gripper's fingers close and grasp the fruit. Then the detaching sequence is initiated to pull the fruit from the branch. A number of different designs for end-effectors have been developed. However, the present design allows the end-effector to approach the fruit at different angles. A close-up of the end-effector is shown in Fig. 11.12. In addition, the CCD camera and ultrasonic sensor are also mounted on the end-effector.

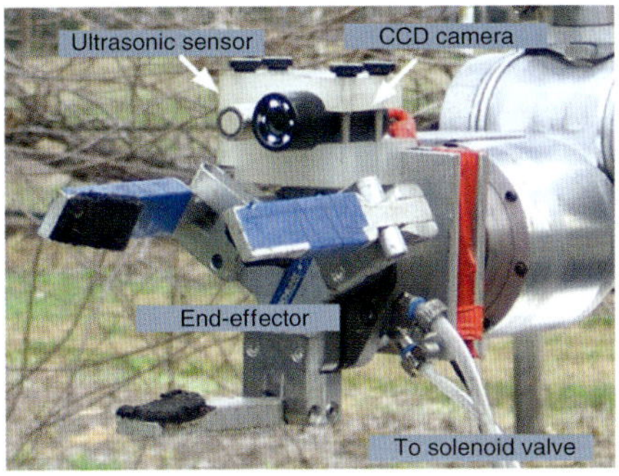

End effector for citrus harvesting

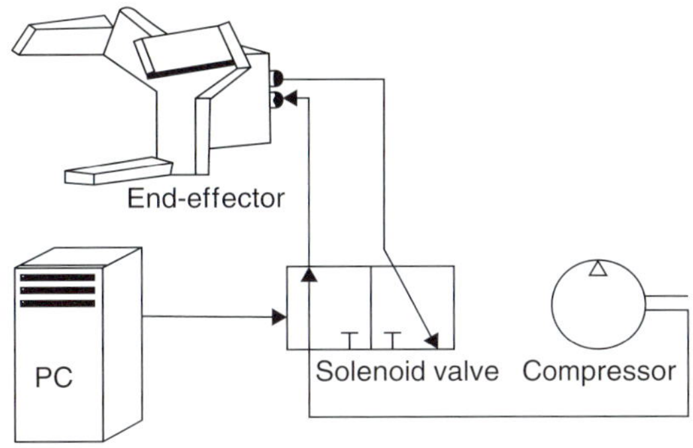

Opening and closing control of end effector

Fig. 11.12. Developed end-effector with CCD camera and ultrasonic sensor for harvesting oranges.

11.7.4 System architecture

Figure 11.13 shows the system architecture of the vision-based robotic harvesting system. This diagram can be used to visualize how the harvesting operation works. The system's flow starts with the recognition and location of the fruits on the tree using data from the machine vision system and the ultrasonic sensor. PC 1 performs the image processing to detect fruits within the canopy and outputs the location of the fruit to PC 2, whose responsibility it is to monitor systems status, execute visual servo control, and provide the interface between the vision system and the manipulator controller. PC 2 passes visual servo-based position control updates to PC 3. PC 3 is the real-time manipulator controller responsible for determining manipulator forward and inverse kinematic, actuator joint torques and position command, and monitors the manipulator status. Once PC 3 has received the position update, it will execute the inverse kinematics to establish the new joint angles required to reach the desired tool frame destination. It also provides feedback to the visual servo control systems of the current manipulator position. The manipulator control on PC 3 and the visual servo control on PC 2 execute in real-time under the INtime Real-Time Operating System (RTOS), while the image processing operates under the Windows operating system. Communication between the three computers is made using the TCP/IP platform.

The robotic harvesting sequence implemented has six steps (Fig. 11.14). In the first step (scene analysis), the robot is moved to its start position and the vision system executes fruit segmentation and localization. In the next step, the target fruit is chosen based on size and distance from image center. Harvesting path optimization algorithms can be implemented here to minimize cycle time and energy consumption. The robot is moved to orient the target fruit in the center of the image by visual servoing, which improves the range estimation accuracy of the ultrasonic sensor. Maintaining the fruit in the image center, the robot approaches the target

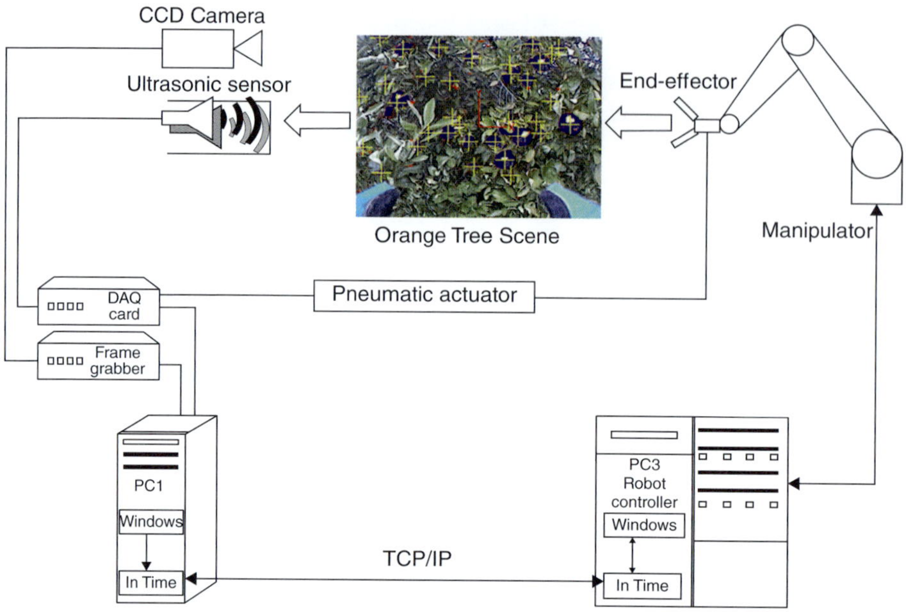

Fig. 11.13. System architecture for robotic citrus harvesting.

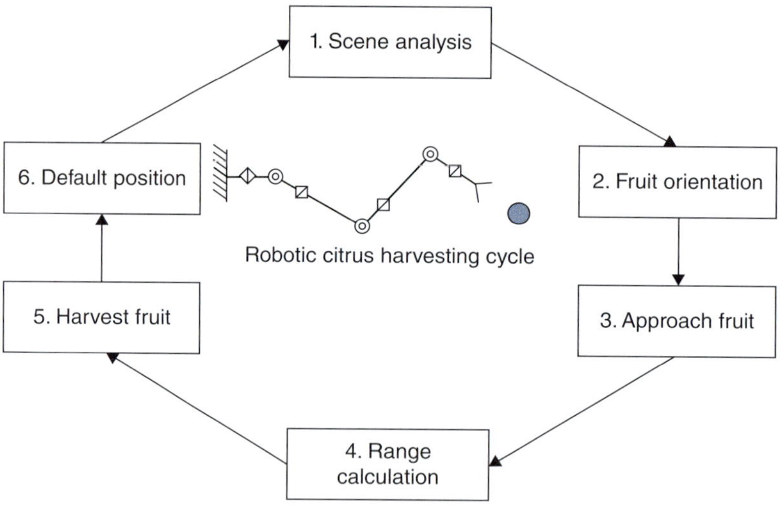

Fig. 11.14. Fruit robotic harvesting sequence.

fruit. Once the robot is at a pre-set distance away from the target, the end of the tool is moved a pre-set distance downwards while continually tracking the target. This technique provides additional target range information through a stereo vision technique. Once the range is estimated, further image processing is executed to determine if the target is a single fruit or a cluster of fruits, through a combination of edge detection and circle detection. The within-cluster top fruit is selected as the target and the robot harvests the fruit. Then the robot returns to its start position to search for a new target.

11.7.5 Fruit detection and harvesting trials

The fruit recognition algorithm was applied to a set of 24 randomly selected images taken from the grove under different lighting conditions. Table 11.1 shows a positive detection rate of 74% when ignoring fruit clusters, while a recognition rate of 95% was achieved using the de-clustering algorithm. The inability to detect the remaining 5% of oranges can be attributed to poor fruit color and occlusion. It is also important to point out that there were no false detections in the tested images.

The robot was tested in an orange grove with the robot positioned near the outer canopy of an orange tree using the macro-positioning system as shown in Fig. 11.9b. Of the 450 harvesting attempts performed, the robot had 357 (79.33%) successful attempts and 93 (20.67%) failed attempts. Table 11.2 summarizes the causes of the failed harvesting attempts. Most of the failed attempts were caused by range estimation. Correct estimation of the range allows the robot to properly position the fruit inside the end-effector. This could be compensated for by adding a fruit proximity sensor in the grip of the end-effector to provide positive feedback that the fruit is in harvesting position before the gripper closes. Several factors contributed to the erroneous range estimation. Firstly, the ultrasonic sensor's analog range signal is rather noisy, due to irregularity in the surface of the tree and the fruit. Although an attempt was made to compensate for this using stereo vision, difficulty in corresponding the features in the image pair can also

Table 11.1. Performance of fruit detection.

	Total number of fruit	Detected	Rate
Positive detections	436	323	74%
Positive detections with clusters	436	415	95%
False detections	436	0	0

Table 11.2. Sources of harvesting failures.

Causes of failure	Number of failed attempts
1. Range estimation with ultrasonic and triangulation	31
2. Grabbing of multiple fruits	21
3. Failure to grab fruit due to occlusion	20
4. Inaccurate fruit center	18
5. Occlusion problem during approach	3

cause erroneous readings. Other causes of the failed attempts included the inability to grab the fruit due to leaf occlusion and clustering, which is partially at least attributed to the end-effector design.

These results have demonstrated the feasibility of robotic fruit harvesting, while also illustrating the challenges that have hindered successful development and commercial adoption.

11.8 Continuing Development and Enhancement Opportunities

As a result of the above case studies, several technological challenge areas continued to persist: (i) maintaining consistent grip on the fruit during the harvesting cycle; (ii) locating fruit hidden in the interior canopy; and (iii) controlling the manipulator in the presence of obstacles and disturbances to the initially predicted fruit position. Consequently, additional research studies have continued, with varied progress achieved, on the topics of robust end-effector development, fruit detection in the canopy interior, and visual servo control in the presence of obstacles and disturbances. A brief overview of additional efforts in fruit detection and end-effector development will be presented in this section, while a more rigorous treatment of advances in visual servo control will be provided.

11.8.1 Fruit detection systems

Development of fruit detection systems has sought to improve the ability to locate fruit in the tree canopy, overcoming some of the difficulties associated with traditional machine vision approaches attempted in the past. Bulanon *et al.* (2009a) showed that fruit visibility in the canopy can be increased by acquiring six or more different perspective views of the canopy. This viewing method is similar to the way human pickers locate fruits in the canopy. Image processing techniques to separate fruits that are in cluster and to detect partially occluded fruits have also been developed, and we have explored novel imaging techniques using sensor fusion with thermal images and multi-spectral imaging (Bulanon *et al.*, 2008, 2009b).

11.8.2 End-effector development

We have conducted tests for citrus peel material properties and harvesting mechanics, which evaluated fruit detachment forces, peel damage criteria, and harvesting mechanics to reduce peel damage. These tests provided valuable information that was used in developing an efficient three-fingered gripper end-effector for harvesting citrus, which had machine vision and optical proximity detection incorporated in the palm of the hand (Flood *et al.*, 2006). Aside from the task of gripping the fruit, the development of the end-effector included the integration of other sensors such as vision, ultrasonic and infrared sensors. In more recent studies, efforts have been directed towards developing a more robust pneumatically actuated end-effector using vacuum gripping, which should be more effective during interior canopy harvesting.

11.9 Novel Approaches in Visual Servo Control Development

Visual servo control refers to closed-loop control of a physical system using image feedback from one or more cameras. Depending on the location of the camera, a vision system can be in a fixed-camera configuration (e.g. a stationary camera viewing a robot) or a camera-in-hand configuration (e.g. a camera held by a robot end-effector). For any camera configuration, the objective of visual servo control is to achieve a desired position and orientation of the system (e.g. a robotic arm) such that the image taken by the camera corresponds to an *a priori* known reference image. The problem of matching two images is equivalent to matching corresponding feature points in each image. Based on how the current image is driven to match the reference image, visual servo control problems can be broadly classified into image-based visual servo control and position-based visual servo control. The former relies solely on 2D image coordinates of the feature points to achieve the control objective, while the latter uses the estimated 3D coordinates of the viewed feature points.

This section will study how image-based visual servo controllers combined with an estimated fruit depth can be used in robotic fruit harvesting. One of the field challenges in robotic harvesting is unknown fruit motion due to exogenous disturbances such as wind gust, robot–tree contact, and canopy unloading. The fruit motion may cause unsuccessful pick cycles and consequently reduce harvesting efficiency. Therefore, we shall introduce advanced robust and adaptive visual servo controllers to compensate for the unknown fruit motion.

11.9.1 Image-based visual servo control

A vision system or a camera is a widely used sensor in robotic harvesting due to its information-rich feedback. Grand D'Esnon (1985) and Grand D'Esnon *et al.* (1987) developed a vision-based three-DOF, hydraulically powered spherical coordinate manipulator, called MAGALI, for golden apple harvesting, where a monocular camera detected a fruit during vertical scan. Subsequently, the telescopic arm translated along the optical beam until it reached the fruit, which was sensed by a photoelectric sensor. Levi *et al.* (1988) investigated a vision-based cylindrical manipulator system for robotic citrus harvesting. Grand D'Esnon *et al.* (1987) and Levi *et al.* (1988) found that harvesting efficiency was susceptible to mechanical backlash, bearing wear, fruit motion etc. due to dead-reckoning during the reaching stage. The Florida Citrus Picking Robot (Harrell *et al.*, 1989, 1990 a, b) aimed to overcome the limitations of dead-reckoning by using a closed-loop camera-in-hand (CIH) configuration along with an ultrasound transducer for fruit range identification. However, concerns regarding stability of the system existed, since the control gains increased by two orders of magnitude as the robot approached a fruit. Spanish and French researchers (Juste and Sevila, 1991; Rabatel *et al.*, 1995) proposed a robotic citrus harvesting system called EUREKA. A Bayesian classifier detected mature fruit from grayscale images captured by a monocular vision system. For the proposed spherical manipulator, the robot motion trajectory was along the straight line between the camera optical center and the fruit. Inadequate fruit-depth information was found to be the cause of the majority of failures. Muscato *et al.* (2005) developed a vision-based citrus harvesting robot prototype called CRAM, where differential image size was used to identify the distance to fruit, thus avoiding additional range measurement sensors. Research in robotic harvesting at large has been focused on the development of robotic manipulators and target classification methods.

Harvesting efficiency is one of the most influential factors in robotic harvesting economics that depends on the stability and performance of closed-loop control systems. However, relatively little attention is paid to control formulation and rigorous stability analysis of harvesting systems. In this section, a cooperative visual servo controller is developed to regulate a robot end-effector to the target fruit location. Lyapunov-based stability analysis guarantees global exponential stability of the closed-loop system such that the desired transient performance can be obtained by appropriately selecting control gains. Similar to Van Henten *et al.* (2002, 2003), a cooperative vision system consisting of a fixed camera and a CIH is incorporated such that the fixed camera provides a global view of a tree canopy, while the CIH, due to proximity, provides high-resolution fruit images.

One of the challenges associated with vision-based harvesting systems is in determining the Euclidean position of a fruit. Apart from using an additional range sensor (Harrell *et al.*, 1989, 1990a, b), the most popular method of determining the Euclidean depth of fruit is by using stereo vision or triangulation (Buemi *et al.*, 1996; Kondo *et al.*, 1996a, b; Recce *et al.*, 1996; Van Henten *et al.*, 2002, 2003). In contrast to these methods, we leverage on a model-based approach designed in Mehta and Burks (2014) to obtain absolute range estimates. The opted range estimation approach is computationally less complex than a stereo vision method, thus making it suitable for the real-time harvesting application. A global view from the fixed camera along with the range estimates can be used to generate a global map of fruit locations from which a target fruit can be selected for harvesting, using a pre-defined criteria. A rotation controller is developed to orient the robot end-effector towards the target fruit such that the target fruit enters the field of view (FOV) of the CIH. This enables harvesting of fruits that were not initially visible to the CIH. Subsequently, the developed visual servo controller regulates the end-effector to the fruit location. For improved dexterity and accuracy, a seven-DOF kinematically redundant electric motor-driven manipulator was selected. The performance of the developed controller was verified by conducting experiments on a synthetic citrus tree with randomly distributed fruits.

11.9.1.1 Euclidean reconstruction

Consider the orthogonal coordinate frames F, F_f and F_b as shown in Fig. 11.15. The time-varying coordinate frame F is attached to a CIH, i.e. a camera held by a robot end-effector. The coordinate frame F_f is attached to a fixed camera, for example, a stationary camera mounted in the workspace of a robot, and the coordinate frame F_b is attached to the stationary base of a robot. $O^* \in \mathbb{R}^3$ denotes the fruit position measured in the base frame F_b. The unknown Euclidean coordinates of the fruit center, $\bar{m}(t)$, $\bar{m}_f \in \mathbb{R}^3$, expressed in terms of F and F_f, respectively, are given as:

$$\bar{m}(t) - \begin{bmatrix} x(t) & y(t) & z(t) \end{bmatrix}^T, \ \bar{m}_f = \begin{bmatrix} x_f & y_f & z_f \end{bmatrix}^T \tag{11.7}$$

where $z(t)$, $z_f \in \mathbb{R}$ denote the unknown depth of the target fruit expressed in F and F_f, respectively.

The Euclidean space is projected on to the image-space, so let $m_i(t)$ and m_{fi} denote the corresponding normalized Euclidean coordinates of the fruit center as:

$$m(t) = \frac{\bar{m}(t)}{z(t)} = \begin{bmatrix} \frac{x(t)}{z(t)} & \frac{y(t)}{z(t)} & 1 \end{bmatrix}^T, \ m_f = \frac{\bar{m}_f}{z_f} = \begin{bmatrix} \frac{x_f}{z_f} & \frac{y_f}{z_f} & 1 \end{bmatrix}^T. \tag{11.8}$$

Assumption 1: *In equation 11.8, it is assumed that the unknown depths* $z(t), z_f > \varepsilon$, *where* $\varepsilon \in \mathbb{R}_{>0}$ *is a constant. This is a standard assumption in visual servo control, which physically means that the target is always in front of the camera.*

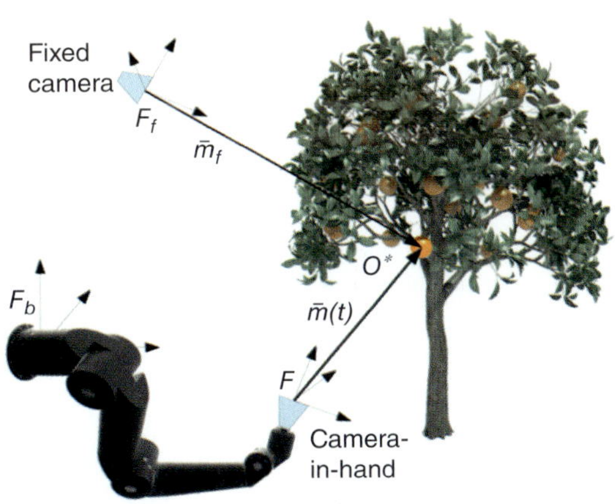

Fig. 11.15. Coordinate frame relationships, where the time-varying frame F is attached to the CIH, F_f corresponds to the fixed camera, and F_b is attached to the stationary base of the robot.

In addition to having normalized task-space coordinates, the target point will also have pixel coordinates acquired by the CIH and the fixed camera. Let $p(t), p_f \in \mathbb{R}^2$ denote the pixel coordinates of the target center expressed in F and F_f, respectively, as:

$$p(t) \triangleq [u(t) \quad v(t)]^T, \; p_f \triangleq \begin{bmatrix} u_f & v_f \end{bmatrix}^T. \tag{11.9}$$

Since the normalized Euclidean coordinates in equation 11.8 cannot be measured directly, a global invertible transformation (i.e. the pinhole camera model) is used to determine the normalized Euclidean coordinates from the corresponding pixel information as:

$$\begin{bmatrix} p^T & 1 \end{bmatrix}^T = Am, \; \begin{bmatrix} p_f^T & 1 \end{bmatrix}^T = A_f m_f. \tag{11.10}$$

where $A, A_f \in \mathbb{R}^{3\times3}$ are the known, constant, invertible, intrinsic camera calibration matrices for the CIH and the fixed camera, respectively.

Leveraging on our efforts in Mehta and Burks (2014), the depth of a fruit can be estimated via perspective transformation by assuming known geometry of the fruit variety as:

$$\hat{z}_f = \frac{f_f(\lambda_{xf} d_{ox} d_{iy} + \lambda_{yf} d_{oy} d_{ix})}{2 d_{ix} d_{iy}} \tag{11.11}$$

$$d_{ix} = 2\left(\frac{A_p}{\pi\sqrt{1-e^2}}\right)^{1/2}, \; d_{iy} = 2\left(\frac{A_p\sqrt{1-e^2}}{\pi}\right)^{1/2} \tag{11.12}$$

where $\hat{z}_f \in \mathbb{R}$ denotes the estimated Euclidean depth of a fruit measured in F_f; d_{ox}, $d_{oy} \in \mathbb{R}$ denote the sample mean major and minor axes, respectively, of an ellipsoidal fruit; d_{ix}, $d_{iy} \in \mathbb{R}$ denote the major and minor axes, respectively, in the image plane; $A_p \in \mathbb{R}$ denotes the area of the fruit in the image plane of the fixed camera (in pixels); $e \in \mathbb{R}$ is the known eccentricity of the ellipse; the constant $f_f \in \mathbb{R}_{>0}$ represents the focal length in pixels for the fixed camera; and λ_{xf}, $\lambda_{yf} \in \mathbb{R}_{>0}$ are the scaling factors in the image x and y directions of the fixed camera, respectively.

Remark 1: *Any inaccuracy in estimating the fruit size Ap affects the major and minor axes, dix and div, respectively. As stated in Remark 1 in Mehta and Burks (2014), it can be shown that* d_{ix}/d_{iy} *is constant and hence the unknown depth ratio* $\hat{z}_f/z_f$*, denoted by* $\gamma_z \in \mathbb{R}_{>0}$*, is also constant.*

Remark 2: *In the presence of partial occlusions or clustered fruit, advanced methods such as perimeter detection and shape analysis techniques (Plebe and Grasso, 2001; Hannan* et al.*, 2009) can be used to directly obtain the image-space diameters dix, diy of the fruit to get* $\hat{z}_f$ *using equation 11.11.*

11.9.1.2 Control objective

The objective is to locate the robot end-effector to the target fruit position for harvesting, i.e. to regulate the CIH coordinate frame F to the target fruit in the sense that $O_{F(t)} \to O^*$, where $O_{F(t)} \to \mathbb{R}^3$ denotes the time-varying position of the frame F measured in F_b. The control objective can be achieved by regulating the time-varying fruit pixel coordinates $p(t)$ to the desired image coordinates, and regulating the end-effector to the desired fruit depth. Hence, mathematically, the control objective can be stated as:

$$p(t) \to p_d,\ p_d = \begin{bmatrix} u_0 & v_0 \end{bmatrix}^T \text{ and } z(t) = z_d \tag{11.13}$$

where $z_d \in \mathbb{R}_{>0}$ denotes the maximum desired depth of the fruit in F, and $u_0, v_0 \in \mathbb{R}$ denote the pixel coordinates of the principal point (i.e. the intersection of an optical axis with the image plane) of the CIH.

11.9.1.3 Controller development

As discussed earlier, the fixed camera can view an entire or part of a tree canopy. Using equations 11.11 and 11.12, the fixed camera can obtain a global fruit map and the corresponding harvesting sequence such as in Edan *et al.* (1991). However, the fruit to be harvested from the sequence may not be visible to the CIH, say, because the CIH is pointing away from the fruit. Therefore, a non-linear rotation controller is developed to orient the CIH such that the target fruit enters its FOV. The rotation controller uses the estimated fruit position obtained by the fixed camera to determine the desired orientation of the CIH. Once the target fruit is visible to the CIH, the translation controller regulates the end-effector to the target fruit using image feedback from the CIH.

11.9.1.4 Rotation controller

In this section, a controller is developed to orient the robot end-effector such that the target fruit enters the FOV of the CIH. Using equations 11.8, 11.11 and 11.12, let the estimated Euclidean position of the fruit in the fixed camera F_f be denoted by $\hat{\bar{m}}_f \in \mathbb{R}^3$. The Euclidean coordinates $\hat{\bar{m}}_f$ can be expressed in the CIH as $\hat{\bar{m}}'(t)$ (see (9) and (12) in Mehta and Burks, 2014). Therefore, the objective is to align $\hat{\bar{m}}'(t)$ along the direction of the camera's optical axis $\begin{bmatrix} 0 & 0 & 1 \end{bmatrix}^T$.

The rotation error $e_\omega(t) \in \mathbb{R}^3$ defined as orientation mismatch to bring the target fruit in the FOV of the CIH can be represented in terms of angle-axis representation as

$$e_\omega = u\theta \tag{11.14}$$

where $u(t) \in \mathbb{R}^3$ represents a unit axis of rotation such that $u(t) = \hat{\bar{m}}'(t) \wedge [0 \quad 0 \quad 1]^T$, and $\theta(t) = \cos^{-1} \hat{\bar{\bar{m}}}'(t), [0 \quad 0 \quad 1]^T \in \mathbb{R}$ denotes the rotation angle about $u(t)$ that brings $\hat{\bar{m}}'(t)$ along the optical axis, such that $0 \leq \theta(t) \leq \pi$. In equation 11.14, $\hat{\bar{\bar{m}}}'(t) \in \mathbb{R}^3$ represents a unit vector along $\hat{\bar{m}}'(t)$.

Based on the rotation error in equation 11.14, the angular velocity $\omega_c(t) \in \mathbb{R}^3$ of the camera can be designed using the following PD controller:

$$\omega_c = -k_{p\omega}(I_3 + k_{d\omega}L_\omega)^{-1}e_\omega \tag{11.15}$$

where $k_{p\omega}, k_{d\omega} \in \mathbb{R}_{>0}$ are the proportional and derivative control gains, respectively. Various loop tuning methods, such as Ziegler-Nichols and manual (trial-and-error), can be adopted to determine the control gains. In equation 11.15, I3 denotes a 3×3 identity matrix, and $L_\omega(t) \in \mathbb{R}^{3\times3}$ is a measurable Jacobian-like function defined as:

$$L_\omega = I_3 - \frac{\theta}{2}[u]_\times + \left(1 - \frac{\text{sinc}(\theta)}{\text{sinc}^2\left(\frac{\theta}{2}\right)}\right)[u]_\times^2 \tag{11.16}$$

where sinc(θ) is the unnormalized sinc function, and $[u]_\times$ is the skew-symmetric matrix of $u(t)$. The determinant of $L_\omega(t)$ is $det(L_\omega) = 1/\text{sinc}^2(\theta/2)$, thus being singular only at $\theta = 2k\pi \forall k \in N_{>0}$, i.e. outside of $0 \leq \theta(t) \leq \pi$.

Theorem 1: *The angular velocity control input in equation 11.15 ensures global exponential regulation of robot end-effector such that the target fruit is in the FOV of the CIH in the sense that:*

$$\|e_\omega(t)\| = \zeta_0 \exp\{-\zeta_1 t\} \tag{11.17}$$

where $\zeta_0 \cdot \zeta_1 \in R$ denote positive bounding constants.
Proof: *see Mehta and Burks* (2014).

11.9.1.5 Translation controller

The objective of the translation controller is to regulate the CIH to the target fruit position. Based on the control objective, the translation errors $e_{v1}(t) \in \mathbb{R}^2$ and $e_{v2}(t) \in \mathbb{R}$ can be defined as:

$$e_{v1} \triangleq p_d - p \tag{11.18}$$

$$e_{v2} \triangleq z_d - \alpha\hat{z} \tag{11.19}$$

The error $e_{v1}(t) \in \mathbb{R}^2$ corresponds to regulating the fruit to the image center, and the error $e_{v2}(t) \in \mathbb{R}$ is designed to regulate the end-effector to the target fruit depth. In equation 11.19, $\hat{z}(t) = [0 \quad 0 \quad 1]\hat{\bar{m}}(t)$ is the estimated depth of the target fruit from the CIH, and $\hat{\bar{m}}(t)$ is the estimated fruit position in the CIH coordinate frame. The estimated fruit depth $\hat{z}(t)$ is assumed to be a continuous function of time. In equation 11.19, $\alpha \in \mathbb{R}_{>0}$ denotes a scaling factor such that $z < \alpha\hat{z} \forall t$. The constant α is selected based on an upper bound on the Euclidean depth estimation error, and, as a rule of thumb, α can be selected arbitrarily high to ensure that the robot reaches the target fruit despite any estimation errors. Since $z < \alpha\hat{z}$, the robot may overshoot the target fruit, and hence the end-effector is equipped with an infrared proximity sensor to stop once the fruit is reached.

Let $v_c(t) \triangleq [v_{cx}(t) \quad v_{cy}(t) \quad v_{cz}(t)]^T$ be the linear velocity of the CIH and define $v_{c1} \triangleq [v_{cx}(t) \quad v_{cy}(t)]^T$. Taking the time derivative of equation 11.19, the linear velocity $v_{cz}(t)$ along the optical axis of the camera can be obtained as:

$$v_{cz} = -k_{pv2}(1+\alpha k_{dv2})^{-1} e_{v2} \qquad (11.20)$$

where $k_{pv2}, k_{dv2} \in \mathbb{R}_{>0}$ are the proportional and derivative control gains, respectively.

Taking the time derivative of the first expression in equation 11.10, the velocity of the CIH can be related to the velocity $\dot{p}(t) \in \mathbb{R}^2$ of the target centroid in the image frame as:

$$\dot{p} = -\frac{1}{z} J_v' v_{c1} + \frac{1}{z} J_v'' v_{cz} \qquad (11.21)$$

where $J_v'(u,v) \in \mathbb{R}^{2\times 2}$ and $J_v''(u,v) \in \mathbb{R}^2$ are measurable image Jacobians. Since no orientation change is required during translation control, the image dynamics in equation 11.21 are obtained considering $\omega_c(t) = 0$.

Using equation 11.21, the velocity $v_{c1} \in \mathbb{R}^2$ along the x- and y-axis of the CIH can be designed as:

$$v_{c1} = \left(I_2 + \frac{k'_{dv1}}{\hat{z}} J_v' \right)^{-1} \left(-k_{pv1} e_{v1} + \frac{k'_{dv1}}{\hat{z}} J_v'' v_{cz} \right) \qquad (11.22)$$

where $k_{pv1}, k'_{dv1} \in \mathbb{R}_{>0}$ are proportional and derivative control gains, respectively, such that $k_{pv1} = k_{pv11} + k_{pv12}$. In equation 11.22, the facts that $\hat{z}/z = \gamma$ and $z \le \alpha\hat{z}$ are used.

Theorem 2: *The translation velocity control input vc1(t) and vcz(t) in equations 11.22 and 11.20 ensures global exponential regulation of robot end-effector to the desired fruit depth in the sense that:*

$$\|e_{v1}(t)\| \le \zeta_2 \exp\{-\zeta_3 t\} \qquad (11.23)$$

$$|e_{v2}(t)| = \zeta_4 \exp\{-\zeta_5 t\} \qquad (11.24)$$

where $\zeta_2, \zeta_3, \zeta_4, \zeta_5 \in \mathbb{R}$ are positive bounding constants.
Proof: *see Mehta and Burks* (2014).

11.9.1.6 Experimental validation

The performance of the developed visual servo controller was demonstrated using a seven-DOF Robotics Research K1207 manipulator (Fig. 11.16). The indoor experiment comprised an artificial citrus tree and two color CCD cameras (KT&C, KPCS20-CP1) with focal length of 4.3 mm and resolution of 640 ×480 that served as the fixed camera and the CIH. The fixed camera was mounted on the stationary base of the manipulator, while the CIH was attached to the robot end-effector (Fig. 11.16). Images from both the cameras were digitized using USB frame grabbers. The image processing workstation (IPW) was used to identify fruits from the captured images using the method described in Hannan *et al.* (2009). The robot control workstation (RCW) hosted a lower-level controller to generate joint torque commands based on the control input from the IPW. Also, the RCW broadcasted the joint position feedback along with the end-effector position and orientation to the IPW using a real-time communication network.

The experiment was repeated several times for different robot and fruit positions. The actual fruit position O^* was measured using forward kinematic analysis by positioning the end-effector at the center of the fruit. Figure 11.17 shows the polar plot of the Euclidean distance error between $F(n)$ and O^*. To assist in visualizing the results, the fruit is shown as an ellipse of axes $\{d_{ox}, d_{ou}\}$. From the fact that the regulation error is less than the radius of the fruit and that the fruit was harvested successfully during each trial, the preliminary results indicate satisfactory performance of the developed visual servo controller.

11.9.2 Robust visual servo control

One of the practical challenges in citrus harvesting is that the fruit may not be stationary. Exogenous disturbances such as wind gusts, fruit detachment forces, canopy unloading, and robot–tree contact may cause an unknown time-varying fruit motion, which aggravates for fruits with longer stems. If not considered during control system development, the fruit motion could result in an unsuccessful pick cycle and reduce the overall harvesting efficiency. Robotic harvesters using open-loop servo control or dead-reckoning (Grand D'Esnon, 1985; Levi *et al.*, 1988; Ceres

(a) (b)

Fig. 11.16. (a) Robotic Research K1207 manipulator with the fixed camera mounted on the stationary base of the robot and (b) the CIH located inside the robot end-effector.

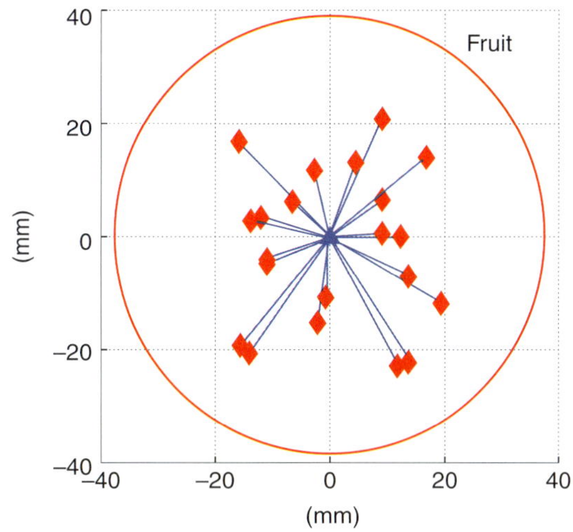

Fig. 11.17. Plot showing Euclidean distance error between the final position of CIH *F*(*n*) (red ◇) and the fruit centroid *O** (blue △) for the obtained 21 observations.

et al., 1998) are highly vulnerable to fruit motion, since fruit position is not updated during the reaching stage. The existing approaches to vision-based robotic harvesting either do not consider fruit motion or rely on high-gain controllers (Mehta and Burks, 2014) in conjunction with or separate from high-frequency camera feedback (Harrell *et al.*, 1989, 1990a, b) for fruit motion compensation. Approaches using high-gain controllers are susceptible to measurement noise as inadvertently noise gets amplified along with the feedback signal, which could lead to high-bandwidth actuation (causing chattering of the end-effector) and system instability. In the field conditions, where a fruit can easily be partially occluded or clustered, additional image processing becomes necessary to robustly identify the fruit to improve fruit detection rates, which limits the rate of image feedback. A passive approach to fruit motion compensation using high-frequency image feedback may not be viable with the need for robust image processing (Muscato *et al.*, 2005) and the desire to process higher-resolution imagery for improved positioning accuracy. Additionally, since disturbance dynamics are not taken into account, passive approaches fail to guarantee stability and performance of the closed-loop harvesting system. Various researchers (Fortuna *et al.*, 1996; Hayashi *et al.*, 2002; Muscato *et al.*, 2005; Bac *et al.*, 2014; Mehta and Burks, 2014) have expressed the need for robust and adaptive control methods to improve harvesting efficiency when uncertainties in fruit detection and tracking arise. In this section, a robust visual servo controller is developed by assuming the knowledge of the bounds on the fruit motion.

11.9.2.1 Control objective

The goal is to locate the CIH coordinate frame to the target fruit position in the presence of unknown fruit motion, which can be achieved by regulating the time-varying fruit image coordinates $p(t)$ to the desired image coordinates, and driving the end-effector to the desired fruit depth. Hence, mathematically, the control objective can be stated as:

$$p(t) \to p_d,\ p_d = [u_0 \quad v_0]^T \text{ and } z(t) = z_d \tag{11.25}$$

where $z_d \in \mathbb{R}_{>0}$ denotes the maximum desired depth of the fruit in F, and $u_0, v_0 \in \mathbb{R}$ denote the pixel coordinates of the principal point (i.e. the intersection of an optical axis with the image plane) of the CIH.

11.9.2.2 Rotation controller

The rotation error in equation 11.14 depends on the camera motion as well as the unknown fruit motion. Therefore, the open-loop error system obtained by taking the time-derivative of equation 11.14 contains two terms as:

$$\dot{e}_\omega = L_\omega \omega_c + d_\omega \tag{11.26}$$

where $L_\omega(t) \in \mathbb{R}^{3\times3}$ is defined in equation 11.26 and $d_\omega \in \mathbb{R}^3$ is the exogenous disturbance as a result of fruit motion. The disturbance is assumed to be bounded such that $\| d_\omega \| \le \gamma_\omega$, where $\gamma_\omega \in \mathbb{R}_{>0}$ and $\|\cdot\|$ denotes the L2 vector norm.

Based on the open-loop error dynamics in equation 11.26 and the subsequent stability analysis, the angular velocity of the camera can be designed as:

$$\omega_c = -k_\omega e_\omega - \frac{e_\omega \gamma_\omega^2}{\| e_\omega \| \gamma_\omega + \varepsilon_\omega} \tag{11.27}$$

where $k_\omega \in \mathbb{R}_{>0}$ is the control gain, and $\varepsilon \in \mathbb{R}_{>0}$ is chosen to be arbitrarily small.

Theorem 1: *The rotation controller orients the end-effector such that the target fruit appears arbitrarily close to the center of the CIH. Formally, the rotation control input developed in equation 11.27 ensures uniformly ultimately bounded regulation of the end-effector in the sense that:*

$$\| e_\omega(t) \| \le \zeta_0 \exp\{\zeta_1 t\} + \zeta_2 \tag{11.28}$$

where ζ_0, ζ_1 and $\zeta_2 \in \mathbb{R}$ denote positive bounding constants.
Proof: *see Mehta and Burks (2016).*

11.9.2.3 Translation controller

Assuming the estimated fruit depth $\hat{z}(t)$ to be a continuous function of time, the open-loop error dynamics for depth regulation can be obtained by taking time-derivative of equation 11.19 as:

$$\dot{e}_{v2} = \alpha \xi v_{cz} + d_z \tag{11.29}$$

where $d_z(t) \in \mathbb{R}$ is the component of fruit motion along the optical axis, such that $|d_z(t)| \le \gamma_d$ for $\gamma_d \in \mathbb{R}_{>0}$, and ξ denotes the constant depth ratio

$\hat{z}/z.\gamma_d$ indicates an upper bound on the fruit velocity along the optical axis due to disturbance. Based on equation 11.29, the linear velocity $v_{cz}(t)$ of the CIH along the optical axis can be designed as:

$$v_{cz} = k_z e_{v2} + \frac{e_{v2}\gamma_d^2}{|e_{v2}|\gamma_d + \varepsilon_z} - w\,\|e_{v1}\|^2 \tag{11.30}$$

where $k_z = k_{z1} + k_{z2} \in \mathbb{R}_{>0}$ is the constant control gain, $\varepsilon_z \in \mathbb{R}_{>0}$ is an arbitrarily small design constant, and $w \in \mathbb{R}_{>0}$ is the user defined weight on the term $\|e_{v1}\|^2$. The control gain k_z determines response of the controller to the input error $e_{v2}(t)$; however, large gains should be avoided in the presence of measurement noise.

To design the linear velocity $v_{c1}(t)$, an open-loop error system can be obtained by taking the time-derivative of equation 11.18 as:

$$\dot{e}_{v1} = -\frac{1}{z}J_v{}'v_c{}' - \frac{1}{z}J_v{}''v_{cz} + d_p. \tag{11.31}$$

Based on the open-loop error system in equation 11.31 and the subsequent stability analysis, the linear control velocity $v_{c1}(t) \in \mathbb{R}^2$ of the CIH can be designed as:

$$v_{c1}(t) = J_v{}'^{-1}\left(\alpha\hat{z}k_p e_{v1} - J_v{}''v_{cz} + \frac{\alpha\hat{z}e_{v1}\gamma_p^2}{\|e_{v1}\|\gamma_p + \varepsilon_p}\right) \tag{11.32}$$

where $k_p \in \mathbb{R}_{>0}$ is the constant control gain, and $\varepsilon_p \in \mathbb{R}_{>0}$ is an arbitrarily small design constant. Similar to k_z, the control gain k_p determines response of the controller to input error $e_{v1}(t)$.

Theorem 2: *The translation controller guarantees that the end-effector is placed arbitrarily close to the target fruit location. Formally, the translation control inputs developed in equations 11.30 and 11.32 ensure uniformly ultimately bounded regulation of the end-effector in the sense that:*

$$\begin{aligned} \|e_{v1}(t)\| &\le \zeta_3 \exp\{\zeta_4 t\} + \zeta_5 \\ |e_{v2}(t)| &\le \zeta_6 \exp\{\zeta_7 t\} + \zeta_8 \end{aligned} \tag{11.33}$$

where $\zeta_3, \zeta_4, \zeta_5, \zeta_6, \zeta_7, \zeta_8 \in \mathbb{R}$ denote positive bounding constants.
Proof: *see Mehta and Burks (2016).*

11.9.2.4 Experimental validation

We used an artificial citrus fruit suspended in the air using a string to simulate a fruit attached to a stem. The fruit was manually perturbed in the experiment and the performance of the controller with and without robust feedback elements was recorded. The displacement of the fruit centroid was ≈ 120 mm. The plot of regulation error using the robust controller is shown in Fig. 11.18, and the image-space trajectory of the fruit is shown in Fig. 11.19.

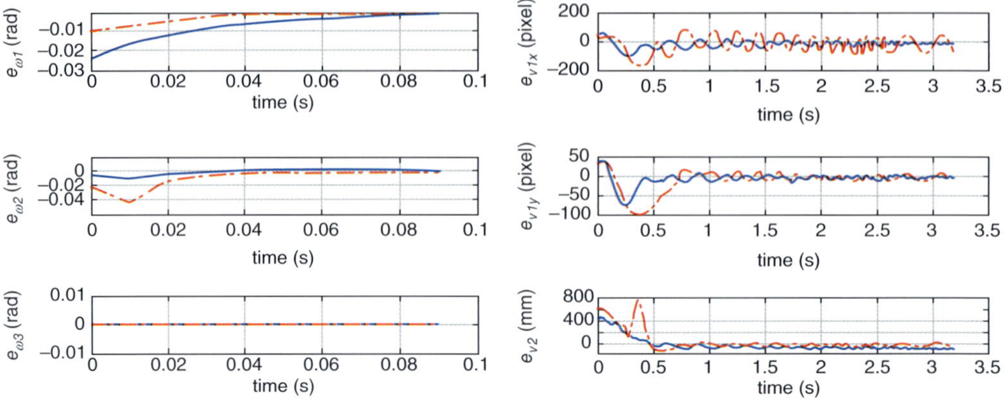

Fig. 11.18. Rotation error $e_\omega(t)=[e_{\omega1}(t)e_{\omega2}(t)e_{\omega3}(t)]^T$ and translation error $e_v(t)=[e_{v1}^T(t)e_{v2}(t)]^T$ where $e_{v1}(t)=[e_{v1x}(t)e_{v1y}(t)]^T$ using the proposed robust controller (blue line) and without robust feedback terms (red dash-dot line).

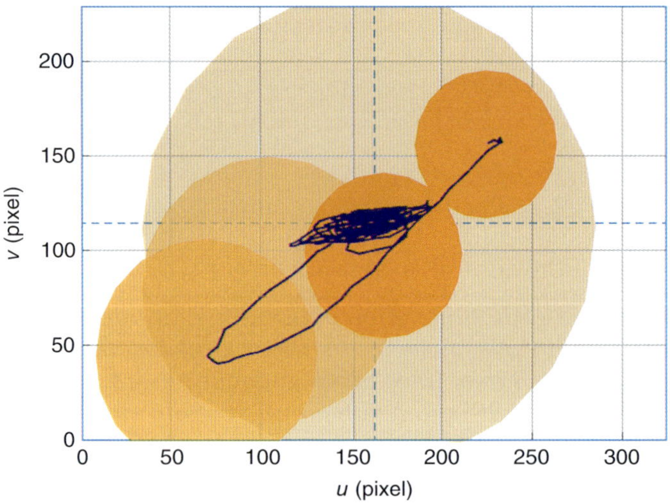

Fig. 11.19. Time-varying fruit position (blue line) and relative fruit size in the image plane of the camera.

11.9.3 Adaptive visual servo control

The robust control approach presented in the previous section compensates for the unknown fruit motion by including robust feedback terms that were designed to upper bound a non-linear disturbance. The designed controller guaranteed that a robot can be regulated arbitrarily close to fruit in the presence of fruit motion (for details, see Mehta *et al.*, 2016). The motivation behind using direct adaptive control approaches is to 'learn' and compensate the fruit motion in real-time. With the gained knowledge of the fruit motion, effective compensation can be offered as against using robust disturbance bounding terms. On the downside, learning requires that the functional form of the fruit motion is known. Instead of considering an arbitrary fruit motion as in Mehta *et al.* (2016), this result assumes that the suspended fruit follows the motion of a simple pendulum, which is a mild assumption for fruits with long stems such as orange. The fruit motion is analyzed in the image plane and along the optical axis of the camera using a second-order spring-mass system. By linearly parameterizing the motion dynamics, an adaptive update law is designed to identify the unknown fruit motion, and the developed adaptive visual servo controller regulates the robot to a target fruit. To account for modeling uncertainties, robust feedback elements are included in the control structure.

11.9.3.1 Fruit motion modeling

A suspended fruit, similar to a simple pendulum, is considered to move in *x*, *y*, and *z* directions with respect to *F*. Using the small angle approximation (i.e. angular displacement $\ll$ 1 rad), a pendulum can be considered a harmonic oscillator. The fruit motion in the image-space is analyzed as a combination of two second-order spring-mass systems (Fig. 11.20).

For a fruit modeled using a spring-mass system as shown in Fig. 11.20 and observed by a stationary camera, its motion can be analyzed by separating the motion in the image plane $(\dot{u}(t), \dot{v}(t))$ from the out-of-plane motion $\dot{z}(t)$ along the optical axis of the camera as:

$$\begin{aligned} \dot{u} &= u_0\omega \cos(\omega t) \\ \dot{v} &= -2v_0\omega \sin(2\omega t) \\ \dot{z} &= z_0\omega \cos(\omega t) \end{aligned} \tag{11.34}$$

where $u_0, v_0 \in \mathbb{R}$ is the unknown amplitude of motion in the image plane, $z_0 \in \mathbb{R}$ is the unknown amplitude of motion along the optical axis, and $\omega \in \mathbb{R}$ is the unknown angular frequency of motion.

Using polynomial approximation of cyclic functions and segregating the unknown parameters u_0, v_0 and ω from the known functions of time,

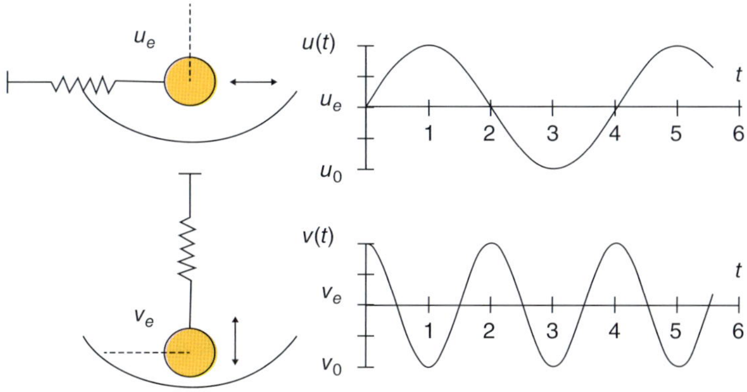

Fig. 11.20. Fruit motion in the image plane is modeled as a spring-mass system.

i.e. after linearly parameterizing the uncertainty, the image dynamics can be obtained as:

$$\dot{p} = -\frac{1}{z} J_{v1} v_c{}' + \frac{1}{z} J_{v2} v_z + J_\omega \omega_c + Y_p \Theta_p + R_p \tag{11.35}$$

where $Y_p(t) \in \mathbb{R}^{2 \times 2n}$ is the regression matrix of known functions of time, $\Theta_p \in \mathbb{R}^{2n}$ is the vector of constant unknown parameters, and $R_p \in \mathbb{R}^2$ is defined as:

$$R_p \triangleq \begin{bmatrix} u_0 \omega R_1 \\ -2v_0 \omega R_2 \end{bmatrix} \tag{11.36}$$

such that $\| R_p \| \leq \gamma_p x$ for any $\gamma_p \in \mathbb{R}_{>0}$. Similarly, the depth dynamics can be obtained as:

$$\dot{z} = -v_z + Y_z \Theta_z + R_z \tag{11.37}$$

where $Y_z(t) \in \mathbb{R}^{1 \times n}$ is a vector of known functions of time, $\Theta_z \in \mathbb{R}^n$ is the vector of constant unknown parameters, and $R_z = z_0 \omega R_1$ such that $|R|_z \leq \gamma_z$ for any $\gamma_z \in \mathbb{R}_{>0}$.

11.9.3.2 Controller development

The regulation error in the image plane $e_p(t) \in \mathbb{R}^2$ and along the optical axis $e_z(t) \in \mathbb{R}$ can be defined as:

$$\begin{aligned} e_p &\triangleq p - p_d \\ e_z &\triangleq z - z_d \end{aligned} \tag{11.38}$$

where the fruit depth $z(t) \in \mathbb{R}$ is assumed to be known (e.g. using triangulation).

Based on the error dynamics and the stability analysis, the velocity of the robot along the optical axis $v_z(t) \in \mathbb{R}$ can be designed as:

$$v_z = k_z e_z + Y_z \hat{\Theta}_z + \frac{e_z \gamma_z^2}{|e_z| \gamma_z + \varepsilon_z} \tag{11.39}$$

where $k_z \in \mathbb{R}_{>0}$ is the control gain, $\hat{\Theta}_z(t) \in \mathbb{R}^n$ is the time-varying estimate of the unknown parameter vector Θ_z, and $\varepsilon_z \in \mathbb{R}_{>0}$ is an arbitrarily small design constant. The estimate $\hat{\Theta}_z(t)$ in equation 11.39 is obtained using the following parameter update law:

$$\dot{\hat{\Theta}}_z = proj\left(\Gamma_z Y_z^T e_z\right) \tag{11.40}$$

where $\Gamma_z \in \mathbb{R}^{n \times n}$ is the positive definite gain matrix, and proj denotes the normal projection algorithm, which ensures that the elements $\hat{\Theta}_{zi}(t) \forall i = 1, 2, \cdots, n$ of $\hat{\Theta}_z(t)$ are bounded as $\underline{\Theta}_{zi} \leq \hat{\Theta}_{zi}(t) \leq \bar{\Theta}_{zi}$, where $\underline{\Theta}_{zi}, \bar{\Theta}_{zi} \in \mathbb{R}$ denote the known constant lower and upper bounds of $\hat{\Theta}_{zi}(t)$, respectively.

Similarly, the camera velocity $v_c'(t) \in \mathbb{R}^2$ in the *xy*-plane of *F* can be obtained as:

$$v_c' = zJ_{v1}^{-1}\left(k_p e_p + \frac{1}{z}J_{v2}v_z + J_\omega \omega_c + Y_p\hat{\Theta}_p + \frac{e_p\gamma_p^2}{\|e_p\|\gamma_p + \varepsilon_p}\right) \quad (11.41)$$

where $\hat{\Theta}_p(t) \in \mathbb{R}^{2n}$ is the estimate of the unknown vector Θ_p that is obtained using:

$$\dot{\hat{\Theta}}_p = proj\left(\Gamma_p Y_p^T e_p\right) \quad (11.42)$$

where $k_p \in \mathbb{R}_{>0}$ is the control gain, $\varepsilon_p \in \mathbb{R}_{>0}$ is an arbitrarily small design constant, and $\Gamma_p \in \mathbb{R}^{2n\times 2n}$ is a positive definite adaptation gain matrix.

Theorem 1: *The adaptive visual servo controller in equations 11.39–11.42 ensures uniformly ultimately bounded regulation of the fruit in the sense that:*

$$\begin{aligned} \|e_p(t)\| &\le \zeta_0 \exp\{-\zeta_1 t\} + \zeta_2 \\ |e_z(t)| &\le \zeta_3 \exp\{-\zeta_4 t\} + \zeta_5 \end{aligned} \quad (11.43)$$

where $\zeta_0, \zeta_1, \zeta_2, \zeta_3, \zeta_4, \zeta_5 \in \mathbb{R}$ *denote positive bounding constant.*
Proof: *see Mehta and Burks (2016b).*

11.9.3.3 Simulation results

A numerical simulation was performed to demonstrate the performance of the adaptive controller. A non-vanishing disturbance was assumed to perturb the fruit, with velocity causing the fruit centroid to oscillate with an amplitude of about 210 mm. The image coordinates of the fruit were assumed to be affected by a zero-mean Gaussian noise of standard deviation one pixel. The performance of the developed adaptive controller was compared with a pure high-gain controller.

Figure 11.21 compares the regulation error using the adaptive and the high-gain controller. For clarity, the transient response of the adaptive controller using a larger time scale is shown in Fig. 11.22. It can be seen from Fig. 11.22 that the adaptive controller adapts to the fruit motion to offer excellent disturbance rejection.

Figure 11.23a shows the image-space trajectory of the fruit during the closed-loop operation using the adaptive controller, and the corresponding image-space trajectory using the high-gain controller is shown in Fig. 11.23b. The adaptive controller regulates the fruit close to the image center (blue line) as shown in Fig. 11.23a, while Fig. 11.23b demonstrates poor performance and potentially unstable closed-loop operation using the high-gain controller.

11.10 Conclusions

The modern era of tree fruit automation began in the 1950s as producers sought to improve harvesting labor productivity in numerous crops, such

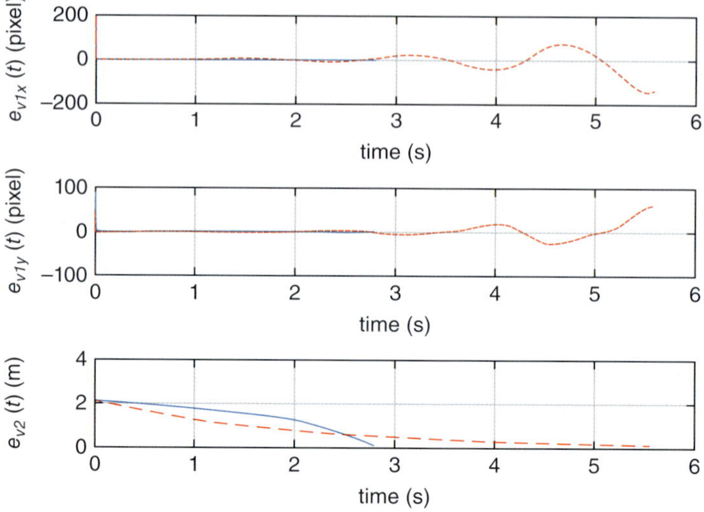

Fig. 11.21. Error comparison between the adaptive controller (blue line) and the high-gain controller (red dotted line).

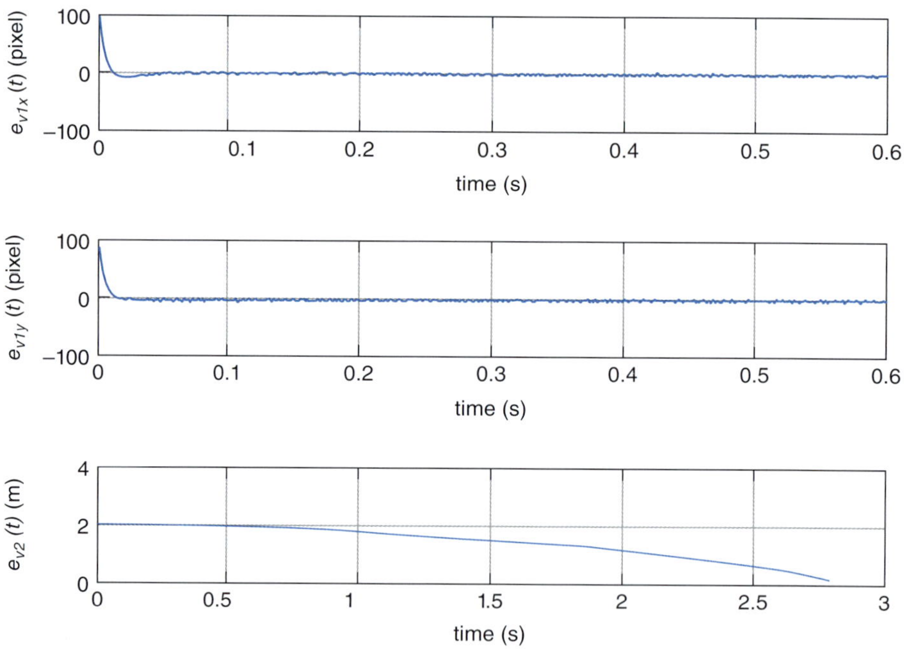

Fig. 11.22. Error $e_v(t) = [e_{v1}^T(t) e_{v2}(t)]^T$ for $d_c = 210$ mm, where $e_{v1}(t) = [e_{v1x}(t) e_{v1y}(t)]^T$ for the proposed adaptive controller.

as potatoes, cabbages and tomatoes. As technologies advanced, development programs for more complicated production tasks began to emerge using mechanization and general automation concepts. A variety of mechanized solutions are commercially available, or currently under development,

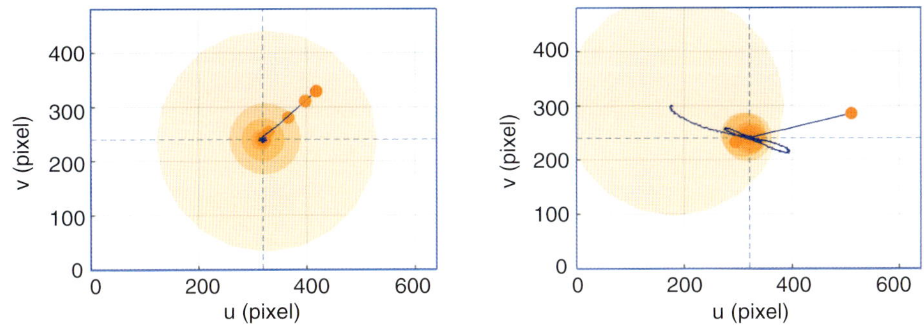

(a) Adaptive Controller (b) High-Gain Controller

Fig. 11.23. Time-varying fruit position (blue line) and relative fruit size in the image plane of the camera.

for production tasks such as harvesting, pruning, hedging, planting, spraying and fertilization. In addition, precision technologies are being explored for crop status and yield-monitoring tasks. Although robotic solutions have been pursued in some domains within tree fruit production over the past several decades, very few have been commercially adopted, for various reasons. As shown in this chapter, most solutions to fruit automation problems are multi-disciplinary in nature. Although there have been significant technology advances, many scientific challenges remain. Viable solutions will require engineers, horticultural scientists, plant breeders, entomologists and pathologists working together who understand crop-specific biological systems and production practices, as well as the machinery, robotics and control issues associated with the automated production systems. Clearly focused multi-disciplinary teams are needed to address the full range of commodity-specific technical issues involved. Although there will be common technology components, such as machine vision, robotic manipulation, vehicle guidance, and so on, each crop application will be specialized, due to the unique nature of the biological system. However, collaboration and technology sharing between commodity groups will offer the benefit of leveraged research and development funds and reduced overall development time for multiple commodities.

References

Annamalai, P. and Lee, W.S. (2003) *Citrus yield mapping system using machine vision*. ASAE Annual International Meeting, ASAE Paper No. 031002. American Society of Agricultural Engineers, St Joseph, Michigan.

Annamalai, P., Lee, W.S. and Burks, T.F. (2004) *Color vision system for estimating citrus yield in real-time*. ASAE Annual International Meeting, ASAE Paper No. 043054. American Society of Agricultural Engineers, St Joseph, Michigan.

Bac, C.W., Van Henten, E.J., Hemming, J. and Edan, Y. (2014) Harvesting robots for high-value crops: state-of-the-art review and challenges ahead. *Journal of Field Robotics* 31(6), 888–911.

Bedford, R., Ceres, J.L., Pons, A.R., Jimenez, J., Martin, M. and Calderon, L. (1998) Design and implementation of an aided fruit-harvesting robot (Agribot). *Industrial Robot* 25(5), 337–346.

Ben-Tal, Y. (1983) Horticultural aspects of mechanical fruit harvesting. *Proceedings of the International Symposium on Fruit, Nut and Vegetable Harvesting Mechanization.* ASAE SP-5, 372–375. American Society of Agricultural Engineers, St Joseph, Michigan.

Benson, E.R., Reid, J.F. and Zhang, Q. (2001) *Machine vision based steering system for agricultural combines.* ASAE Paper No. 01-1159. American Society of Agricultural Engineers, St Joseph, Michigan.

Berenstein, R., Shahar, O.B., Shapiro, A. and Edan, Y. (2010) Grape clusters and foliage detection algorithms for autonomous selective vineyard sprayer. *Intelligent Service Robotics* 3(4), 233–243.

Buemi, F., Massa, M., Sandini, G. and Costi, G. (1996) The agrobot project. *Advances in Space Research* 18(1), 185–189.

Bulanon, D.M., Burks, T.F. and Alchanatis, V. (2008) Study on temporal variation in citrus canopy using thermal imaging for citrus fruit detection. *Biosystems Engineering* 101, 161–171.

Bulanon, D.M., Kataoka, T., Zhang, S., Ota, Y. and Hiroma, T. (2001) *Optimal thresholding for the automatic recognition of apple fruits.* ASAE Paper No. 01-3133. American Society of Agricultural Engineers, St Joseph, Michigan.

Bulanon, D.M., Burks, T.F. and Alchanatis, V. (2009a) Fruit visibility analysis for robotic citrus harvesting. *Transactions of the ASABE* 52(1), 277–283.

Bulanon, D.M., Burks, T.F. and Alchanatis, V. (2009b) Image fusion of visible and thermal images for fruit detection. *Biosystems Engineering* 103(2009), 12–22.

Burks, T.F., Shearer, S.A. and Payne, F.A. (2000) Classification of weed species color texture features and discriminant analysis. *Transactions of the ASAE* 43(2), 441–448.

Campoy, J., Gonzalez-Mora, J. and Dima, C. (2010) *Advanced sensing for tree canopy modeling and precision spraying.* ASABE Annual International Meeting, ASABE Paper No. 1009470. American Society of Agricultural and Biological Engineers, St Joseph, Michigan.

Carew, J. (1969) *The Prospects for Fruit and Vegetable Mechanization: Horticulture Outlook.* Report No. 16, Rural Manpower Center. Michigan State University, East Lansing, Michigan, pp. 79–83.

Cargill, B.F. (1983) Harvesting high density red tart cherries. *Proceedings of the International Symposium on Fruit, Nut and Vegetable Harvesting Mechanization*, ASAE Paper No. SP-5, 195–200. American Society of Agricultural Engineers, St Joseph, Michigan.

Cavalieri, S. and Plebe, A. (1996) Manipulator adaptive control by neural networks in an orange picking robot. *International Journal of Neural Systems* 7(6), 735–755.

Ceres, R., Pons, J.L., Jimenez, A.R., Martin, J.M. and Calderon, L. (1998) Design and implementation of an aided fruit-harvesting robot (Agribot). *Industrial Robot* 25(5), 337–346.

Chen, Y., Zhu, H., Ozkan, H.E., Derksen, R.C. and Krause, C.R. (2011) *An experimental variable-rate sprayer for nursery and orchard applications.* ASABE Annual International Meeting, Paper No. 1110497. American Society of Agricultural and Biological Engineers, St Joseph, Michigan.

Chinchuluun, R., Lee, W.S. and Burks, T.F. (2006) Machine vision-based citrus yield mapping system. *Proceedings of the Florida State Horticultural Society* 119, 142–147.

Davis, V. (1969) *Mechanization of Fruit and Pod Vegetables Introduction.* Report No. 16, Rural Manpower Center. Michigan State University, East Lansing, Michigan, pp. 205–213.

Dawson, J.R., Hooper, A.W. and Ambler, B. (1976) A continuous weigher for agricultural use. *Journal of Agricultural Engineering Research* 21(4), 389–397.

Edan, Y., Flash, T., Peiper, U., Shmulevich, I. and Sarig, Y. (1991) Near-minimum-time task planning for fruitpicking robots. *IEEE Transactions on Robotics and Automation* 7(1), 48–56.

Flood, S. (2006) Design of a robotic citrus harvesting end effector and force control model using physical properties and harvesting motion tests. PhD Dissertation, University of Florida, Gainesville, Florida.

Flood, S.J., Burks, T.F. and Teixeira, A.A. (2006) Physical properties of oranges in response to applied gripping forces for robotic harvesting. *Transactions of the ASABE* 49(2), 341–346.

Fortuna, L., Muscato, G., Nunnari, G., Pandolfo, A. and Plebe, A. (1996) Application of neural control in agriculture: an orange picking robot. *Acta Horticulturae* 406, 441–450.

Fujiura, T. (1997) Agricultural robots using 3D vision sensor. *Robot* 117, 32–38.

Grand D'Esnon, A. (1985) Robotic harvesting of apples. *Proceedings of the Agri-Mation*, 1, 210–214. ASAE Paper No. 1-85. American Society of Agricultural Engineers, St Joseph, Michigan.

Grand D'Esnon, A., Rabatel, G., Pellenc, R., Journeau, A. and Aldon, M.J. (1987) *MAGALI: a self-propelled robot to pick apples*. ASAE Paper No. 87-1037. American Society of Agricultural Engineers, St Joseph, Michigan.

Han, S. and Burks, T.F. (2010) *Multiple layered hierarchical feature tracking for grove scene*. 2010 ASABE Annual International Meeting, Paper No. 1008886. American Society of Agricultural and Biological Engineers, St Joseph, Michigan.

Hannan, M.W., Burks, T.F. and Bulanon, D.M. (2009) A machine vision algorithm combining adaptive segmentation and shape analysis for orange fruit detection. *Agricultural Engineering International: CIGR Journal* 11, 1–17.

Harrell, R.C., Adsit, P.D., Pool, T.A. and Hoffman, R. (1988) *The Florida robotic grove-lab*. ASAE Paper No. 88-1578. American Society of Agricultural Engineers, St Joseph, Michigan.

Harrell, R.C., Slaughter, D.C. and Adsit, P.D. (1989) A fruit-tracking system for robotic harvesting. *Machine Vision and Applications* 2(2), 69–80.

Harrell, R.C., Adsit, P.D., Munilla, R.D. and Slaughter, D.C. (1990a) Robotic picking of citrus. *Robotica* 8(04), 269–278.

Harrell, R.C., Adsit, P.D., Pool, T.A. and Hoffman, R. (1990b) The Florida robotic grove-lab. *Transactions of the ASAE* 33(2), 391–399.

Hayashi, S., Ganno, K., Ishii, Y. and Tanaka, I. (2002) Robotic harvesting system for eggplants. *Japan Agricultural Research Quarterly* 36, 163–168.

Iida, M. and Burks, T.F. (2002) Ultrasonic sensor development for automatic steering control of orchard tractor. *Proceedings of the Conference Automation Technology for Off-Road Equipment*, ASAE Publ. 701P0502, 221–229. American Society of Agricultural Engineers, St Joseph, Michigan.

Ito, N. (1990) Agricultural robots in Japan. *Proceedings of IEEE International Workshop on IRS*, 3–6 July 1990, 249–253. IEEE, Piscataway, New Jersey.

Jiminez, A.R., Ceres, R. and Pons, J.L. (2000) Vision system based on a laser range-finder applied to robotic fruit harvesting. *Machine Vision and Applications* 11(6), 321–329.

Juste, F. and Sevila, F. (1991) Citrus: an European project to study the robotic harvesting of oranges. *Proceedings of the 2nd Workshop on Robotics in Agriculture and Food Industry (IARP), Genoa*, 187–196.

Juste, F., Gracia, C., Molto, E., Ibanez, R. and Castillo, S. (1988) Fruit bearing zones and physical properties of citrus for mechanical harvesting. *Citriculture: Proceedings of the Sixth International Congress, Tel Aviv, Israel*, 1801–1809.

Juste, F., Fornés, I., Plá, F. and Sevila., F. (1992) An approach to robotic harvesting of citrus in Spain. *International Society of Citriculture* 3, 1014–1018.

Jutras, P.J. and Kretchman, D.W. (1962) A topping machine for Florida citrus groves. *Proceedings of the Florida State Horticultural Society* 75, 32–35.

Kim, D.G. (2011) Detection of citrus diseases using computer vision techniques. PhD Dissertation, University of Florida, Gainesville, Florida.

Kodagoda, K.R.S., Wijesoma, W.S. and Teoh, E.K. (2002) Fuzzy speed and steering control of an AGV. Institute of electrical and electronics engineers. *Transactions of Control Systems Technology* 10(1), 112–120.

Kondo, N. and Ting, K.C. (1998) Design and Control of Manipulators. In: *Robotics for Bioproduction Systems*. American Society of Agricultural Engineers, St Joseph, Michigan, pp. 31–63.

Kondo, N., Monta, M. and Fujiura, T. (1996a) Basic constitution of a robot for agricultural use. *Advanced Robotics* (104), 339–353.

Kondo, N., Nishitsuji, Y., Ling, P.P. and Ting, K.C.R. (1996b) Visual feedback guided robotic cherry tomato harvesting. *Transactions of the ASAE* 39(6), 233–2338.

Lapushner, D., Frankel, R. and Edelman, E. (1983) Genetical and cultural aspects for mechanical harvested fresh market tomatoes. *Proceedings of the International Symposium on Fruit, Nut and Vegetable Harvesting Mechanization* 5(12), 404–407.

Lee, W.S. and Slaughter, D.C. (2004) Recognition of partially occluded plant leaves using a modified watershed algorithm. *Transactions of the ASAE* 47(4), 1269–1280.

Levi, P., Falla, A. and Pappalardo, R. (1988) Image controlled robotics applied to citrus fruit harvesting. In: Guttropf, W. (ed.) *Advanced Sensor Technology: Proceedings of the 7th International Conference on Robot Vision and Sensory Controls* (RoViSeC-7), Zurich, 2–4 February 1988. IFS Publications, Bedford, UK.

Mehta, S. and Burks, T. (2014) Vision-based control of robotic manipulator for citrus harvesting. *Computers and Electronics in Agriculture* 102, 146–158.

Mehta, S., MacKunis, W. and Burks, T. (2016) Robust visual servo control in the presence of fruit motion for robotic citrus harvesting. *Computers and Electronics in Agriculture* 123, 362–375.

Mehta, S., and Burks, T. (2016) Adaptive visual servo control of robotic harvesting systems. *IFAC-PapersOnLine*, 49(16), 287–292.

Mishra, A.R., Eshani, R., Lee, W.S. and Albrigo, G. (2007) *Spectral characteristics of citrus greening (Huanglongbing)*. ASABE Annual Meeting, Paper No. 073056. American Society of Agricultural and Biological Engineers, St Joseph, Michigan.

Mizushima, A., Noguchi, N., Ishii, K. and Terao, H. (2002) Automatic navigation of the agricultural vehicle by the geomagnetic direction sensor and gyroscope. *Proceedings of the Conference Automation Technology for Off-Road Equipment*. ASAE Publ. 701P0502. 204–211. American Society of Agricultural Engineers, St Joseph, Michigan.

Monselise, S.P. and Goldschmidt, E.E. (1982) Alternate bearing in fruit trees. *Horticultural Reviews* 4, 155–158.

Morris, J.R. (1983) Effects of mechanical harvesting on the quality of small fruits and grapes. *Proceedings of the International Symposium on Fruit, Nut and Vegetable Harvesting Mechanization* 5(12), 332–348.

Muscato, G., Prestifilippo, M., Abbate, N. and Rizzuto, I. (2005) A prototype of an orange picking robot: past history, the new robot, and experimental results. *Industrial Robot* 32(2), 128–138.

Nagasaka, Y., Umeda, N. and Kanetai, Y. (2002) National agr. res. center, Japan: automated rice transplanter with GPS and FOG. *Proceedings of the Conference Automation Technology for Off-Road Equipment*. ASAE Publ. 701P0502, 190–195. American Society of Agricultural Engineers, St Joseph, Michigan.

Noguchi, N., Kise, M., Ishii, K. and Terao, H. (2002) Field automation using robot tractor. *Proceedings of the Conference Automation Technology for Off-Road Equipment*. ASAE Publ. 701P0502, 239–245. American Society of Agricultural Engineers, St Joseph, Michigan.

O'Conner, M., Elkaim, G. and Parkinson, B. (1995) *Kinematic GPS for closed-loop control of form and construction vehicles*. Paper No. 10NGPS-95, presented at ION GPS-95, Palm Springs, California, 12–15 September 1995. Stanford University, Stanford, California.

Parrish, E.A. and Goksel, A. (1997) Pictorial pattern recognition applied to fruit harvesting. *Transactions of the ASAE* 20(5), 822–827.

Peterson, D.L., Bennedsen, B.S., Anger, W.C. and Wolford, S.D. (1999) A systems approach to robotic bulk harvesting of apples. *Transactions of the ASAE* 42(4), 871–876.

Plebe, A. and Grasso, G. (2001) Localization of spherical fruits for robotic harvesting. *Machine Vision and Applications* 13(2), 70–79.

Pool, T.A. and Harrell, R.C. (1991) An end-effector for robotic removal of citrus from the tree. *Transactions of the ASAE* 34(2), 373–378.

Pydipati, R., Burks, T.F. and Lee, W.S. (2006) Identification of citrus disease using color texture features and discriminant analysis. *Computers and Electronics in Agriculture* 52, 49–59.

Qiu, H., Zhang, Q. and Reid, J. (2001) Fuzzy control of electrohydraulic steering systems for agricultural vehicles. *Transactions of the ASAE* 44(6), 1397–1402.

Rabatel, G., Bourely, A., Sevila, F. and Juste, F. (1995) Robotic harvesting of citrus: state-of-art and development of the French Spanish Eureka project. *Proceedings of the International Conference on Harvest and Post Harvest Technology for Fresh Fruits and Vegetables*, Guanajunto, Mexico, 232–239.

Recce, M., Taylor, J., Plebe, A. and Tropiano, G. (1996) Vision and neural control for an orange harvesting robot. *Proceedings of the International Workshop on Neural Networks for Identification, Control, Robotics, and Signal/Image Processing, 1996*, pp. 467–475.

Rohrbach, R.P. (1983) Mechanized maintenance of blueberry quality. *Proceedings of the International Symposium on Fruit, Nut and Vegetable Harvesting Mechanization*, SP-5, 134–140.

Rosa, U.A., Cheetancheri, K.G., Gliever, C.J., Lee, S.H., Thompson, J. and Slaughter, D.C. (2008) An electro-mechanical limb shaker for fruit thinning. *Computers and Electronics in Agriculture* 61(2), 213–221.

Sankaran, S., Ehsani, R. and Etxeberria, E. (2010) Mid-infrared spectroscopy for detection of huanglongbing (greening) in citrus leaves. *Talanta* 83(2), 574–581.

Sansavini, S. (1978) Mechanical pruning of fruit tree. *Acta Horticulturae* 65, 183–197.

Sarig, Y. (1993) Robotics of fruit harvesting: a state-of-the-art review. *Journal of Agricultural Engineering Research* 54(4), 265–280.

Schertz, C. and Brown, G. (1968) Basic consideration in mechanizing citrus harvesting. *Transactions of the American Society of Agricultural Engineers* 11, 343–346.

Schupp, J.R., Baugher, T.A., Miller, S.S., Harsh, R.M. and Lesser, K.M. (2008) Mechanical thinning of peach and apple trees reduces labor input and increases fruit size. *HortTechnology* 18(4), 660–670.

Senoo, S., Mino, M. and Funabiki, S. (1992) Steering control of automated guided vehicle for steering energy saving by fuzzy reasoning. *Proceedings of 1992 Institute of Electrical and Electronics Engineers Industry Applications Society Annual Meeting*, 1712–1716.

Sims, W.L. (1969) *Cultural Practices for Fruit Vegetables*. Report No.16, Rural Manpower Center. Michigan State University, East Lansing, Michigan, pp. 225–237.

Sivaraman, B. (2006) Design and development of a robot manipulator for citrus harvesting. PhD Dissertation, University of Florida, Gainesville, Florida.

Slaughter, D. and Harrell, R.C. (1989) Discriminating fruit for robotic harvest using color in natural outdoor scenes. *Transactions of the ASAE* 32(2), 757–763.

Subramanian, V., Burks, T.F. and Arroyo, A.A. (2006) Machine vision and laser radar-based vehicle guidance systems for citrus grove navigation. *Computers and Electronics in Agriculture* 53, 130–143.

Subramanian, V., Burks, T.F. and Dixon, W.E. (2009) Sensor fusion using fuzzy logic enhanced Kalman filter for autonomous vehicle guidance in citrus groves. *Transactions of the ASABE* 52(5), 1–12.

Swanson, M., Dima, C. and Stentz, A. (2010) *A multi-modal system for yield prediction in citrus trees.* ASABE Annual International Meeting, Paper No. 1009474. American Society of Agricultural and Biological Engineers, St Joseph, Michigan.

Tucker, D.P.H., Wheaton, T.A. and Muraro, R.P. (1994) *Citrus Tree Pruning Principles and Practices.* Horticultural Sciences Department, Florida Cooperative Extension Service, IFAS, Gainesville, Florida.

Tumbo, S.D., Salyani, M., Whitney, J.D., Wheaton, T.A. and Miller, W.M. (2002) Investigation of laser and ultrasonic ranging sensors for measurements of citrus canopy volume. *Applied Engineering in Agriculture* 18, 367–372.

Tutle, E.G. (1985) *Robotic Fruit Harvester.* US Patent No. 4,532,757. Washington, DC.

Upadhyaya, S.K., Mir, S.S. and Leroy, O.G. (2006) *Development of an impact type electronic weighing system for processing tomatoes.* ASABE Annual International Meeting, Paper No. 061190. American Society of Agricultural and Biological Engineers, St Joseph, Michigan.

Van Henten, E., Hemming, J., Van Tuijl, B., Kornet, J., Meuleman, J., Bontsema, J. and Van Os, E. (2002) An autonomous robot for harvesting cucumbers in greenhouses. *Autonomous Robots* 13(3), 241–258.

Van Henten, E., Hemming, J., Van Tuijl, B., Kornet, J., Meuleman, J. and Bontsema, J. (2003) Collision-free motion planning for a cucumber picking robot. *Biosystems Engineering* 86 (2), 135–144.

VRC (2011) Grape Pruning. Available at: http://visionrobotics.com/vrc/index.php?option=com_zoom&Itemid=26&catid=6.

Wangler, R.J., Flower, K.L., McConnerll, R.E. and Robert, E. (1992) *Object Sensor and Method for Use in Controlling an Agricultural Sprayer.* Patent No. 5278423. Orlando, Florida.

Wei, J. and Salyani, M. (2005) Development of a laser scanner for measuring tree canopy characteristics: phase 2. Foliage desity measurement. *Transactions of the ASAE* 48, 1595–1601.

Whitney, J.D., Miller, W.M., Wheaton, T.A., Salyani, M. and Schueller, J.K. (1999) Precision farming applications in Florida citrus. *Applied Engineering in Agriculture* 15(5), 399–403.

Whitney, J.D., Ling, Q., Miller, W.M. and Wheaton, T.A. (2001) A DGPS yield monitoring system for Florida citrus. *Applied Engineering in Agriculture* 17(2), 115–119.

Wolf, I. and Alper, Y. (1983) Mechanization of Paprika Harvest. *Proceedings of the International Symposium on Fruit, Nut and Vegetable Harvesting Mechanization, ASAE*, 265–275.

Zarchan, P. and Musoff, H. (eds) (2005) *Fundamentals of Kalman Filtering: A Practical Approach*, 2nd edn. Progress in Astronautics and Aeronautics series, Volume 208. American Institute of Aeronautics and Astronautics (AIAA), Reston, Virginia.

Zhang, Q., Reid, J. and Noguchi, N. (1999) Automatic guidance control for agricultural tractor using redundant sensor. *Journal of Commercial Vehicles* 108, 27–31.

Zocca, A. (1983) Shaking harvesting trials with market apples. *Proceedings of the International Symposium on Fruit, Nut and Vegetable Harvesting Mechanization* 5(12), 65–72.

Index

Note: Page numbers in **bold** type refer to **figures**
Page numbers in *italic* type refer to *tables*